AF362060

Environmental Hydraulics Research

Environmental Hydraulics Research

Editors

Helena M. Ramos
Armando Carravetta
Aonghus McNabola
Kemi Adeyeye

MDPI • Basel • Beijing • Wuhan • Barcelona • Belgrade • Manchester • Tokyo • Cluj • Tianjin

Editors
Helena M. Ramos
University of Lisbon
Portugal

Armando Carravetta
University Federico II of Naples
Italy

Aonghus McNabola
Structural & Environmental Engineering
Ireland

Kemi Adeyeye
University of Bath
UK

Editorial Office
MDPI
St. Alban-Anlage 66
4052 Basel, Switzerland

This is a reprint of articles from the Special Issue published online in the open access journal *Water* (ISSN 2073-4441) (available at: https://www.mdpi.com/journal/water/special_issues/ Environmental_Hydraulics_Research).

For citation purposes, cite each article independently as indicated on the article page online and as indicated below:

LastName, A.A.; LastName, B.B.; LastName, C.C. Article Title. *Journal Name* **Year**, *Volume Number*, Page Range.

ISBN 978-3-0365-0808-5 (Hbk)
ISBN 978-3-0365-0809-2 (PDF)

Contents

About the Editors

Helena M. Ramos is Professor at Instituto Superior Técnico—IST (the Engineering Faculty from University of Lisbon) and is a member of CERIS in the Civil Engineering Department. She has participated in 30 scientific projects, and she has more than 5750 citations, an h-index of 40, and an i-10 of 124. She has been a member of editorial teams and a reviewer of different scientific journals. She has more than 400 publications, with 180 being in international journals with referee, more than 180 in Conferences, 22 book chapters, and 5 books: In Small Hydropower Plants 2000 and Pump as Turbines 2018, Bombas operando como Turbinas 2020, New Challenges in Water Systems MDPI, amongst other documents of technical and scientific disclosure. She has received 3 international scientific recognitions. She is an expert in transients, hydropower, pumped storage and hybrid renewable solutions, water–energy nexus, leakage detection, and hydrodynamics. For more detail: http://scholar.google.pt/citations?sortby=pubdate&hl=pt-PT&user=9jTHk6oAAAAJ   ; http://orcid.org/0000-0002-9028-9711.

Armando Carravetta is Full Professor in Hydraulics at the Department of Civil, Architecture and Environmental Engineering, University of Naples, Federico II, IT. He undertakes research in technical innovation for water systems, energy efficiency and resilience of pumping systems, energy recovery by PAT technology, and fluid dynamics of slurry flows. He represents the Italian Association of Pump Manufacturers in the Europump Lot 28 Working Group for the implementation of the European standards on Ecodesign. He is one of the partners in the ongoing EU Interreg project, Reducing Energy Dependency in Atlantic Area Water Networks (REDAWN).

Aonghus McNabola is a Professor in Energy and the Environment at the Dept of Civil, Structural and Environmental Engineering, TCD. His research interests lie in the field of environmental fluid dynamics, where he has applied this expertise in air pollution, energy efficiency, and/or water services sectors. He has been active in this field of research for over 15 years. Prof. McNabola currently leads a group of 8 PhD students and 3 postdoctoral researchers and is involved, primarily as the lead partner and principal investigator, in a number of national and international collaborative projects, funded by Horizon 2020, INTERREG ERDF, SEAI, and the EPA. Prof. McNabola has an h-index of 25 and an i-10 index of 45. He has generated funding of over 5.25 million from national and primarily EU sources in over 20 funded research projects since 2008.

Kemi Adeyeye currently works at the Department Architecture and Civil Engineering, University of Bath. Kemi does research in Architecture & Planning, Architectural Engineering, Sustainability, Resilience, and Environmental Engineering. She has participated in 3 funding scientific projects, and has 72 publications, including edited book and journal articles.

Editorial

Environmental Hydraulics Research

Helena M. Ramos [1],*, Armando Carravetta [2], Aonghus Mc Nabola [3] and Kemi Adeyeye [4]

[1] Department of Civil Engineering and Architecture, CERIS, Instituto Superior Técnico, University of Lisbon, 1049–001 Lisbon, Portugal

[2] Department of Civil, Architecture and Environmental Engineering, University Federico II of Naples, 80125 Naples, Italy; armando.carravetta@unina.it

[3] Department of Civil, Structural and Environmental Engineering, Trinity College Dublin, D02 PN40 Dublin, Ireland; amcnabol@tcd.ie

[4] Department of Architecture & Civil Engineering, University of Bath, Claverton Down, Bath BA2 7AY, UK; K.Adeyeye@bath.ac.uk

* Correspondence: hramos.ist@gmail.com or helena.ramos@tecnico.ulisboa.pt

Received: 29 September 2020; Accepted: 30 September 2020; Published: 1 October 2020

Abstract: Environmental hydraulics research includes the different domains of hydrodynamics, such as the investigation and implementation of the physical and experimental applications, and research into the quantity, quality, modelling and simulation of the attributes associated with flowing water. This topic is studied both from a technical and environmental point of view, with the objective of protecting and enhancing the quality of the environment. It is a cross-disciplinary field of study which comprises open channel/river flows and pressurised systems, combining, among others, new technological, social, and environmental hydraulic challenges. It provides researchers and engineers working in water-related fields with available information, new concepts and tools, new design solutions, eco-friendly technologies, and the advanced materials necessary to address the increasing challenges of ensuring a sustainable water environment—that is, a water environment effectively managed and adequated for generations to come by promoting the adaptation, flexibility, integration and sustainability of recognised environmental solutions. Using advanced numerical and physical models in field experiments, and tests in different types of laboratory set-ups, specialists in environmental hydraulics produce the best analyses, concepts, techniques, tools, and solutions to environmental hydraulic problems, as well as in relation to the water, energy and environmental nexus.

Keywords: new hydraulic concepts; sustainable developments; CFD models; water systems efficiency; hydropower systems; eco-design; environmentally-friendly solutions; hydrologic and ecologic challenges; hydraulic structures; free surface flows; pressurised flows; soil structure; groundwater; erosion and energy dissipaters; hydrodynamics

---

1. Introduction

Environmental hydraulics research includes the different domains of hydrodynamics, such as the investigation and implementation of the physical and experimental applications, and research into the quantity, quality, modelling and simulation of the characteristics associated with water flow. Using advanced numerical and physical models in field experiments, and tests in different types of laboratory set-ups, specialists in environmental hydraulics present the best analyses, new design concepts, apply different nature-friendly techniques, use advanced simulation tools, and obtain friendly solutions to environmental hydraulic problems as well as for water, soil, civil works, energy dissipaters and to produce devices and complex hydraulic phenomena under an integrated environmental nexus.

This Special Issue addresses environmental hydraulics as an environmental continuum of water, air and soil, focussing on the motion of liquid at the macroscopic scale [1]. It considers groundwater, hydraulic structures, spillways, stormwater infrastructures, erosion problems, river and lake interactions, energy dissipaters, river bed and scour processes, water–energy potential, as well as the water, soil and ecosystem issues pertaining to hydrological activities and events, water quality and all hydraulic activities and investigations within natural and non-natural water domains. Of note in this Special Issue is the essential role of pumped-storage hydropower systems as a flexible renewable energy source, especially when integrated with wind and solar energy sources [2]. Hydropower is one of the oldest energy sources developed by humankind and continues to also be one of the most prominent ones in terms of cost, as well as social and environmental impacts. However, the use of hydropower requires a more flexible, efficient, environmentally and socially friendly and acceptable approach to complement wind and solar energy production, in particular by (i) increasing hydropower production through the implementation of new environmentally friendly multipurpose hydropower schemes and by using the hidden potential in existing water infrastructures, (ii) increasing the flexibility of generation from existing hydropower plants by adaptation and optimization of infrastructures and equipment combined with innovative solutions for the mitigation of environmental impacts and (iii) increasing storage by the construction of new reservoirs, which have to ensure not only flexible energy supply but also support the food and water nexus [3–6]. Therefore, a hybrid system comprised of several renewable energy resources of a complementary nature working alongside other sources in an integrated and flexible way is of utmost importance in the electric production system and energy management.

2. Main Subjects Analysed

This book comprises research papers which address the different approaches, advanced studies and practice tools for environmental hydraulics. The works presented in this Special Issue cover the following main themes, as described below:

1. In cold regions, hydraulic conductivity is a critical parameter for determining the water flow in frozen soil. Previous studies have shown that hydraulic conductivity hinges on the pore structure, which is often depicted as the pore size and porosity. However, these two parameters do not sufficiently represent the pore structure [7,8]. To enhance the characterization ability of the pore structure, one study introduces fractal theory to investigate the influence of pore structure on hydraulic conductivity. Phosphorus is a major cause of lake eutrophication. Understanding the characteristics regarding the release of phosphorus from sediments under hydrodynamic conditions is crucial for the regulation of lake water quality.

2. Water wells play an increasingly important role in providing water for the civilian population all over the world. The number of boreholes is increasing not only due to population growth, but also due to a decrease in the productivity of the existing assets. Like other engineering structures, wells are subject to ageing processes resulting in degradation, which is observed as a reduction in hydraulic efficiency throughout their lifespan. Hydraulic conductivity hinges on the pore structure, which is often depicted as the pore size and porosity. Unlike "sustainable yield", for the concept of efficiency, compensation is not a constant-rate or constant-head method. The "sustainable efficiency" concept requires the regulation of both flow rate and drawdown. The authors in [9,10] investigate the problems associated with the drilling of groundwater wells.

3. Lakes provide valuable ecosystem services for riparian communities and play an important role in sustaining ecological security and sustainable development. Hence, the effects of sediment suspension on the release characteristics of phosphorus from sediment are analysed.

4. Reusing stormwater is more critical in water-scarce regions and regions where rainfall patterns and rainfall frequencies are changing [5]. Several regions are trying to reuse stormwater as a sustainable method of water resource management; thus, timely research into the feasibility of reusing stormwater from a water quality perspective is presented in [11]. The Soil Conservation Service (SCS) curve number model is used as the basis of an integrated hydrological–CFD model to generate flow rates from the watershed draining into rain ponds due to storms and estimating

bacterial levels in storm water ponds. An experimental investigation of the erosion characteristics of fine-grained cohesive sediments is used to model the transport of fine sediment and the associated nutrients in a river system. The erosion experimental data allowed the identification of the critical shear stress for erosion. Urban stormwater, as an important environmental problem, requires us to find green infrastructure solutions, which provide water treatment and retention at the same time. Floating treatment wetlands, which are porous patches that continue down from the free surface with a gap between the patch and bed, are innovative instruments for nutrient management in lakes, ponds, and slow-flowing waters.

5. Lakes are important for global ecological balance and provide rich biological and social resources, but they are sensitive to climate change and anthropogenic activities in which river–reservoir erosion and the water level can be affected by an existing dam operation. Fish protection and downstream migration measures are also considered essential in the hydraulic design, construction or retrofitting and operation of hydropower plants. Since downstream migrating fish, particularly juveniles, tend to swim within the main current, they will consequently pass the turbines if appropriate measures are not put in place at the water intake [12]. Different types of spur dikes, such as rock fill, permeable, w-shaped rock fill, and w-shaped permeable are evaluated using flume experiments for spur dike hydrodynamics and fish aggregation effects. The results could provide theoretical support for habitat heterogeneity research and the ecologically optimal design of spur dikes. The effects of changes in the angle of the pool impact plate, plunging depth, and discharge upon the dynamic pressure caused by ski jump buckets are investigated in the laboratory. Due to dynamic pressures resulting from the flow in hydraulic structures, the riverbed is frequently affected by scouring. Moreover, many lakes are connected to rivers to form complicated river–lake systems [13–15]. As a safe device in a dam, the understanding of jet features is crucial in designing the pool and determining the plunging rate [16]. Recently, many researchers and engineers have also studied the ecological effects of spur dikes through model tests, numerical simulations and on-site observations. The results in [17–19] show that spur dikes can enhance river habitat heterogeneity and improve the suitability of fish habitats under low and medium water flows. In hydropower, the requirements for fish protection at hydro power plants have led to a significant decrease in the bar spacing at trash racks, as well as the need for an inclined or angled design to improve the guidance effect using a fish-friendly trash rack.

6. In a water system where an excess of internal energy needs to be dissipated effectively, a specific dissipater reduces the downstream flow speed, smoothly connects the downstream flow, and avoids the erosion of the river channel in a well-defined water conveyance project described in [20–22]. On the other hand, the toothed internal energy dissipater is a new type of dissipater that controls the excess flow energy. However, in the open-channel control algorithm, good feedforward controllers will reduce the transition time of the canal and improve its performance. Hybrid pumped hydro storage energy solutions with wind and PV integration present further improvements in flexibility, reliability and in the associated energy costs. A technique based on a multi-criteria evaluation, for a sustainable environmental solution based on renewable source integration, is presented, with the results demonstrating that, technically, pumped hydro storage with wind and PV is an ideal solution to achieve energy autonomy and to increase flexibility and reliability. A hydrostatic pressure machine is also analysed using CFD modelling with detailed characterization and evaluation, with promising results [23]. Another study recently examined a technology based on a transient flow-induced compressed air energy storage system, which takes advantage of a compressed air vessel to combine pumping and hydropower solutions.

The importance of environments in rivers and pipe systems are not overstated. They provide water, energy and food, safety and better controlled systems. They can cause major damage through flooding and erosion, and they are the means to avoid water scarcity or a lack of quality. Therefore, engineering needs a solid understanding of the mechanics governing phenomena. This book provides readers with necessary knowledge about eco-design developments and troubleshoots a variety of environmental hydraulic engineering works. It analyses the fascinating hydraulics of rivers, including the interactions between sediment (e.g., sand, gravel) and water flow, erosion and scour problems, engineering intervention in river and lakes, in fish protection infrastructures, in storm water sustainable management, in eco-friendly solutions, in flood modelling and mitigation, hydropower renewable integration systems, embankments, dams or dikes and spillways and jet effects in relation to hydraulic–ecological effects. Environmental issues are presented in the wider context of civil engineering and environmental hydraulics via computer-based studies, and laboratory or field results.

Author Contributions: H.M.R. conceived and led the development of this Special Issue and this editorial; A.C., A.M.N. and K.A. contributed substantially to the writing. All authors have read and agreed to the published version of the manuscript.

Acknowledgments: The authors of this editorial, who served as guest editors of this Special Issue, wish to thank the journal editors, all authors submitting papers to this Special Issue, and the many referees who contributed to paper revision and the improvement of all published articles.

Conflicts of Interest: The authors declare no conflict of interest.

References

1. Singh, V.P.; Hager, W.H. What is Environmental Hydraulics. In *Environmental Hydraulics*; Springer: Dordrecht, The Netherlands, 1996; pp. 1–5.
2. Vieira, F.; Ramos, H.M. Hybrid solution and pump-storage optimization in water supply system effciency: A case study. *Energy Policy* **2008**, *36*, 4142–4148. [CrossRef]
3. Carravetta, S.A.; Ramos, H.M.; Houreh, S.D. *Pumps as Turbines, Fundamentals and Applications*; Springer International Publishing: Berlin/Heidelberg, Germany, 2018.
4. Jaramillo-Duque, Á.; Castronuovo, E.D.; Sánchez, I.; Usaola, J. Optimal operation of a pumped-storage hydro plant that compensates the imbalances of a wind power producer. *Electr. Power Syst. Res.* **2011**, *81*, 1767–1777. [CrossRef]
5. International Hydropower Association (IHA). *The World's Water Battery: Pumped Hydropower Storage and the Clean Energy Transition*; IHA Working Paper; International Hydropower Association: London, UK, 2018.
6. International Hydropower Association (IHA). *Pumped Storage Hydropower Has 'Crucial Role' in Europe's Energy Strategy*; IHA Working Paper; International Hydropower Association: London, UK, 2020.
7. Tao, G.L.; Zhu, X.L.; Hu, Q.Z.; Zhuang, X.S.; He, J.; Chen, Y. Critical pore-size phenomenon and intrinsic fractal characteristic of clay in the process of compression. *Rock Soil Mech.* **2019**, *40*, 81–90.
8. Xiao, B.; Fan, J.; Ding, F. A fractal analytical model for the permeabilities of fibrous gas diffusion layer in proton exchange membrane fuel cells. *Electrochim. Acta* **2014**, *134*, 222–231. [CrossRef]
9. Walton, W.C. Ground-Water Hydraulics as an Aid to Geologic Interpretation. *Ohio J. Sci.* **1955**, *55*, 13–20.
10. Dufresne, D.P. Developing Better Regional Groundwater Flow Models with Effective Use of Step-Drawdown Test Results. *Fla. Water Resour. J.* **2011**, *63*, 36–40.
11. Ahilan, S.; Guan, M.; Wright, N.; Sleigh, A.; Allen, D.; Arthur, S.; Haynes, H.; Krivtsov, V. Modelling the long-term suspended sedimentological effects on stormwater pond performance in an urban catchment. *J. Hydrol.* **2019**, *571*, 805–818. [CrossRef]
12. Clevenot, L.; Carré, C.; Pech, P. A Review of the factors that determine whether stormwater ponds are ecological traps and/or high-quality breeding sites for amphibians. *Front. Ecol. Evol.* **2018**, *6*, 40. [CrossRef]
13. Böttcher, H.; Gabl, R.; Ritsch, S.; Aufleger, M. Experimental study of head loss through an angled fish protection system. In Proceedings of the 4th IAHR Europe Congress, Liege, Belgium, 27–29 July 2016; pp. 637–642.
14. Yao, J.; Zhang, Q.; Li, Y.L.; Li, M.F. Hydrological evidence and cause of seasonal low water levels in a large river-lake system: Poyang Lake, China. *Hydrol. Res.* **2016**, *47*, 24–39. [CrossRef]
15. Pu, J.H.; Huang, Y.F.; Shao, S.D.; Hussain, K. Three-Gorges Dam Fine Sediment Pollutant Transport: Turbulence SPH Model Simulation of Multi-Fluid Flows. *J. Appl. Fluid Mech.* **2016**, *9*, 1–10. [CrossRef]
16. Huang, G.X.; Zhou, J.J.; Lin, B.L.; Xu, X.F.; Zhang, S.H. Modelling flow in the middle and lower Yangtze River, China. *Water Manag.* **2016**, *170*, 298–309. [CrossRef]

17. Mahmoud, H.; Kriaa, W.; Mhiri, H.; Le Palec, G.; Bournot, P. Numerical analysis of recirculation bubble sizes of turbulent co-flowing jet. *Eng. Appl. Comput. Fluid.* **2012**, *6*, 58–73. [CrossRef]
18. Chakravarti, A.; Jain, R.K.; Kothyari, U.C. Scour under submerged circular vertical jets in cohesionless sediments. *ISH J. Hydraul. Eng.* **2014**, *20*, 32–37. [CrossRef]
19. Yin, Z.; Shi, B.; Zhao, L.; Sun, D. Numerical simulation of plug energy dissipater flow. *Adv. Water Sci.* **2008**, *1*, 89–93.
20. Wang, Z.; Wu, B.; Wang, G. Fluvial processes and morphological response in the Yellow and Weihe Rivers to closure and operation of Sanmenxia Dam. *Geomorphology* **2007**, *91*, 65–79. [CrossRef]
21. Lian, L.; Wang, W.; Tian, Z. Characteristics of energy dissipation and cavitation for complex plug in discharge tunnel. *J. Hydroelectr. Eng.* **2012**, *2*, 62–70.
22. Rydlewicz, W.; Rydlewicz, M.; Pałczyński, T. Experimental investigation of the influence of an orifice plate on the pressure pulsation amplitude in the pulsating flow in a straight pipe. *Mech. Syst. Signal Process.* **2019**, *117*, 634–652. [CrossRef]
23. Senior, J.; Saenger, N.; Müller, G. New hydropower converters for very low head differences. *J. Hydraul. Res.* **2010**, *48*, 703–714. [CrossRef]

Article

A Fractal Model of Hydraulic Conductivity for Saturated Frozen Soil

Lei Chen [1,2], Dongqing Li [1], Feng Ming [1,*], Xiangyang Shi [1,2] and Xin Chen [1,2]

[1] State Key Laboratory of Frozen Soil Engineering, Northwest Institute of Eco-Environment and Resources, Chinese Academy of Sciences, Lanzhou 730000, China; ChenLei8@lzb.ac.cn (L.C.); dqli@lzb.ac.cn (D.L.); shixiangyang2008@163.com (X.S.); chenxcumt@sina.com (X.C.)

[2] University of Chinese Academy of Sciences, Beijing 100049, China

* Correspondence: mingfeng05@lzb.ac.cn; Tel.: +86-93-1496-7469

Received: 10 December 2018; Accepted: 18 February 2019; Published: 21 February 2019

Abstract: In cold regions, hydraulic conductivity is a critical parameter for determining the water flow in frozen soil. Previous studies have shown that hydraulic conductivity hinges on the pore structure, which is often depicted as the pore size and porosity. However, these two parameters do not sufficiently represent the pore structure. To enhance the characterization ability of the pore structure, this study introduced fractal theory to investigate the influence of pore structure on hydraulic conductivity. In this study, the pores were conceptualized as a bundle of tortuous capillaries with different radii and the cumulative pore size distribution of the capillaries was considered to satisfy the fractal law. Using the Hagen-Poiseuille equation, a fractal capillary bundle model of hydraulic conductivity for saturated frozen soil was developed. The model validity was evaluated using experimental data and by comparison with previous models. The results showed that the model performed well for frozen soil. The model showed that hydraulic conductivity was related to the maximum pore size, pore size dimension, porosity and tortuosity. Of all these parameters, pore size played a key role in affecting hydraulic conductivity. The pore size dimension was found to decrease linearly with temperature, the maximum pore size decreased with temperature and the tortuosity increased with temperature. The model could be used to predict the hydraulic conductivity of frozen soil, revealing the mechanism of change in hydraulic conductivity with temperature. In addition, the pore size distribution was approximately estimated using the soil freezing curve, making this method could be an alternative to the mercury intrusion test, which has difficult maneuverability and high costs. Darcy's law is valid in saturated frozen silt, clayed silt and clay, but may not be valid in saturated frozen sand and unsaturated frozen soil.

Keywords: frozen soil; soil freezing curve; hydraulic conductivity; fractal model; Darcy's law

1. Introduction

Hydrology in cold regions is a rapidly progressing research field, representing an important sub-discipline of hydrology. Recently, research interest in cold region hydrology has been spurred by global warming-induced hydrological and ecological changes to cold regions, such as permafrost degradation, glacial recession and surface runoff shrinkage [1]. Cold region hydrology includes permafrost hydrology, glacier hydrology and snow hydrology, of which permafrost hydrology is the broader scientific field owing to permafrost's wide distribution, covering about a quarter of the land surface of the world [2]. Frozen soil is a special type of soil-water system composed of soil particles, liquid water, ice and gas. The emergence and existence of ice changes the driving force of water migration, leading to changes in migration direction, migration distance, migration velocity and migration amount; these produce further changes in the water distribution in frozen soil. Previously, frozen soil has often been conceptualized as an impermeable barrier that inhibits infiltration and promotes surface and near-surface runoff; however, researchers have since investigated the hydraulic conductivity of frozen soil and found that frozen soil is not a complete aquiclude, as liquid water could migrate in the frozen soil. Thus, water migration in frozen soil is one of the main processes in permafrost hydrology. Due to the free energy of soil particles and water-air interfaces, some water in frozen soil remains unfrozen at sub-zero temperatures [3]—defined as unfrozen water. The unfrozen water will flow along a temperature gradient or pressure gradient, resulting in water and solute redistribution [4], frost heaving [5], and salt expansion [6]. Such damage causes problems during the construction of engineering ground structures in cold regions, including roads, railways, pipelines, buildings and dams. In frozen soil, at least in high-temperature frozen soil, the unfrozen water transported in the frozen soil is always assumed to follow Darcy's law:

$$V = k_\mathrm{f} \frac{\Delta H}{L} \tag{1}$$

where V is Darcy's flux (m/s); k_f is the hydraulic conductivity of frozen soil (m/s); ΔH is the water head loss between two cross sections (m), $\Delta H = H_1 - H_2$, where H is the total head (m), defined as the sum of the elevation head Z and the pressure head h_w; and L is the distance of the seepage path between these two cross sections (m). As Equation (1) shows, hydraulic conductivity is one of the key parameters for determining water flow in frozen soil.

To further predict water redistribution, frost heave and salt expansion, several hydrology and hydrogeology models had been developed by coupling a groundwater flow equation with heat transfer equations. Researchers applying these mathematical models are faced with difficulties due to a lack of precise values of hydraulic conductivity. The main reason for this deficiency is that measuring the hydraulic conductivity of frozen soil remains difficult and a complete expression for hydraulic conductivity is not yet available. This problem leads to a general assumption being made that the hydraulic conductivity of frozen soil is, in most cases, zero. However, it has been found in most of the studies that water flow still occurs in frozen soil, especially in high-temperature frozen soil, making this assumption unreasonable. The main objective of this paper is to provide a solution to this problem by presenting a new mathematical model for estimating the hydraulic conductivity of saturated frozen soil with a limited set of data.

2. Background

Many attempts have been undertaken to determine the hydraulic conductivity of frozen soil. Burt and Williams [7] measured saturated hydraulic conductivity using a permeameter. In this study, water was used as the fluid. To prevent the water from melting the frozen soil, lactose was added to ensure that the fluid was in thermodynamic equilibrium with the water in the soil. In addition, the lactose had restricted entry to the soil through the use of a dialysis membrane. In other studies, to solve the problem of the water melting the frozen soil, different fluids other than water have been used to measure hydraulic conductivity; for example, Horiguchi and Miller [8] used distilled water; Wiggert et al. [9] used dacane, octane, heptane and NaCl brine; McCauley et al. [10] used diesel; and Seyfried and Murdock [11] used air. All of these studies tended to overestimate hydraulic conductivity because the fluids could flow due to air-filled porosity, making it difficult to obtain an accurate hydraulic conductivity measuring by such experiments alone.

Many mathematical models for predicting the hydraulic conductivity of frozen soil have since been recommended. These mathematical models include empirical models and statistical models. For empirical models, there are three main types. The first type is determined by the assumption that the hydraulic conductivity of frozen soil is a function of temperature. For the convenient application in numerical models, Horiguchi and Miller [12] gave an empirical formula, $k_\mathrm{f} = C \times T^D$, where, k_f is the hydraulic conductivity of frozen soil (m/s), T is the temperature (°C), and C and D are constant fitting parameters. Nixon [13] proposed a similar temperature-dependent empirical formula as follows, $k_\mathrm{f} = k_0/(-T)^\delta$, where, k_0 is the hydraulic conductivity at -1 °C and δ is the slope of the

$k_f - T$ relationship in a log-log coordinate; however, these models must be determined using some type of experimental data, thus limiting their applicability. The second type of empirical model is determined by the assumption that the ice in saturated frozen soil and the air in unsaturated unfrozen soil plays a similar role in the process of water seepage when there is the same water content; that is the hydraulic conductivity of saturated frozen soil is closed to that of unsaturated melted soil. A drawback of this model is that it always overrates the hydraulic conductivity of frozen soil. The third model type is determined by the fact that the second model overestimates the hydraulic conductivity of frozen soil, resulting in an ice impedance being introduced. Jame and Norum [14] introduced an impedance factor, $k_f = k_u/10^{-E\theta_i}$, where, k_u is the hydraulic conductivity of unfrozen soil with the same liquid water content (m/s), E is an empirical constant and, θ_i is the volumetric content of ice (m^3/m^3). Based on agreement with experimental data from Jame and Norum [14] and Burt and Williams [7], Taylor and Luthin [15] gave a specific impedance factor, $k_f = k_u/10^{10\theta_i}$. Mao et al. [16] presented another impedance factor form to express hydraulic conductivity, $k_f = k_u \times (1 - \theta_i)^3$; however, many researchers have criticized this model because of the arbitrary choice of impedance factor. Shang et al. [17] suggested that the impedance factor would be better determined using $I = k_f/k_u$; however, this method is still limited by a paucity of the measured data and, moreover, the impedance factor is not a constant.

Meanwhile, several statistical models have also been proposed, of which there are two main types. The first type of statistical model is determined by the capillary bundle model and the Hagen-Poiseuille equation. Watanabe and Flury [18] developed a new capillary bundle model, in which the ice is assumed to form in the centre of the capillaries, leaving a circular annulus open for liquid water flow. On this premise, a newly developed hydraulic conductivity model was suggested, as seen in Equation (2):

$$k_f = \frac{\gamma_w \pi}{8\mu\tau} \sum_{J=k+1}^{M} n_J \left[R_J^4 - r_{i_J}^4 + \frac{\left(R_J^2 - r_{i_J}^2\right)^2}{\ln\left(r_{i_J}/R_J\right)} \right] \quad R_J - d(T) \geq r_i(T)$$

$$k_f = \frac{\gamma_w \pi}{8\mu\tau} \sum_{J=k+1}^{M} n_J R_J^4 \quad R_J - d(T) < r_i(T)$$

$$(2)$$

where, n_J is the number of capillaries of radius R_J per unit area; r_{i_J} is the radius of cylindrical ice (m); $d(T)$ is the thickness of the water film (m); μ is the dynamic viscosity of the fluid (kg/m·s); τ is tortuosity; M is the number of different capillary size classes and $k = 0, 1, 2, \cdots$, is an index for each decrease of water content. In this model, the soil water characteristic curve is used to determine the hydraulic conductivity of frozen soil, but the soil water characteristic curve is the constitutive equation in the unfrozen soil. Similarly, Weigert and Schmidt [19] proposed another model for predicting unsaturated hydraulic conductivity of partly frozen soil. The model considered that the effective saturated hydraulic conductivity of partly frozen soil equals the difference between that of saturated melted soil and unsaturated melted soil. Thus, the hydraulic conductivity of saturated melted soil equals the sum of the hydraulic conductivity of each class of capillaries, which is proportional to $r_x^2 \Delta\theta_x$; therefore, the model was determined, as shown in Equation (3):

$$k_f = k_s - k_u$$

$$k_u = \sum_{J=k+1}^{M} k_s \frac{r_x^2 \Delta\theta_x}{r_1^2 \Delta\theta_1 + r_2^2 \Delta\theta_2 + r_3^2 \Delta\theta_3 + \cdots + r_J^2 \Delta\theta_J}$$

$$(3)$$

where, k_u and k_s represent unsaturated and saturated hydraulic conductivity, respectively (m/s); r_x is the mean radius (m); and $\Delta\theta_x$ is the share of the pore class x of the total volumetric water content (cm^3/cm^3). The second type of statistical model is determined by the hydraulic conductivity model for melted soil and the Clausius-Clapeyron equation for describing the phase change process between ice and water. The hydraulic conductivity of frozen soil is converted by the hydraulic conductivity of unsaturated melted soil and the Clausius-Clapeyron equation. Azmatch [20] and Tarnawski [21] present their models to predict the hydraulic conductivity of frozen soil based on different hydraulic conductivity models of melted soils, although this method tends to overestimate hydraulic conductivity [3].

Given the problems with the above methods, there is great need for a new approach to estimating hydraulic conductivity. Since fractal theory (see Section 3) was first created, this method has been applied to different areas to determine the permeability of porous material; for example, Yu [22,23] proposed a fractal permeability model for fabrics and bi-dispersed porous media, predicting the corresponding permeability well. Yao [24] investigated the fractal characteristics of coals using the mercury porosimetry method, and analysed the influence of the fractal dimension of pores on the permeability of coals. Pia [25] presented an intermingled fractal units model based on fractal theory for the purpose of predicting the permeability of porous rocks. Xu [26] used fractal geometry to describe soil pores and obtained the fractal dimension of pores from porosimetric measurements. A permeability function was eventually put forward to predict the permeability of soils. The fractal theory used by this study to estimate the permeability performed well for the chosen materials. The pore size distribution of frozen soil was confirmed as satisfying the fractal characteristics [27–29] required; therefore, we attempted to apply the fractal theory to determine its hydraulic conductivity.

In the present study, a fractal model for hydraulic conductivity of saturated frozen soil was presented. A method for determining the model parameters was also proposed. Model performance was subsequently evaluated by comparing predictions with experimental data from the existing literature and other models. Furthermore, the validity of Darcy's law in saturated frozen soil was explained.

3. Fractal Model for Hydraulic Conductivity of Frozen Soil

3.1. Fractal Theroy

In classical Euclidean geometry, the dimensions of ordered objects are the integers; for example, the dimensions of points, straight lines, planes, and volumes are 0, 1, 2 and 3, respectively. However, for disordered objects such as irregular lines, planes or volumes, Euclidean geometry is not applicable. To solve this problem, fractal geometry has been introduced. In fractal geometry, the dimension of fractal objects is not an integer, but a fraction. Normally, $1 < D < 2$ in two-dimensional space and $2 < D < 3$ in three-dimensional space. Fractal theory emerges based on fractal geometry. Fractal theory describes and studies the objects from the perspective of fractal dimensions and its mathematical methods. The fractal theory broke out of the traditional barriers of one-dimensional lines, two-dimensional surfaces, three-dimensional spaces, bringing it closer to a description of the real properties and states of complex systems, more consistent with the diversity and complexity of objects.

3.2. The Fractal Capillary Bundle Model of Frozen Soil

The real size or geometric shape of pores are difficult to describe, meaning pores are sometimes visualized as circular capillary tubes [30]. A capillary bundle model has consequently been proposed and frequently applied [31–33]. Although the model is different from real soils, it contains many of the same properties and so can be used to analyse the water flow in frozen soil [18]. As illustrated in Figure 1, the pores are conceptualised as an assembly of tortuous capillary tubes.

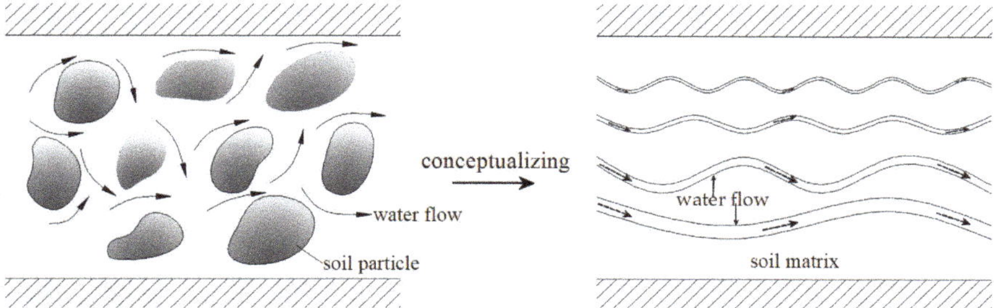

Figure 1. Schematic of capillary bundle model of frozen soil.

To develop the fractal capillary bundle model, we assumed the frozen soil was homogeneous and isotropic; the solid matrix was incompressible; the pores could be treated as capillary tubes with different radii; the capillaries were initially filled with liquid water and the water flow in the different capillary tubes remain independent; The freezing of pore water started from macrospores and the freezing temperature of pore water reduced with the decrease of pore diameter. Figure 2 gives a schematic of the freezing process in capillary tubes, where the fully saturated tubes are filled with liquid water in the unfrozen state. As the temperature decreases, the liquid water in larger tubes are frozen into ice and a thin liquid film remains. The radii of ice filled tubes and the thickness of the liquid film both depend on temperature.

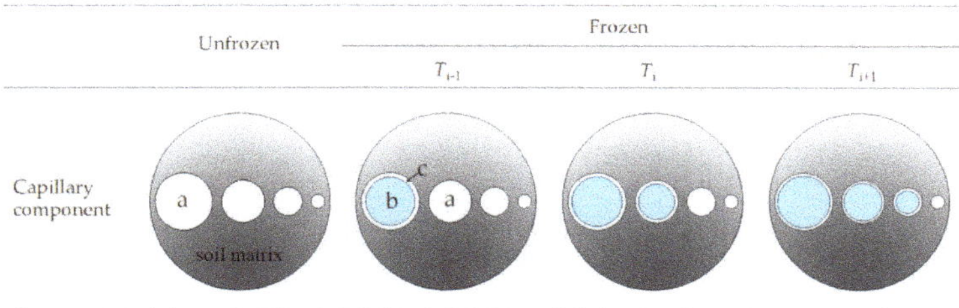

ᵃWater filled capillary tubes; ᵇIce filled capillary tubes; ᶜFilm of water

Figure 2. Schematic of freezing process of capillary tubes.

3.3. Derivation of the Fractal Model to Determine the Hydraulic Conductivity of Frozen Soil

The cumulative pore size distribution follows the fractal scaling law [34]; therefore the cumulative number of pores satisfied:

$$N(\geq r) = \left(\frac{r_{\max}}{r}\right)^D \tag{4}$$

where N is the cumulative number of pores with a radius greater than or equal to r; $r_{\max}$ is the maximum radius (m) and, D is the pore size dimension, $1 < D < 2$. Differentiating Equation (4) with respect to r yields:

$$-dN(\geq r) = Dr_{\max}{}^D r^{-(D+1)}dr \tag{5}$$

Equation (5) gives the pore number between the pore size r and $r + dr$. In Equation (5), $-dN > 0$, which indicates that pore number decreases with an increase in pore size.

The soil pore system is assumed to be a bundle of capillary tubes with different radii; therefore, the Hagen-Poiseuille equation can be used to describe the water flow in the soil, as given by [22,35]:

$$q = r^4 \frac{\pi}{8} \frac{\rho_w g}{\eta} \frac{\Delta H}{L_e} \tag{6}$$

where q is the water flow rate in a single capillary (m^3/s); r and L_e are the radius and length of the capillary, respectively (m); η is the dynamic viscosity of the fluid (Pa · s); ΔH is the hydraulic gradient; ρ_w is the density of water (kg/m^3); and g is gravitational acceleration (m/s^2). The total flow rate Q (m^3/s) can be obtained by integrating the single flow rate, q, from the minimum pore radius to the maximum radius with the aid of Equation (6):

$$Q = -\int_{r_{\min}}^{r_{\max}} q\,dN = \frac{\pi\rho_w g}{8\eta}\frac{\Delta H}{L_e}\frac{D}{4-D}r_{\max}{}^4\left[1 - \left(\frac{r_{\min}}{r_{\max}}\right)^{4-D}\right] \tag{7}$$

Since $1 < D < 2$, the exponent $4 - D > 2$, $r_{\min}/r_{\max} < 10^{-2}$; therefore, $(r_{\min}/r_{\max})^{4-D} \ll 1$. Then Equation (7) is reduced to:

$$Q = \frac{\pi\rho_w g}{8\eta}\frac{\Delta H}{L_e}\frac{D}{4-D}r_{\max}{}^4 \tag{8}$$

The average velocity J_a (m/s) through the capillaries can be obtained by dividing Q by the total pore area A_p (m^2):

$$J_a = \frac{\pi\rho_w g}{8\eta A_p}\frac{\Delta H}{L_e}\frac{D}{4-D}r_{\max}{}^4 \tag{9}$$

The total pore area in Equation (9) can be expressed as [36]:

$$A_p = \int_{r_{\min}}^{r_{\max}} \pi r^2(-dN) = \frac{\pi D(1-\varepsilon)r_{\max}{}^2}{2-D} \tag{10}$$

where ε is the porosity. Inserting Equation (10) into Equation (9) yields:

$$J_a = \frac{\rho_w g}{8\eta}\frac{\Delta H}{L_e}\frac{2-D}{4-D}\frac{r_{\max}{}^2}{1-\varepsilon} \tag{11}$$

The Darcy flux J_T can be determined by using the Dupuit-Forchheimer equation:

$$J_T = \varepsilon J_a = \frac{\rho_w g}{8\eta}\frac{\Delta H}{L_e}\frac{\varepsilon}{1-\varepsilon}\frac{2-D}{4-D}r_{\max}{}^2 \tag{12}$$

where τ is a tortuosity, which is the ratio between the length of soil column L (m) and the length of the capillary L_e (m). Comparing Equation (12) with Darcy's law, $J_T = k\frac{\Delta H}{L}$, results in the expression for the hydraulic conductivity as follows:

$$k = \frac{\rho_w g}{8\eta\tau}\frac{\varepsilon}{1-\varepsilon}\frac{2-D}{4-D}r_{\max}{}^2 \tag{13}$$

where k is the hydraulic conductivity. For a straight capillary bundle fractal model ($\tau = 1$), the hydraulic conductivity is reduced to:

$$k = \frac{\rho_w g}{8\eta}\frac{\varepsilon}{1-\varepsilon}\frac{2-D}{4-D}r_{\max}{}^2 \tag{14}$$

It can be found from Equation (13) that for water flow through a given soil sample, the hydraulic conductivity is related to the maximum pore size, pore size dimension, and tortuosity. Of these parameters, maximum pore size plays a key role in affecting hydraulic conductivity. These parameters reflect the pore structure characteristic of the soil, and each of them has a clear physical meaning, which reveals a significant advantage of the model.

In the model, ice is assumed to form first in the largest water-filled tubes upon freezing based on the Gibbs-Thomson effect. The Gibbs-Thomson effect describes how a depression in the freezing point is inversely proportional to pore size. Due to the free energy of the capillary wall, some water in the vicinity of the capillary wall remains unfrozen at subzero temperatures; hence why the frozen capillary is considered to be composed of an ice column and a thin liquid film in the vicinity of the tube wall. As the solid phase is assumed to be incompressible, the radius of the capillary cannot be affected by the frost heave of the water in the capillary. Therefore, the radius of the capillary is measured as the sum of the radius of the ice column and the thickness of the thin liquid film. The radius of the ice column can be assessed by the Gibbs-Thomson equation [37]:

$$r_i{}^i = \frac{2\sigma_{sl}}{L_f \cdot \rho_i} \frac{T_m}{T_m - T^i} \tag{15}$$

where, $r_i{}^i$ (m) is the critical radius of ice column at T^i (°C); σ_{sl} is the ice-liquid water interfacial free energy, $\gamma = 0.0818$ J/m^2; L_f is latent heat of fusion, $L = 334.56$ kJ/kg; ρ_i is the density of ice, $\rho_i = 917$ kg/m^3; and T_m is the freezing temperature of bulk water given in Kelvin, $T_m = 273.15$ K. The thickness of the liquid water film can be assessed by [18]:

$$d^i = \left[-\frac{A_H}{6\pi\rho_i L} \frac{T_m}{T_m - T^i} \right]^{1/3} \tag{16}$$

where d^i is the film thickness of unfrozen water (m) and, A_H is the Hamaker constant, $A_H = -10^{-19.5}$ J. The soil freezing curve (SFC) describes how the volumetric content of unfrozen water decreases with a decrease in the sub-zero temperature. The number of capillaries of radius r^i can be determined using the SFC. As shown in Figure 3, the SFC is divided into several different spaced water-content intervals of width $\Delta\theta_u{}^i$, according to the experimental points, and the temperature $T(\theta_u{}^i)$ associated with the decrease in $\theta_u{}^i$ can be determined. All capillaries of radius $r \geq r^i$ are assumed to be frozen when $T = T^i$. Thus, the number of capillaries N^i frozen at T^i, is equivalent to the increase in ice content in real soil, $\Delta V_i{}^i / \pi r_i{}^i L_e$, where $\Delta V_i{}^i$ is the increment of ice volume at T^i. $\Delta V_i{}^i$ can be calculated from the SFC, then the cumulative number of capillaries $N(r \geq r^i)$ is given by summing the number of capillaries N^i of radius r^i when $r \geq r^i$.

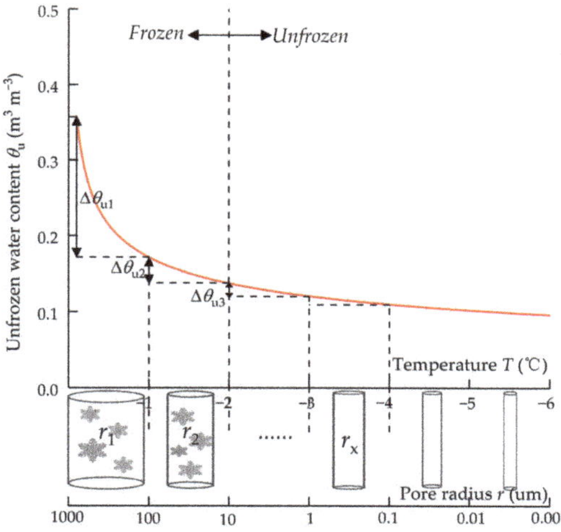

Figure 3. Scheme of a physical approach to determine the pore size distribution of frozen soil.

The pore size dimension was determined by the accumulative pore size distribution. The cumulative pore size distribution of fractal capillaries follows the fractal power law; therefore, taking the logarithm of both sides of Equation (4) gives:

$$D = \frac{\lg N}{\lg(r_{\max}/r)} \tag{17}$$

As seen in Equation (17), the pore size dimension was the slope of the accumulative pore size distribution in the double logarithmic coordinates, which was approximately a straight line. For frozen soil, the accumulative pore size distribution changed with temperature, leading to pore size dimension changes with temperature. The pore size dimension of frozen soil at different subzero temperatures needed to be determined using the accumulative pore size distribution at the corresponding temperature, the calculation of which had already been described above. In Equation (13), the dynamic viscosity of the water can be given as [18]

$$\eta = \eta_0 \exp\left(\frac{c}{T}\right) \tag{18}$$

where $\eta_0 = 9.62 \times 10^{-7}$ Pa · s and $c = 2046$ K. Equation (18) yields $\mu = 10^{-3}$ Pa · s when $T = 21.5\,^\circ$C. The tortuosity can be assessed using the following relationship [34,35,38]

$$\tau = 1 + 0.41 \ln(1/\varepsilon) \tag{19}$$

In this model, the porosity is the ratio between the total unfrozen volume of capillaries and the volume of the soil column. As the unfrozen capillaries are assumed to be completely filled with unfrozen water, so the volume of unfrozen water can be conceived as the volume of the capillaries; thus, the porosity is equivalent to the volumetric content of unfrozen water. To illustrate the fractal hydraulic conductivity modeling process well, a flow diagram is given in Figure 4.

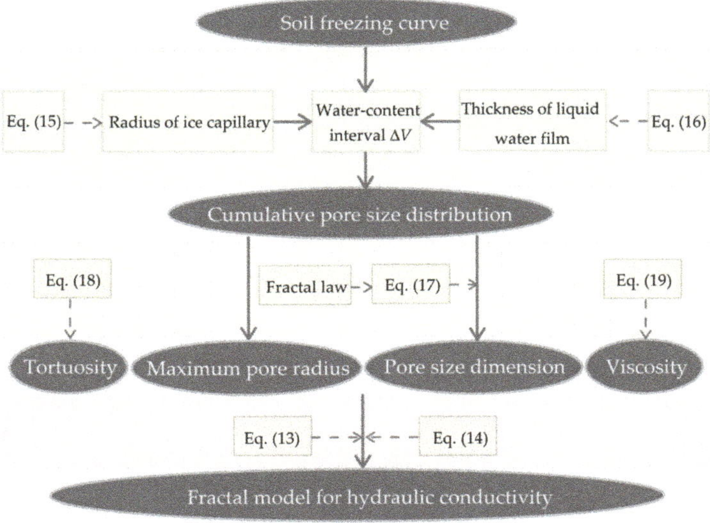

Figure 4. Flow diagram of fractal model building.

4. Methods

The present model was applied to existing literature on the measured hydraulic conductivity of saturated frozen soil to test its validity. Some existing studies did measure saturated conductivity; however, only a few simultaneously report the SFC, initial volumetric content and the dimension of the soil column, which are required by the model. Thus, model testing was limited by the paucity of the literature.

Burt and Williams [7] and Tokoro et al. [39] measured saturated hydraulic conductivity using self-developed experimental apparatus; their apparatus allows water to flow as a fluid. A summary of the physical parameters of the soil samples is given in Table 1.

Table 1. Physical parameters of the soil samples.

Soil Type	Length (cm)	Diameter (cm)	Dry Density (g/m^3)	Initial Water Content (m^3 m^{-3})
Silt [7]	3	3.8	1.52	0.50
clayey silt [7]	3	3.8	3.8	0.33
Silt [39]	3	7	7	0.38

Figure 5 presents the SFCs of the three soil samples. As shown, the freezing curve decreases dramatically near 0 °C, after which the freezing curve became gradual. This result was because the proportion of capillaries with small radii increased as temperature decreased.

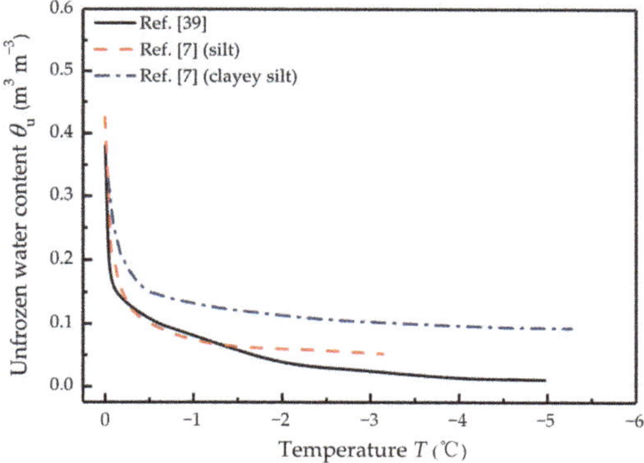

Figure 5. Soil freezing curve.

5. Results

5.1. Cumulative Pore Size Distribution Curve

Here, the experimental data from Tokoro et al. [39] were used as an example to describe the calculation process for finding the hydraulic conductivity of frozen soil. Figure 6 presents the cumulative pore size distribution estimated by the SFC, with the solid line representing the cumulative volume of pores and the dashed line representing the cumulative number of pores. The number of pores with small radii was markedly higher than the number with large radii. As the temperature decreased, the maximum radius decreased which led to changes in the pore size distribution curve. Additionally, the cumulative pore size distribution (the dashed line plotted in the double logarithmic coordinates) is approximately a straight line, the slope of which was the pore size dimension.

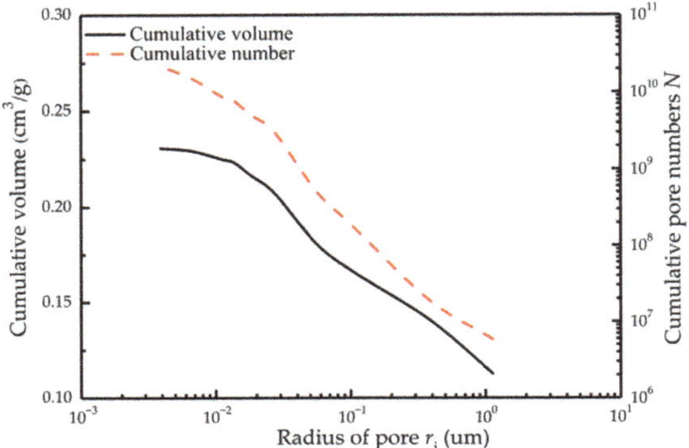

Figure 6. Pore size distribution determined by the soil freezing curve.

5.2. Maximum Pore Size, the Pore Size Dimension and Tortuosity

Figure 7a shows the relationship between the maximum pore radius and temperature, which was calculated by Equations (15) and (16). The maximum pore radius decreased with temperature. Figure 7b gives the relationship between the tortuosity and temperature, obtained using Equation (19). The tortuosity increased with temperature. Figure 7c presents the relationship between the pore size dimension and temperature, which was computed from the cumulative pore size distribution using Equation (17). The pore size dimension decreased linearly with temperature. As shown in the figure, the three parameters all varied with temperature, which caused the hydraulic conductivity to be affected by the temperature according to Equation (13).

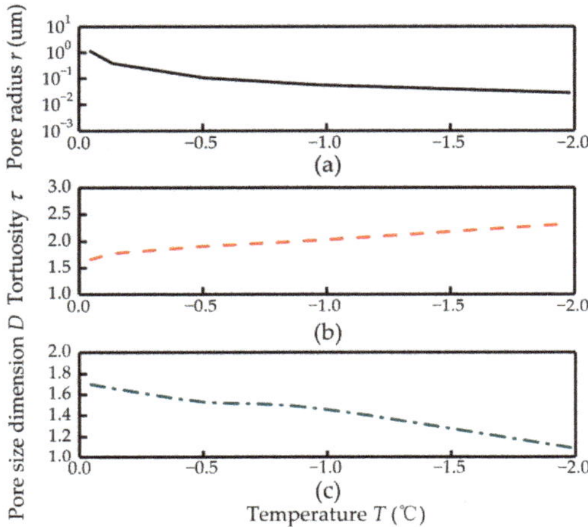

Figure 7. Variation of pore size dimension (**a**), maximum pore radius (**b**) and tortuosity (**c**)'with temperature.

5.3. Fractal Values of Hydraulic Conductivity

Figure 8a presents the hydraulic conductivity predicted by the present model for different pore size dimensions from Equations (13) and (14). Figure 8b shows the hydraulic conductivity predicted by the present model for different maximum pore radii from Equations (13) and (14). The hydraulic conductivity changed dramatically with the decrease in pore size dimension and maximum pore radius. This phenomenon can be explained as: the decrease of pore size and pore size dimension causing a decrease in the number of capillaries, as well as the increase in the tortuosity causing an increase in water flow paths, leading to a decrease in flow rate. Of all these influencing factors, pore size plays a role in affecting hydraulic conductivity.

In the same way, the hydraulic conductivities of the three soil samples were predicted by the present model for different temperature and volumetric water content from Equation (13). The results are shown in Figure 8c,d. From Figure 8c, the hydraulic conductivity changed dramatically within $-1\ ^\circ$C, with similar results observed by Burt et al. [7] and Miller et al. [8]. These results can be explained by the fact that the decrease in temperature caused a decrease of the pore size and pore size dimension and increase of the tortuosity as shown in Figure 7, leading to a decrease in the hydraulic conductivity. Figure 8d shows the relationship between the hydraulic conductivity and volumetric water content, calculated from Equation (13). Figure 8d shows that the hydraulic conductivity increases with an increased in unfrozen water content.

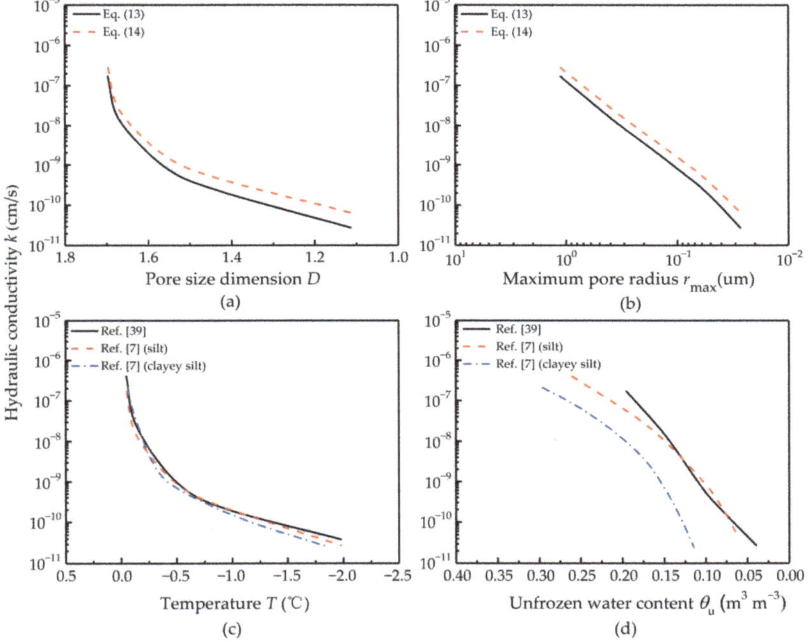

Figure 8. The effect of pore size dimension (**a**), maximum pore radius (**b**), temperature (**c**) and unfrozen water content (**d**) on hydraulic conductivity.

6. Discussion

6.1. Model Application

Figure 9 presents the comparison between the fractal model and the experimental data, with the symbols representing the experimental data, and the dashed line and dashed-dot line representing the predictions by the present models calculated from Equations (13) and (14), respectively. In both cases, the present model was in close agreement with the experimental data. From Figure 9a, using Equation (13), the present model shows good agreement with the experimental data except for the data point at 0.25 $^\circ$C; however, using Equation (14), the present model overrated the experimental data. Figure 9b shows that using Equation (14), the present model showed good agreement with the experimental data while using Equation (13), the present model underestimate the experimental data. Figure 9c indicates the present model showed a slight difference from the experimental data; however, the behavior of the experimental data is approximately captured by the present model. Overall, the present model based on fractal theory showed close agreement with the experimental data. Thus, the validity of the present model was verified.

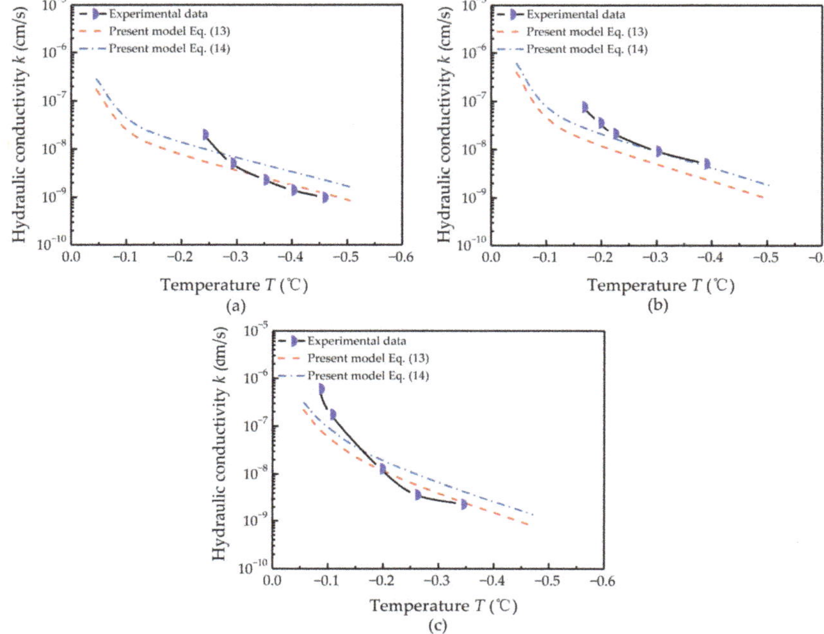

Figure 9. Comparison of the model results and experimental data. (**a**) Tokoro et al. [39] (silt); (**b**) Burt and Williams [7] (silt); (**c**) Burt and Williams [7] (clayey silt).

The differences between the predictions and experimental data can be translated into quantifiable terms by means of the mean squared error, MSE, which is an indicator of the overall magnitude of the residuals. The MSE is computed as follows

$$\text{MSE} = \frac{1}{N}\sum_{i=1}^{N}\left[\ln(k_{e,i}) - \ln(k_{p,i})\right]^2 \tag{20}$$

where N is the number of experimental data points; $k_{e,i}$ is the ith measured hydraulic conductivity (m/s) and $k_{p,i}$ is the ith predicted hydraulic conductivity (m/s). Table 2 summarises the MSE for the three data sets. As shown, the present model–using Equations (13) and (14)—made the best prediction for the Tokoro's [39] silt data set and Burt and Williams' [7] silt data set, respectively.

Table 2. Model prediction statistics for various data sets.

Reference	Present Model	Mean Squared Error	Squared Error				
			$-0.25\,°\text{C}$	$-0.3\,°\text{C}$	$-0.35\,°\text{C}$	$-0.4\,°\text{C}$	$-0.45\,°\text{C}$
Tokoro et al. [39]	Equation (13)	0.34	1.62	0.08	0	0.001	0.01
	Equation (14)	0.41	0.55	0.08	0.33	0.54	0.55
Burt and Williams (silt) [7]	Equation (13)	0.91	2.23	1.02	0.55	0.30	0.47
	Equation (14)	0.21	0.84	0.21	0.02	0	0
Burt and Williams (clayey silt) [7]	Equation (13)	0.98	3.50	1.19	0	0.22	0
	Equation (14)	0.78	1.98	0.41	0.16	0.97	0.40

As an additional test, the present fractal model was further compared with past models. The following three models were considered: the Nixon model [13], Mao et al. model [16] and Jame and Norum model [14]. Here, the experimental data from the study by Tokoro [39] were used as an example. In the Nixon model, k_0 and δ were obtained using the $k_f - T$ relationship in a log-log coordinate. The value of k_0 and δ were 2.0×10^{-10} cm/s and -2.273, respectively. In the Mao et al. model and the Jame and Norum model, the value of k_u was $8.1 \times 10^{-3} \times \theta_u^{6.21}$ cm/s and θ_i values were calculated from the SFC, $\theta_i = \theta_w \rho_w / \rho_i$. In the Jame and Norum model, the impedance factor was determined by the empirical constant E. To avoid an arbitrary choice of impedance factor, the empirical constant E was determined using the method proposed by Shang et al. [17]. The experimental data at $-0.25\,°C$, $-0.3\,°C$ and $-0.35\,°C$ were used to determine the empirical constants, respectively. The corresponding empirical constants E were -0.48, -2.82 and -3.66, respectively.

Figure 10 presents a visual comparison of various models for the silt data set from Tokoro [39]. The inferences on model performance can be drawn from model predictions for the experimental data. As shown, the Nixon model predicted the experimental data reasonably. This is ascribed to the fact that the model was determined from experimental data. Generally, the Mao et al. model overrated the experimental data. For the Jame and Norum model, when $E = -0.48$; the model deviated greatly from the experimental data; however, a significant improvement was observed when $E = -2.82$ and $E = -3.66$. It was found that the Jame and Norum model changed dramatically with the impedance factor. This shortcoming leads to a difficulty in accurately predicting the experimental data. Using Equation (13) in the present model showed much closer agreement with the experimental data. Moreover, the present model using Equation (13) is close to the Nixon model. Using Equation (13) in the present model with Equation (13) overrated the experimental data overall; this may be attributed to the fact that the water flow path in the soil was tortuous.

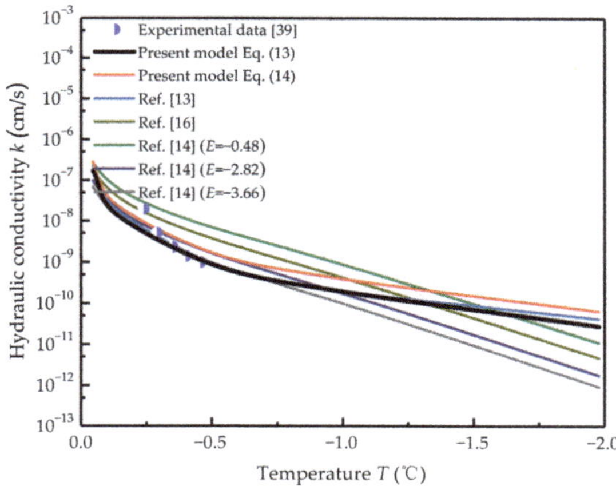

Figure 10. Comparison of the experimental data with different models for Tokoro [39] silt.

Table 3 summarises the MSE for the various models used to evaluate the model performance. As shown, the MSE value for the Nixon model is lowest, at 0.33, followed by the present model with Equation (13) where the MSE was 0.34. In contrast, the present model with Equation (13) is close to the Nixon model, which was determined by the experimental data. Using Equation (13) in the present model, the squared error at $-0.25\,°C$ was much higher than that at the other temperatures. This difference can be attributed to experimental error and model error. The latter error may be related to the cracks caused by frost heave. Cracks are preferential flow paths, which result in an increase in conductivities. This influence of frost heave on the pore structure was not considered in the fractal model; however, on the whole, the difference was small. The MSE value for the Mao et al. model and Jame and Norum model ($E = -0.48$) was approximately 4 and 9 times larger, respectively, than for the present model with Equation (13). The MSE for the Jame and Norum model ($E = -2.82$ and $E = -3.66$) was close to that of the present model using Equation (14).

Table 3. Comparison results between various models.

Model Type	Mean Squared Error	Squared Error				
		$-0.25\,°C$	$-0.3\,°C$	$-0.35\,°C$	$-0.4\,°C$	$-0.45\,°C$
Present model Equation (13)	0.34	1.62	0.08	0.001	0.009	0.01
Present model Equation (14)	0.41	0.55	0.08	0.33	0.54	0.55
Nixon model [13]	0.33	1.45	0.05	0.01	0.06	0.06
Mao et al. model [16]	1.37	0.03	0.80	1.55	2.14	2.32
Jame and Norum model [14] $E = -0.48$	2.96	0.14	2.19	3.47	4.36	4.66
Jame and Norum model [14] $E = -2.82$	0.40	0.84	0.02	0.22	0.46	0.48
Jame and Norum model [14] $E = -3.66$	0.39	1.75	0.12	0.0007	0.03	0.04

In contrast, the fractal model provided a good agreement with the hydraulic conductivity data overall. Moreover, comparison with other models revealed the following advantages of the fractal model: fewer parameters; parameters are easy to obtain; each of the parameters has a clear physical meaning; and no measured conductivity is required in the model. Additionally, the fractal model can explain the reason for hydraulic conductivity changing with temperature; however, the influence of frost heave on the pore structure of frozen soil is not considered in the model, which may result in underestimating of the experimental data.

The present model has some limitations. Firstly, the model may be not valid in sandy soils, as these contain massive large pores that will freeze once temperature is lower than zero degrees Celsius, according to the SFC and Gibbs-Thomson equation. Second, the model does not consider the influence of the difference in SFC determined by the different methods on the hydraulic conductivity of saturated frozen soil. We expect this effect would not be pronounced when estimating the hydraulic conductivity, as hydraulic conductivity largely depends on pore radius. Third, the model does not consider the influence of frost heave on the pore structure of frozen soil, which may result in underestimating the experimental data. We expect this effect to be less pronounced in high-temperature frozen soil, where only slight frost heave occurs. Fourth, the model does not consider the influence of closed pores on hydraulic conductivity, which will result in overrating the hydraulic conductivity. Lastly, the model is only valid for soils fulfilling the fractal laws.

6.2. Comparison of the Cumulative Pore Size Distribution Determined by the SFC and Mercury Intrusion Porosimetry (MIP)

The SFC was used to determine pore size distribution in the present study. To verify this method, MIP was compared with this method. You et al. [40] and Xiao [41] measured the pore size distribution of silty clay using MIP. Figure 11 represents the SFC of the two soil samples. The solid line represents the freezing curve of You et al. [40], and the dashed line represent that of Xiao [41].

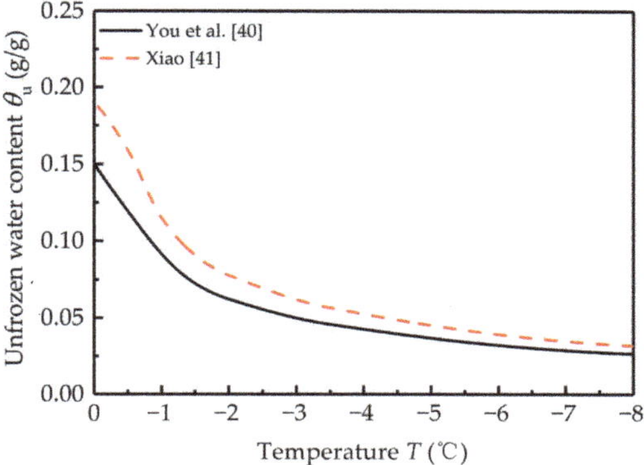

Figure 11. Soil freezing curves of the soils.

Figure 12 shows the comparison of pore size distribution determined by the SFC and MIP. The solid lines represent the pore size distribution determined by MIP and the dashed lines represent the pore size distributions determined by the soil freezing curve. It can be seen that the pore size distribution determined by the SFC captured the distinctive feature of the pore size distribution determined by MIP well; however, this method overrated the smaller pores measured by MIP. This result could be ascribed to the fact that the isolated pores—those that had no communication with the exterior of the sample could not be measured by MIP in any event, regardless of the pressure used [42], However, the isolated pores could be filled with water, which is considered in the pore size distribution determined by the SFC. Fagerlund [37] used a similar method to accurately predict the pore size distribution in the range 0.02 μm~0.5 μm for a certain sand-lime brick as shown in Figure 12c.

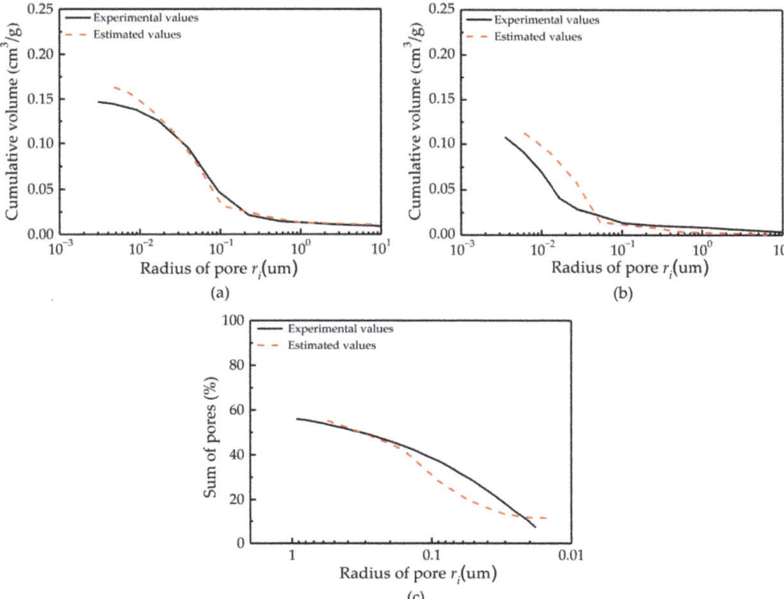

Figure 12. Comparison of the pore size distribution determined by SFC and by MIP. (**a**) Xiao [41]; (**b**) You et al. [40]; (**c**) Fagerlund [42].

6.3. Vvalidity of Darcy's Law

The present model was determined by Darcy's law. Although Darcy's law is widely applied to frozen soil, the validity of Darcy's law in frozen soil remains controversial. Several previous studies have shown that Darcy's law is applicable in frozen soil, at least in high-temperature frozen soil; for example, Burt and William [7] found a linear relationship between the hydraulic gradient and discharge for clayey silt within −0.5 °C using a laboratory experiment. Horiguchi and Miller [8] found the flux linearly changed with hydraulic conductivity for silt at different temperature within −0.15 °C. Tokoro [39] observed the same experimental phenomenon for silt within −0.5 °C.

To further account for the validity of Darcy's law, a theoretical proof based on the capillary bundle model and Reynolds number was proposed. The Reynolds number R_e is always used to justify the flow state of the fluid in the capillaries

$$R_e = \frac{2r_{cr}\rho v}{\mu} \tag{21}$$

where r_{cr} is the critical pore radius (m); ρ is the density of the fluid (kg/m^3); and v is the flow velocity (m/s). The flow velocity can be obtained by dividing water flow rate by the cross area of a single capillary

$$v = q/\pi r^2 = \frac{r^2}{8}\frac{\rho_w g}{\eta}\frac{\Delta H}{L_e} \tag{22}$$

Substituting Equation (22) into Equation (21), the critical pore radius r_{max} for laminar flow was obtained as

$$r_{max} = \sqrt[3]{\frac{4R_e\mu^2\tau}{\rho_w{}^2 gi}} \tag{23}$$

The unfrozen water film is much smaller than the radius of capillaries; therefore, the geometry of ice in frozen soil is assumed to be identical to that of air in drying soil. Thus, the relationship between the matric potential of frozen soil and temperature can be estimated using the generalised form of the Clausius-Clapeyron equation, treating ice pressure the same as gauge pressure

$$h = \frac{L_f}{g}\ln\frac{T_m - \Delta T}{T_m} \tag{24}$$

where $|h|$ is the hydraulic head (m) and ΔT is the freezing temperature depression (°C). Figure 13 presents the $|h|-\Delta T$ relationship obtained using Equation (24).

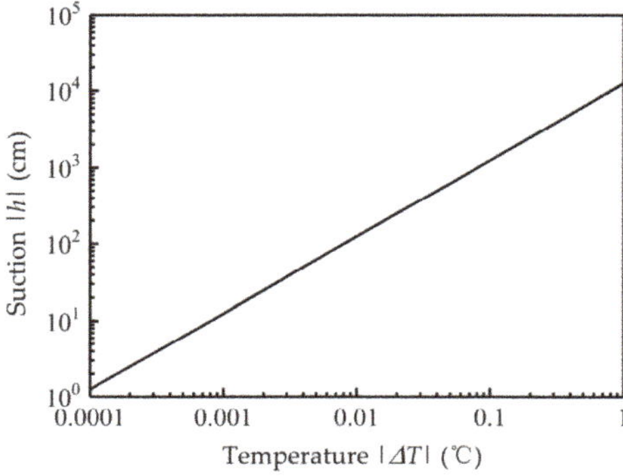

Figure 13. Relationship between the hydraulic head and temperature for frozen soil estimated using the Clausius-Clapeyron equation.

The hydraulic gradient i is

$$i = \frac{|h|}{L} \tag{25}$$

where, L is the length of soil column (m). Inserting Equation (19) into Equation (20) gives the expression for the hydraulic gradient

$$i = \frac{L_f}{Lg} \ln \frac{T_m - \Delta T}{T_m} \tag{26}$$

Substituting Equation (26) into Equation (23) gives the expression for the critical pore radius

$$r_{max} = \sqrt[3]{\frac{4R_e \mu^2 \tau L}{\rho_w{}^2} \frac{1}{L_f \ln \frac{T_m - \Delta T}{T_m}}} \tag{27}$$

With $R_e = 2300$, $g = 9.81$ m/s^2, $\mu = 1.792 \times 10^{-3}$ Pa $\cdot$ s, $\rho_w = 998$ kg/m^3, $L = 334.56$ KJ/kg, $T_m = 273.15$ K, $\Delta T = -1\,^\circ$C $L = 3$ cm and, $\tau = 2$, the critical pore radius is obtained:

$$r_{cr} = 2.95 \times 10^{-4} \sqrt[3]{L_e} \tag{28}$$

Taking $L = 3$ cm and $\tau = 2$ for example, the critical radius obtained is $r_{cr} = 116$ μm. That is, for a soil column with a length of 3 cm and tortuosity of 2, the water flow will be turbulent in pores larger than 116 μm at temperature differences of $-1\,^\circ$C.

In the further test, the critical pore radius was compared with the pore size of soils determined by MIP from You [40], Xiao [41], Tao [43], Penumadu [44], and Juang [45]. As seen in Figure 14, the silt, silty clay, clay and Kaolin clay soils all had barely any pores with a radius larger than 116 μm. For sandy soils except for sand C (70% sand + 30% clay), they all had several pores with a radius larger than 116 μm. Thus, the validity of Darcy's law is reduced in sandy soils with larger pores. The critical radius increased with the increased length of soil sample and tortuosity, but decreased with the temperature difference increase. This method only presents the validity of Darcy's law for saturated frozen soil; for unsaturated frozen soil, the application of Darcy's law is reduced owing to the existence of the matrix potential.

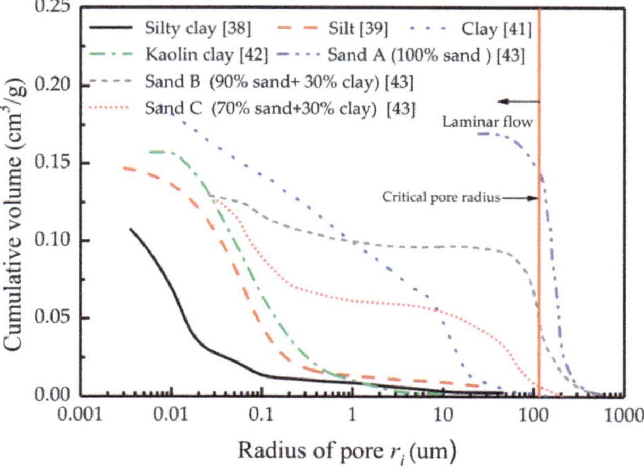

Figure 14. Pore size distribution of the soils.

7. Conclusions

In the present study, based on fractal theory and Hagen-Poiseuille's law, a fractal model of hydraulic conductivity for saturated frozen soil was developed. The model was verified using experimental results. The following conclusions can be drawn:

1. The model provides a good agreement with the hydraulic conductivity data. Compared with other models, this model is simpler to use owing to its simple requirements of the SFC, initial volumetric content of water and the dimensions of the soil column.
2. Pore size played a key role in affecting the hydraulic conductivity of frozen soil. The pore size dimension decreased linearly with temperature, the maximum pore size decreased with temperature and the tortuosity increased with temperature. Changes in these parameters with temperature can explain why the hydraulic conductivity of frozen soil changes with temperature.
3. The pore size distribution is approximately estimated by the SFC, making this method a possible alternative to MIP.
4. Darcy's law is valid in the saturated frozen silt, clayed silt and clay, but maybe not valid in saturated frozen sand and unsaturated frozen soil.

Author Contributions: Conceptualization, L.C. and F.M.; Formal analysis, L.C.; Methodology, L.C. and F.M.; Writing–original draft, L.C.; Writing–review & editing, D.L., F.M., X.S. and X.C.

Funding: This work was funded by the National Nature Science Foundation of China (No. 41701060 and No. 41271080), the funding of Key Research Program of Frontier Science of Chinese Academy of Sciences (QYZDY-SSW-DQC015), and the funding of the State Key Laboratory of Frozen Soil Engineering (No. SKLFSEZT17).

Acknowledgments: The manuscript was further improved by the insightful comments of anonymous reviewers from Water, which is greatly appreciate.

Conflicts of Interest: The authors declare no conflict of interest.

References

1. Kurylyk, B.L.; Watanabe, K. The mathematical representation of freezing and thawing processes in variably saturated, non-deformable soils. *Adv. Water Resour.* **2013**, *60*, 160–177. [CrossRef]
2. Walvoord, M.A.; Kurylyk, B.L. Hydrologic Impacts of Thawing Permafrost—A Review. *Vadose Zone J.* **2016**, *15*. [CrossRef]
3. Watanabe, K.; Osada, Y. Comparison of hydraulic conductivity in frozen saturated and unfrozen unsaturated soils. *Vadose Zone J.* **2016**, *15*. [CrossRef]
4. Ren, R.; Ma, J.; Cheng, Q.; Zheng, L.; Guo, X.; Sun, X. Modeling coupled water and heat transport in the root zone of winter wheat under non-isothermal conditions. *Water* **2017**, *9*, 290. [CrossRef]
5. Wettlaufer, J.S.; Worster, M.G. Premelting dynamics. *Annu. Rev. Fluid Mech.* **2006**, *38*, 427–452. [CrossRef]
6. Trusheim, F. Mechanism of salt migration in northern Germany. *AAPG Bull.* **1960**, *44*, 1519–1540. [CrossRef]
7. Burt, T.P.; Williams, P.J. Hydraulic conductivity in frozen soils. *Earth Surf. Process.* **1976**, *1*, 349–360. [CrossRef]
8. Horiguchi, K.; Miller, R.D. Experimental studies with frozen soil in an ice sandwich permeameter. *Cold Reg. Sci. Technol.* **1980**, *3*, 177–183. [CrossRef]
9. Wiggert, D.C.; Andersland, O.B.; Davies, S.H. Movement of liquid contaminants in partially saturated frozen granular soils. *Cold Reg. Sci. Technol.* **1997**, *25*, 111–117. [CrossRef]
10. McCauley, C.A.; White, D.M.; Lilly, M.R.; Nyman, D.M. A comparison of hydraulic conductivities, permeabilities and infiltration rates in frozen and unfrozen soils. *Cold Reg. Sci. Technol.* **2002**, *34*, 117–125. [CrossRef]
11. Seyfried, M.S.; Murdock, M.D. Use of air permeability to estimate infiltrability of frozen soil. *J. Hydrol.* **1997**, *202*, 95–107. [CrossRef]
12. Horigudhi, K.; Miller, R.D. Hydraulic conductivity functions of frozen materials. In Proceedings of the 4th International Conference on Permafrost; National Academy Press: Washington, DC, USA, 1983; pp. 504–508.
13. Nixon, J.F. Discrete ice lens theory for frost heave in soils. *Can. Geotech. J.* **1991**, *28*, 843–859. [CrossRef]

14. Jame, Y.W.; Norum, D.I. Heat and mass transfer in freezing unsaturated porous media. *Water Resour. Res.* **1980**, *16*, 811–819. [CrossRef]

15. Taylor, G.S.; Luthin, J.N. A model for coupled heat and moisture transfer during soil freezing. *Can. Geotech. J.* **1978**, *15*, 548–555. [CrossRef]

16. Mao, L.; Wang, C.Y.; Tabuchi, Y. A multiphase model for cold start of polymer electrolyte fuel cells. *J. Electrochem. Soc.* **2007**, *154*, 341–351. [CrossRef]

17. Shang, S.H.; Lei, Z.D.; Yang, S.X. Numerical simulation improvement of coupled moisture and heat transfer during soil freezing. *J. Tsinghua Univ.* **1997**, *37*, 62–64. [CrossRef]

18. Watanabe, K.; Flury, M. Capillary bundle model of hydraulic conductivity for frozen soil. *Water Resour. Res.* **2008**, *44*, 1–9. [CrossRef]

19. Weigert, A.; Schmidt, J. Water transport under winter conditions. *CATENA* **2005**, *64*, 193–208. [CrossRef]

20. Azmatch, T.F.; Sego, D.C.; Arenson, L.U.; Biggar, K.W. Using soil freezing characteristic curve to estimate the hydraulic conductivity function of partially frozen soils. *Cold Reg. Sci. Technol.* **2012**, *83–84*, 103–109. [CrossRef]

21. Tarnawski, V.R.; Wagner, B. On the prediction of hydraulic conductivity of frozen soils. *Can. Geotech. J.* **1996**, *33*, 176–180. [CrossRef]

22. Yu, B.M.; Cheng, P. A fractal model for permeability of bi-dispersed porous media. *Int. J. Heat Mass Transf.* **2002**, *45*, 2983–2993. [CrossRef]

23. Yu, B.M.; Lee, L.J.; Cao, H.Q. A fractal in-plane permeability model for fabrics. *Polym. Compos.* **2002**, *23*, 201–221. [CrossRef]

24. Yao, Y.; Liu, D.; Tang, D.; Tang, S.; Huang, W.; Liu, Z.; Che, Y. Fractal characterization of seepage-pores of coals from China: An investigation on permeability of coals. *Comput. Geosci.* **2009**, *35*, 1159–1166. [CrossRef]

25. Pia, G.; Sanna, U. An intermingled fractal units model and method to predict permeability in porous rock. *Int. J. Eng. Sci.* **2014**, *75*, 31–39. [CrossRef]

26. Xu, Y.F.; Sun, D.A. A fractal model for soil pores and its application to determination of water permeability. *Physics A* **2002**, *316*, 56–64. [CrossRef]

27. Katz, A.J.; Thompson, A.H. Fractal sandstone pores: Implications for conductivity and pore formation. *Phys. Rev. Lett.* **1985**, *54*, 1325–1328. [CrossRef] [PubMed]

28. Krohn, C.E.; Thompson, A.H. Fractal sandstone pores: Automated measurements using scanning-electron-microscope images. *Phys. Rev. B* **1986**, *33*, 6366–6374. [CrossRef]

29. Adler, P.M. *Porous Media: Geometry and Transports*; Butterworth-Heinemann: Boston, MA, USA, 1992; pp. 265–267.

30. Jacob, B. *Dynamics of Fluids in Porous Media*; American Elsevier Publishing Company: New York, NY, USA, 1972; p. 41.

31. Cai, J.; Perfect, E.; Cheng, C.L.; Hu, X. Generalized modeling of spontaneous imbibition based on Hagen-Poiseuille flow in tortuous capillaries with variably shaped apertures. *Langmuir* **2014**, *30*, 5142–5151. [CrossRef] [PubMed]

32. Ge, X.M.; Fan, Y.R.; Xing, D.H.; Chen, J.Y.; Cong, Y.H.; Liu, L.L. Predicting the relative permeability of water phase based on theory of coupled electricity-seepage and capillary bundle model. *Open Pet. Eng. J.* **2015**, *11*, 344–349. [CrossRef]

33. Jiang, L.L.; Liu, Y.; Teng, Y.; Zhao, J.F.; Zhang, Y.; Yang, M.J.; Song, Y.C. Permeability estimation of porous media by using an improved capillary bundle model based on micro-CT derived pore geometries. *Heat Mass Transf.* **2016**, *53*, 49–58. [CrossRef]

34. Xiao, B.Q.; Tu, X.; Ren, W.; Wang, Z.C. Modeling for hydraulic permeability and Kozeny–Carman constant of porous nanofibers using a fractal approach. *Fractals* **2015**, *23*, 1550029. [CrossRef]

35. Yun, M.J.; Yu, B.M.; Cai, J.C. Analysis of seepage characters in fractal porous media. *Int. J. Heat Mass Transf.* **2009**, *52*, 3272–3278. [CrossRef]

36. Xiao, B.; Fan, J.; Ding, F. A fractal analytical model for the permeabilities of fibrous gas diffusion layer in proton exchange membrane fuel cells. *Electrochim. Acta* **2014**, *134*, 222–231. [CrossRef]

37. Fagerlund, G. Determination of pore-size distribution from freezing-point depression. *Mater. Constr.* **1973**, *6*, 215–225. [CrossRef]

38. Comiti, J.; Renaud, M. A new model for determining mean structure parameters of fixed beds from pressure drop measurements: Application to beds packed with parallelepipedal particles. *Chem. Eng. Sci.* **1989**, *44*, 1539–1545. [CrossRef]

39. Tokoro, T.; Ishikawa, T.; Akagawa, S. A method for permeability measurement of frozen soil using an ice lens inhibition technique. *Jpn. Geotech. J.* **2010**, *5*, 603–613. [CrossRef]

40. You, Z.M.; Lai, Y.M.; Zhang, M.Y.; Liu, E.L. Quantitative analysis for the effect of microstructure on the mechanical strength of frozen silty clay with different contents of sodium sulfate. *Environ. Earth Sci.* **2017**, *76*, 143. [CrossRef]

41. Xiao, Z.A. Study on the Water and Salt Migration Process and Deformation Mechanism of Saline Freezing Soil. Ph.D. Thesis, Northwest Institute of Eco-Environment and Resources, Chinese Academy of Sciences, Lanzhou, China, 2017.

42. Diamond, S. A critical comparison of mercury porosimetry and capillary condensation pore size distributions of portland cement pastes. *Cem. Concr. Res.* **1971**, *1*, 531–545. [CrossRef]

43. Tao, G.L.; Zhu, X.L.; Hu, Q.Z.; Zhuang, X.S.; He, J.; Chen, Y. Critical pore-size phenomenon and intrinsic fractal characteristic of clay in the process of compression. *Rock Soil Mech.* **2018**, *1*, 1–9. [CrossRef]

44. Penumadu, D.; Dean, J. Compressibility effect in evaluating the pore-size distribution of kaolin clay using mercury intrusion porosimetry. *Can. Geotech. J.* **2000**, *37*, 393–405. [CrossRef]

45. Juang, C.H.; Holtz, R.D. Fabric, Pore Size Distribution, and Permeability of Sandy Soils. *J. Geotech. Eng.* **1986**, *112*, 855–868. [CrossRef]

Article

Numerical Simulation of Phosphorus Release with Sediment Suspension under Hydrodynamic Condition in Mochou Lake, China

Yu Bai [1], Jinhua Gao [2,*] and Tianyi Zhang [3]

[1] School of Water Resources and Hydropower Engineering, Wuhan University, Wuhan 430072, China; 2016102060037@whu.edu.cn
[2] School of Water Resources and Environment Engineering, Changchun Institute of Technology, Changchun 130012, China
[3] Jilin Institute of Hydraulic Science, Changchun 130000, China; 15927044022@sina.com
* Correspondence: zgsherry@sina.com

Received: 25 December 2018; Accepted: 18 February 2019; Published: 21 February 2019

Abstract: Phosphorus is a major cause of lake eutrophication. Understanding the characteristics regarding the release of phosphorus from sediments under hydrodynamic conditions is critical for the regulation of lake water quality. In this work, the effects of sediment suspension on the release characteristics of phosphorus from sediment were investigated under different hydrodynamic conditions. The experimental results showed that in the experimental process, the phosphorus was at first released quickly into the overlying water but then slowed down. Furthermore, the process of dissolved phosphorus (DP) release under hydrodynamic conditions with and without sediment suspension was simulated using a lattice Boltzmann method. The simulation showed satisfying results.

Keywords: dissolved phosphorus; hydrodynamic condition; Lattice Boltzmann method; release characteristics

1. Introduction

Sewage, industrial discharge, and agricultural runoff contribute to most sources of phosphorus released into shallow lakes, and most of the phosphorus is accumulated in lake sediments, which come back to overlying water under certain conditions [1,2]. Investigations on large and shallow lakes [3,4] indicated that wind waves are critical to sediment resuspension processes. Dynamic behaviors of suspended sediments and wind–wave effects have been studied in past decades and has shown that the suspended sediment has a strong effect on the release of phosphorus [5]. Phosphorus is a major factor that leads to the eutrophication of lakes as it can be absorbed by the vegetation of a lake, and studies have shown that strong wave disturbance can double the phosphorus concentration in the overlying water of lakes [6,7]. Therefore, understanding the release characteristics of the phosphorus from sediments under hydrodynamic conditions is critical for the regulation of lake water quality.

Two methods were proposed to investigate the influence of hydrodynamic forces on the release of phosphorus from sediment. One is to measure the concentration of phosphorus in natural lakes under natural hydrodynamic conditions, and then to analyze the effect of the hydrodynamic force on the phosphorus release from the sediment. The other is to take sediment samples from lakes and then simulate the hydrodynamic conditions as natural conditions using laboratory instruments [8]. In terms of the research on natural lakes, Nilsson et al. measured the phosphorus concentration in the overlying water of sediment under different hydrodynamic conditions in recirculating channel and found that the change of hydrodynamic force could significantly affect the phosphorus concentration in overlying water [9]. Previous research has shown that sediment concentration has a significant correlation with nutrients [10]. Wang et al. [11] found a positive correlation between total phosphorus (TP) and total suspended sediment (TSS) concentrations. Moreover, numerical methods have been developed to simulate the phosphorus concentrations in lakes and rivers [12]. Some of these methods consider the processes of adsorption–desorption and the release from bed sediment [13].

In recent years, the lattice Boltzmann method (LBM) has been widely applied to the study of diffusion. Wolf-Gladrow [14] gave the basic equations and the parameter selection method of LBM for solving diffusion problems. Jiaung et al. [15] used the LBM to simulate the process of heat conduction, where the thermal diffusivity was controlled by changing the relaxation time. Chen showed that the LBM is also applicable to simulate the diffusion problem in porous media with uniform and heterogeneous porosities, such as lake sediment [16].

In this study, laboratory experiments were conducted to investigate the release characteristics of phosphorus from sediment in shallow lakes. Experiments with and without sediment suspension were designed. The main objectives were as follows: (1) to investigate the release characteristics of the TP and dissolved phosphorus (DP) from the sediment with and without sediment suspension, and (2) to simulate the DP release with and without sediment suspension using LBM.

2. Methodology

2.1. Laboratory Experiments

The experiment was divided into two parts: a phosphorus release experiment with sediment suspension and a phosphorus release experiment without sediment suspension. The difference between them was the diameter of sediment, which, depending on its size, would either suspend or not under a hydrodynamic condition.

2.1.1. Experimental Device

The structure of the experimental device is shown in Figure 1a,b. A Plexiglas container with a diameter of 30 cm and a height of 50 cm was mounted with a variable speed motor that ran the propeller to adjust the hydrodynamic conditions in the device.

2.1.2. Experimental Sediment

The sediment was collected from Site 1 and Site 2 (Figure 2), respectively. Due to the previous lake slope protection project, a large amount of large diameter sediment was filled in the shore of Mochou Lake in 2015, the sediment diameter of Site 1 along the shore of Mochou Lake is larger than that of Site 2. The diameter of the sediment at Site 2 (D2) ranged from 0.15 cm to 0.23 cm, with a median diameter of 0.18 cm. The sediment was too coarse to suspend, even at the highest propeller speed (300 rad/min). The sediment diameter (D1) of Site 1 is given in Table 1. The diameter of sediment in Site 1 is small enough to have it suspend in a hydrodynamic condition.

Table 1. Grain size distribution of the sediment.

Grain Diameter (D1, μm)	Percentage (%)
D1 < 4	11.25
4 < D1 < 16	29.47
16 < D1 < 32	20.53
32 < D1 < 64	17.15
64 < D1 < 128	14.37
D1 > 128	7.23

Mochou Lake has a history of 1500 years, with an average water depth of 2.5 m, a lake surface area of 0.3 km^2, and a maximum water depth of 4 m. On 21 July 2018, we used a Petersen grab to dig a surface sediment at a depth of 0 cm to 10 cm in Mochou Lake. This lake is frequently influenced by winds, with wind-induced currents transporting dissolved matters (e.g., nutrients). There are two main parts of the sampling device: a sampling grab and a pulling rope. The sampling grab is made of a high-quality alloy material, as shown in Figure 1c. The one-time sampling sediment volume was 1–5 L, and the diameter of the sampling grab was 18 cm.

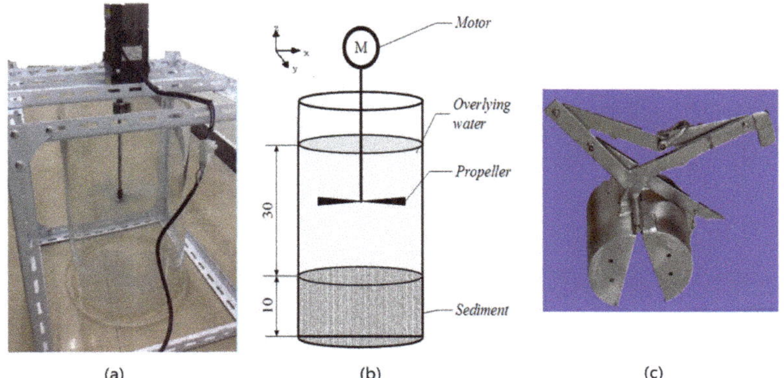

Figure 1. The structure of experimental device and sampling device. (**a**) structure of the experimental device; (**b**) Sketch of the experimental device; (**c**) sampling grab.

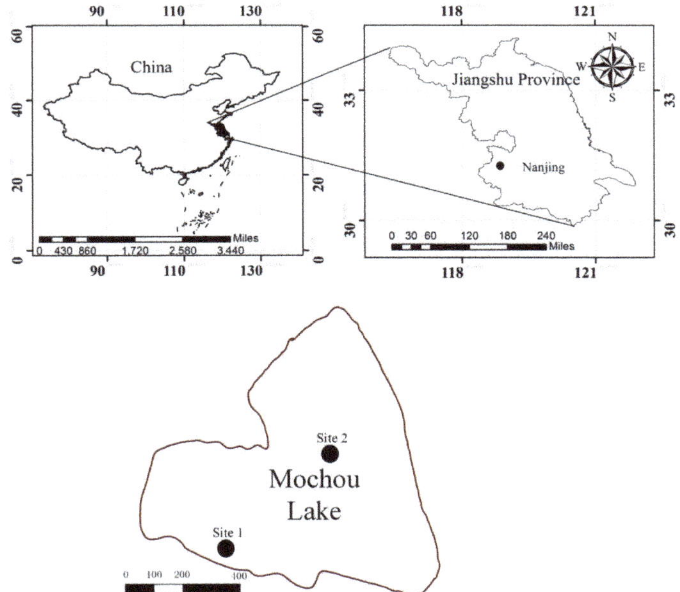

Figure 2. Sample location and structure of the experimental device.

2.1.3. The Experimental Method

Before the experiment, the sediment samples were fully stirred, and the thickness was controlled at 10 cm. A total of 20 L of deionized water was slowly added to the Plexiglas container, and the container was placed for 24 h prior to the experiment. The speed of the propeller for the various steps was set to 100 rad/min, 200 rad/min, and 300 rad/min, respectively. The experimental parameters are shown in Table 2.

Three 100 g sediment samples were taken from the sediment bed before and after each experiment run. The TP concentration of the sediment was determined by Chinese environmental standard HJ632-2011: Soil Determination of Total Phosphorus by alkali fusion-Mo-Sb anti-spectrophotometry method.

For each experiment run, three 30 mL water samples were extracted from each experimental run for each hour of total experimental time. The DP and TP in the water was determined using the Mo-Sb anti-spectrophotometry method of Yuan [17], the samples for TP determination needed to be digested in advance, and the samples for DP determination needed to be filtered without being digested. In addition, the particulate phosphorus (PP) was estimated as the difference between the TP and the DP (i.e., PP = TP − DP). The TSS concentration was calculated using a drying method. The beaker with 20 mL sampling water was placed in an oven at 115 °C till all the water evaporated. The calculated equation was as follows:

$$S = \frac{W_1 - W_2}{V} \tag{1}$$

where W_1 is the weight after drying, W_2 is the weight of the beaker, and V is the volume of sampling water.

Table 2. Experimental parameters.

Run	Sediment Diameter	Propeller Speed (rad/min)
1	D1	100
2	D1	200
3	D1	300
4	D2	100
5	D2	200
6	D2	300

2.1.4. Selection of Propeller Speeds Based on Field Data

At Site 1, the flow velocity (u_z), positioned in the water 0.5 m above the bed, was measured using a LS1206B intelligent portable velocimeter which was made by Nanjing Shunlaida Measurement and Control Equipment Co., Ltd. (Nanjing, China). The velocity measurement error was less than 1.5%, and each sampling time was 5 min. The wind speed (u_w), positioned 0.2 m above the water surface, was measured using an AR866 handheld thermosensitive anemometer made by Suzhou R.B.T Measurement and Control Technology Co., Ltd. (Suzhou, China). The wind speed measurement error was less than 1%, the measurement range was from 0.3 m/s to 30 m/s, and each sampling time was 5 min. The data was measured at 10:00–12:00 am, 2:00–4:00 pm on 22, 24 and 27 July 2018, respectively. Three groups of corresponding velocity and wind speed per hour were measured at Site 1. A statistical analysis of the field data (Figure 3) indicates that the occurrences of the u_z were less than 0.25 m/s and that the u_w was less than 8 m/s.

Based on the similarity principle, the rotational speeds are selected so that the Froude number in the laboratory condition (Fr_m) is equal to that in field conditions (Fr_p) [18].

$$Fr_m = Fr_p \tag{2}$$

Under laboratory conditions,

$$Fr_m = \frac{u_m}{\sqrt{gh_m}} \tag{3}$$

where u_m is the tangential speed in the laboratory experiment, h_m is the height above the sediment surface in the laboratory experiment (herein, h_m = 0.25 m), and g is the gravitational acceleration.

Under the field condition

$$Fr_p = \frac{u_z}{\sqrt{gh_z}} \tag{4}$$

where h_z is the height above the sediment surface in the field (herein, h_z = 0.5 m) under laboratory conditions.

From the general relation between the tangential and angular speeds, the laboratory experimental rotational speed is

$$u_m = \frac{P_s}{60} r \tag{5}$$

where P_s is the propeller speed, and r is the radius of the stirring rod (herein, r = 0.045 m).

This study examined the relationship between u_m and u_w that corresponds to the dynamic condition caused by the propeller speed to the wind speed. When measuring, the maximum wind speed can be less than 8 m/s, but sometimes the maximum wind speed in Nanjing can be larger than 8 m/s. The field wind speed varied from 4.65 m/s to 12.74 m/s, and P_s varied from 0.075 to 0.225 (Table 3). As a result, the propeller speeds were set to 100, 200, and 300 rad/min in the laboratory experiment to simulate the prevailing wind speed of 4.65 m/s to 12.74 m/s in the lake. Huang et al. [18] examined the shear stress caused by lake currents, because they are considered to have similar dynamic effects to the stirring rod on the sediment resuspension, and because the dynamic simulator can produce shear stresses of water currents in the laboratory. The propeller speeds were selected to simulate the prevailing bottom flow velocities of 0 m/s to 0.08 m/s (calculated by Equation (6)), as in Taihu Lake. As a result, the blade stirrer operated at propeller speeds of 0, 100 rad/min, 200 rad/min, 300 rad/min, and 400 rad/min in the laboratory experiment. We used the same method to examine which propeller speeds to select to simulate the prevailing bottom flow velocities of 0 m/s to 0.04 m/s in the lake. As a result, the blade stirrer operated at the propeller speeds of 0, 100 rad/min, 200 rad/min, and 300 rad/min in this laboratory experiment.

The bottom boundary velocity is computed as

$$u_b = \frac{\kappa u_z}{ln\left(\frac{z}{z_0}\right)} \tag{6}$$

where κ is the von Kármán's constant, and z_0 is the bottom boundary roughness, (we assume this based on existing literature [1,8], z_0 = 0.02 m).

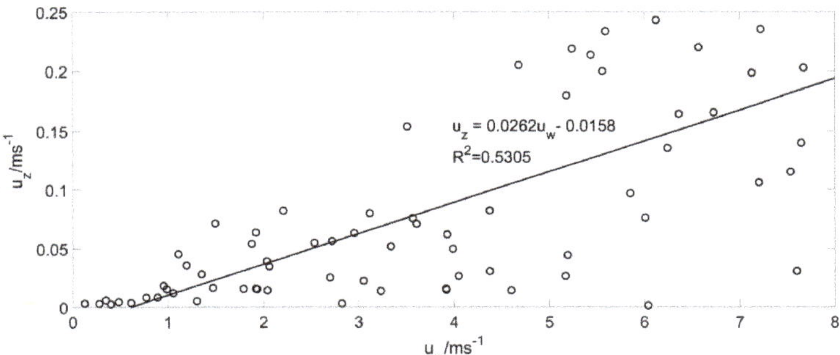

Figure 3. The relation between field flow velocity and wind speed.

Table 3. The contrast table of wind velocity and rotational speeds computed.

Laboratory Experiment		Mochou Lake	
P_s (rad/min)	u_m (m/s)	u_z (m/s)	u_w (m/s)
100	0.075	0.1061	4.65
200	0.15	0.2121	8.7
300	0.225	0.3182	12.74

Note: u_m was calculated by Equation (4); u_w was calculated by the Equation in Figure 3; u_z was calculated by Equations (1)–(3).

2.2. Simulation Method

Generally, the nutrient release in lakes undergoes the following process: the nutrient is released from sediment, the resuspended sediment is desorbed, and then it is diffused in overlying water. For sediment release, it follows the process of diffusion, adsorption, and desorption.

2.2.1. LBM

LBM includes two steps: migration and collision. In the migration step, the particles move in a certain direction and at a certain speed to the adjacent nodes of the grid. The migration process is expressed as:

$$f_k(x,y,t+\Delta t) = f_k(x,y,t) + f_k'(x,y,t)$$ (7)

where f_k is the particle distribution function in terms of a discrete particle along the direction k; f_k' is the value of a particle before a migration along the direction k.

Theoretically, the particle collision process is very complex and difficult to solve. Bhatnagar et al. proposed a simple BGK collision operator for a discretized LBM equation, which can be expressed as:

$$f_k'(x,y,t) = -\omega f_k(x,y,t) + \omega f_k^{eq}(x,y,t)$$ (8)

where f_k^{eq} is the distribution of the equilibrium function along the direction k, and ω is the relaxation frequency.

Bringing Equation (6) into Equation (5), we have:

$$f_k(x,y,t+\Delta t) = f_k(x,y,t)[1-\omega] + \omega f_k^{eq}(x,y,t)$$ (9)

In LBM, the solution region is divided into many lattices. In this paper, two dimensions with a nine-direction (D2Q4) lattice configuration were used (Figure 4).

$$f_k^{eq}(x,z,t) = w_k T(x,z,t)$$ (10)

where f_k is the particle distribution function in terms of a discrete particle along the direction k, f_k^{eq} is the equilibrium distribution function along the direction k; $w_1 = w_2 = w_3 = w_4 = 0.25$, and T is the nutrients concentration.

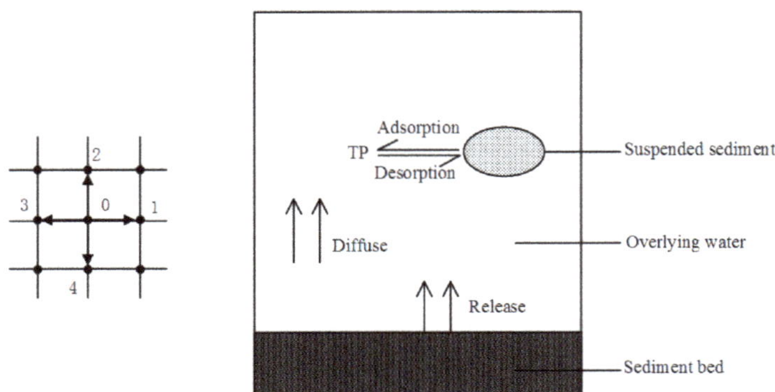

Figure 4. Nutrient migration process in the experimental device.

2.2.2. Phosphorus Release from Sediment

The phosphorus release in lakes includes the phosphorus release from sediment and the phosphorus diffusion in overlying water. For sediment release, it follows the process of diffusion, adsorption, and desorption. The diffusion rate can be measured by the nutrient concentration gradient in the pore water of sediment with Fick's first law, and the adsorption and desorption can be defined by a source item [19].

The phosphorus release from sediment can be expressed as:

$$\varphi \frac{\partial C}{\partial t} = \varphi D_{zs} \frac{\partial^2 c_d}{\partial z^2} - \rho_b \frac{\partial c_s}{\partial t} \ (for \ z \ < \ 0) \tag{11}$$

where c_d is the DP concentration in water (mg/L), t is time (s), φ is the porosity of sediment, z is vertical axis originated (at the sediment-water interface, $z = 0$), D_{zs} is the diffusion coefficient of phosphorus in sediment (m^2/s), $\rho_b \frac{\partial c_s}{\partial t}$ is a source term for phosphorus adsorption and desorption by the sediment bed [20], c_s is the quantity of phosphorus adsorption in the sediment bed (mg/kg) and ρ_b is the density of sediment (kg/m^3).

The Lagergren first-order (LFO) equation is commonly used for describing the adsorption and desorption and for their kinetics research, which is expressed as [21]:

$$\rho_b \frac{\partial c_s}{\partial t} = \rho_b b_1 (c_s - c_{se}) \tag{12}$$

where b_1 is the first-order rate constant: absorption efficiency of sediment bed (g/(Ls)), and c_{se} is the sediment contamination level (mg/kg).

Yuan et al. [22] assumed that the desorption amount of the sediment bed are equal to the amount added to the solution. Then, they modified the LFO equation as:

$$\rho_b \frac{\partial c_s}{\partial t} = \rho_b b_1 (c_d - c_e) \tag{13}$$

where c_e is the equilibrium concentration of TP in water (mg/L) [23].

The modified LFO equation only considers the constant hydrodynamic condition where c_e remains unchanged in a constant condition. Therefore, in an airtight container without phosphorus input, the concentration of phosphorus in overlying water and sediment would be constant values. However, under the action of shear velocity, c_e will vary with the change of the hydrodynamic condition (or the adsorption rate decreases and c_e increases with the increase of shear velocity [24]).

Here the second term on the right of Equation (1) is modified as:

$$\rho_b \frac{\partial c_s}{\partial t} = \rho_b b_1 (c_d - c_e) \tag{14}$$

D_{zs} can be expressed as:

$$D_{zs} = \varphi^{m-1} D_{zm} \tag{15}$$

where D_{zm} is the molecular diffusion coefficient in water (m^2/s) and it varies with the targeting solution; $m = 3$ is a constant [25].

The source term can be added in the LBM function [26], and the LBM function can be described as:

$$f_k(x,z,t+\Delta t) = f_k(x,z,t)(1 - \omega_p) + \omega_p f_k^{eq}(x,z,t) - \frac{\Delta t w_k \rho_b b_1}{\varphi}(c_d(x,z,t) - c_e) \tag{16}$$

where Δt is the time step and ω_p is the relaxation frequency in pore water;

The relationship between the diffusion coefficient and ω_p can be obtained by Chapman-Enskog expansion [26].

$$D_{zs} = \frac{\Delta x^2}{p \Delta t}\left(\frac{1}{\omega_p} - 0.5\right) \tag{17}$$

where p presents the dimension of the model. ($p = 1$ presents one dimension, $p = 2$ presents two dimensions, and $p = 3$ presents three dimensions).

In this paper, we adopted a two-dimension model, and ω can be obtained as follows:

$$\omega_p = 1/\left(\frac{2\Delta t D_{zs}}{\Delta x^2} + 0.5\right) \tag{18}$$

2.2.3. Nutrient Diffusion in Overlying Water

In overlying water, the formulations can be simply described as a diffusion process, as the phosphorus release from TSS and the biochemical reactions are negligible. The governing equations can be expressed as [27]:

$$\frac{\partial c_d}{\partial t} = (D_{zt} + D_{zm})\frac{\partial^2 c_d}{\partial z^2} - S\frac{\partial c_p}{\partial t} \quad (\text{for } z \geq 0) \tag{19}$$

$$\frac{D_{zt}}{v} = \left(A\frac{zu_*}{v}\right)^n \tag{20}$$

where D_{zt} is the turbulent diffusion coefficient (m^2/s), A is the area of water–sediment interface (m^2), v is the kinematic viscosity of water (m^2/s), and $n = 3$ is a constant [28]; $S\frac{\partial c_p}{\partial t}$ is a source term for phosphorus adsorption and desorption by TSS, S is the TSS concentration in the overlying water, and c_p is the quantity of phosphorus adsorption by TSS in the overlying water.

We assumed that the desorption amount of TSS is equal to the amount added to the solution. Then, $S\frac{\partial c_p}{\partial t}$ can be modified as:

$$S\frac{\partial c_p}{\partial t} = b_2 S(c_d - c_e) \tag{21}$$

where b_2 is the first-order rate constant, that is, the absorption efficiency of TSS ($\text{g}/(\text{Ls})$).

The LBM function and ω_w can be obtained with the same method as with the sediment:

$$f_k(x, z, t + \Delta t) = f_k(x, z, t)(1 - \omega_w) + \omega_w f_k^{eq}(x, z, t) - \Delta t w_k b_2 S(c_d(x, z, t) - c_e) \tag{22}$$

$$\omega_w = 1/\left(\frac{2\Delta t(D_{zt} + D_{zm})}{\Delta x^2} + 0.5\right) \tag{23}$$

where u_* is the shear velocity, and the u_* generated by the propeller speed is given by Chandler [29]:

$$u_* = 8.67 \times 10^{-5} P_s - 3.27 \times 10^{-3} \tag{24}$$

2.2.4. Nutrient Desorption from Resuspended Sediment

In this research, the sediment concentration in the overlying water was assumed to be uniform. Here, the dynamic release amount caused by sediment suspension is much larger than the constant release amount caused by diffusion [30], while the vertical velocity component is relatively smaller than the convection velocity component [31]. Therefore, only the effects of sediment re-suspension and sedimentation are considered, and Equation (4) could be simplified into:

$$\frac{dS}{dt} = (S_r - S_s)/H \tag{25}$$

$$S_r = \begin{cases} k(u_* - u_c) & u_* > u_c \\ 0 & u_* < u_c \end{cases} \tag{26}$$

$$S_s = \omega_d S \tag{27}$$

The settling velocity of sediment particles was obtained by Song et al. [32]:

$$\omega_d = \frac{v}{D}D_*^3\left(38.1 + 0.93D_*^{12/7}\right)^{-7/8} \tag{28}$$

$$D_* = \left(\frac{\Delta g}{v^2}\right)^{1/3} D \tag{29}$$

where D is the particle diameter and D_* is the dimensionless particle diameter. $\Delta = \frac{\rho_s}{\rho} - 1$, where ρ_s and ρ represent the density of particles and the density of the fluid, respectively. $K = 9 \times 10^{-3}$ (kg/m^3); S_s is the deposition flux; S_r is the re-suspension flux; u_c is the critical velocity of sediment, which can be expressed as:

$$u_c = \sqrt{d_{cr}\Delta g D_*} \tag{30}$$

Van Rijn [21] related the dimensionless particle parameter to a critical sediment mobility parameter (d_{cr}), which is elaborated in Table 4.

Table 4. Variation of d_{cr} with D_*.

Dimensionless Particle Parameter	Critical Sediment Mobility Parameter
$D_* \leq 4$	$d_{cr} = 0.24(D_*)^{-1}$
$4 < D_* \leq 10$	$d_{cr} = 0.14(D_*)^{-0.64}$
$10 < D_* \leq 20$	$d_{cr} = 0.04(D_*)^{-0.1}$
$20 < D_* \leq 150$	$d_{cr} = 0.13(D_*)^{0.29}$
$D_* > 150$	$d_{cr} = 0.055$

The LBM function for sediment in overlying water can be described as follows,

$$S_{(t+\Delta t)} = \frac{S_{r(t)} - S_{s(t)}}{H} + S_{(t)} \tag{31}$$

2.2.5. Boundary Conditions

By applying LBM, the free surface of overlying water can be defined as the thermal insulating boundary. The nutrient concentration gradient of the free surface is 0, and the container wall is defined as rebound boundary. Please refer to Mohamad [26] for an in-depth explanation of the boundary method.

2.2.6. Model Parameters

The parameters in this model are given in Table 5.

Table 5. Parameters in the model.

Run No.	1	2	3	4	5	6
P_s (rad/min)	100	200	300	100	200	300
Δx (m)	0.01	0.01	0.01	0.01	0.01	0.01
Δy (m)	0.01	0.01	0.01	0.01	0.01	0.01
Δt (s)	1	1	1	1	1	1
u_* (m/s)	0.0054	0.0141	0.0227	0.0054	0.0141	0.0227
φ	0.49	0.49	0.49	0.4	0.4	0.4
ω_p	1.67	1.67	1.67	1.77	1.77	1.77
ω_w	1.67	1.57	1.34	1.77	1.67	1.42
ρ_b (kg/m^3)	1400	1400	1400	1650	1650	1650
b_1 (g/(Ls))	2.2×10^{-4}	2.2×10^{-4}	2.2×10^{-4}	1.9×10^{-4}	1.9×10^{-4}	1.9×10^{-4}
b_2 (g/(Ls))	2.1×10^{-4}	2.1×10^{-4}	2.1×10^{-4}	1.8×10^{-4}	1.8×10^{-4}	1.8×10^{-4}

To conclude the above contents, the sediment and phosphorus transport can be calculated using LBM, as shown in Figure 5.

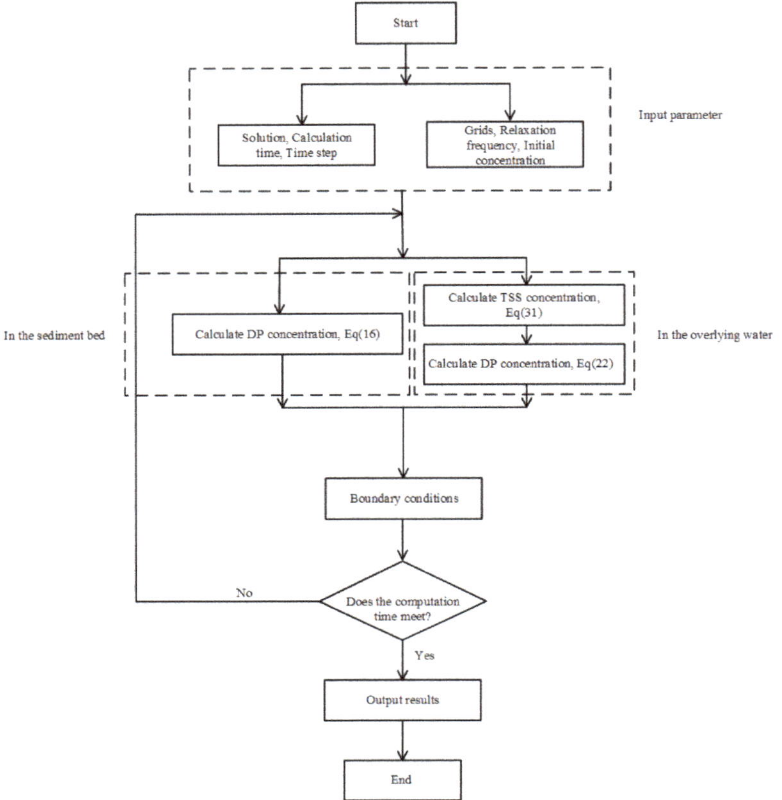

Figure 5. Flow chart of 2-D LBM.

3. Experimental Results

For Run 4–6, there is very little sediment in the samples, and all of the measured TSS concentration are less than 0.02 g/L, which means that the phosphorus absorbed by TSS in the overlying water can be ignored. Therefore, only the TSS concentration of Run 1–3 are listed in Table 6. The TSS concentration is higher with the higher propeller speed.

The TP concentrations are shown in Table 7; the TP concentrations of Run 1 to Run 6 ranged from 0.059 mg/L to 0.119 mg/L, 0.063 mg/L to 0.191 mg/L, 0.065 mg/L to 0.24 mg/L, 0.031 mg/L to 0.128 mg/L, 0.04 mg/L to 0.143 mg/L, and 0.031 mg/L to 0.151 mg/L, respectively, for the 8 experimental hours. The TP concentration in the water increased quickly in the first four hours, then slowed down in the last four hours. The DP concentrations are shown in Table 8; the DP concentrations of Run 1 to Run 6 ranged from 0.048 mg/L to 0.104 mg/L, 0.034 mg/L to 0.163 mg/L, 0.034 mg/L to 0.207 mg/L, 0.037 mg/L to 0.123 mg/L, 0.039 mg/L to 0.136 mg/L, and 0.03 mg/L to 0.146 mg/L, respectively, for the 8 experimental hours. The DP concentration in the water increased quickly in the first four hours, then slowed down in the last four hours.

Table 6. The TSS concentration under different hydrodynamic conditions (g/L).

Run No.	0 h	1 h	2 h	3 h	4 h	5 h	6 h	7 h	8 h
1	0	3.46 ± 0.52	7.12 ± 0.36	8.22 ± 0.57	11.05 ± 0.32	12.07 ± 1.01	12.63 ± 0.92	11.33 ± 0.75	11.1 ± 0.38
2	0	4.29 ± 0.41	9.24 ± 1.03	11.42 ± 0.79	14.26 ± 0.85	18.76 ± 0.71	18.15 ± 0.69	17.11 ± 2.07	16.84 ± 1.13
3	0	5.45 ± 0.39	10.13 ± 0.77	15.53 ± 1.02	19.37 ± 0.94	22.12 ± 1.51	21.57 ± 2.21	22.08 ± 1.49	22.34 ± 1.64

Note: Data in the table are the Mean values ± standard deviation, the same as below.

Table 7. The TP concentration under different hydrodynamic conditions (mg/L).

Run No.	0 h	1 h	2 h	3 h	4 h	5 h	6 h	7 h	8 h
1	0.059 ± 0.004	0.074 ± 0.005	0.091 ± 0.007	0.105 ± 0.005	0.11 ± 0.011	0.111 ± 0.009	0.114 ± 0.021	0.117 ± 0.004	0.119 ± 0.015
2	0.063 ± 0.005	0.099 ± 0.007	0.131 ± 0.005	0.156 ± 0.003	0.165 ± 0.012	0.173 ± 0.011	0.179 ± 0.015	0.185 ± 0.013	0.191 ± 0.005
3	0.065 ± 0.005	0.119 ± 0.004	0.158 ± 0.011	0.191 ± 0.014	0.204 ± 0.004	0.215 ± 0.003	0.224 ± 0.008	0.233 ± 0.007	0.24 ± 0.021
4	0.031 ± 0.003	0.069 ± 0.008	0.087 ± 0.04	0.097 ± 0.012	0.117 ± 0.009	0.12 ± 0.008	0.121 ± 0.011	0.127 ± 0.016	0.128 ± 0.014
5	0.04 ± 0.004	0.081 ± 0.005	0.102 ± 0.07	0.117 ± 0.017	0.139 ± 0.005	0.14 ± 0.012	0.143 ± 0.032	0.143 ± 0.008	0.143 ± 0.005
6	0.031 ± 0.002	0.087 ± 0.003	0.113 ± 0.02	0.133 ± 0.004	0.137 ± 0.012	0.145 ± 0.025	0.148 ± 0.005	0.151 ± 0.007	0.15 ± 0.016

Table 8. The DP concentration under different hydrodynamic conditions (mg/L).

Run No.	0 h	1 h	2 h	3 h	4 h	5 h	6 h	7 h	8 h
1	0.048 ± 0.003	0.057 ± 0.011	0.056 ± 0.004	0.08 ± 0.014	0.109 ± 0.003	0.101 ± 0.022	0.102 ± 0.007	0.105 ± 0.012	0.104 ± 0.007
2	0.034 ± 0.005	0.078 ± 0.009	0.11 ± 0.012	0.143 ± 0.015	0.140 ± 0.014	0.149 ± 0.007	0.152 ± 0.013	0.159 ± 0.012	0.163 ± 0.011
3	0.034 ± 0.008	0.092 ± 0.007	0.096 ± 0.013	0.174 ± 0.007	0.188 ± 0.003	0.194 ± 0.006	0.201 ± 0.011	0.211 ± 0.007	0.207 ± 0.012
4	0.031 ± 0.006	0.068 ± 0.005	0.087 ± 0.007	0.093 ± 0.005	0.082 ± 0.007	0.106 ± 0.013	0.115 ± 0.008	0.123 ± 0.011	0.123 ± 0.008
5	0.039 ± 0.004	0.08 ± 0.007	0.098 ± 0.005	0.112 ± 0.005	0.098 ± 0.019	0.135 ± 0.017	0.137 ± 0.014	0.14 ± 0.006	0.136 ± 0.009
6	0.03 ± 0.005	0.085 ± 0.008	0.110 ± 0.003	0.128 ± 0.006	0.134 ± 0.008	0.141 ± 0.004	0.139 ± 0.009	0.146 ± 0.009	0.146 ± 0.011

The variations of TP concentration in the sediment before and after experiment are shown in Figure 6. We can see that the initial TP concentration of Run 1–3 is higher than Run 4–6. After the experiment TP, decreased by 18.5%, 29.7%, 38.8%, 36.7, 46.2%, and 48.9%, from Run 1 to Run 6, respectively.

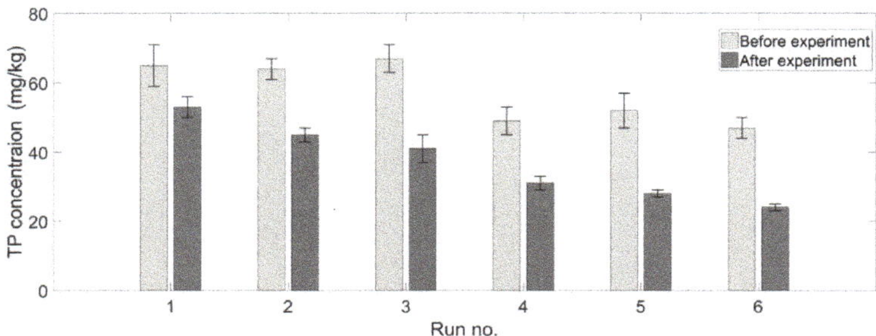

Figure 6. Variations of TP concentration in the sediment before and after the experiment.

4. Simulation Results

The simulation result of TSS concentration is shown in Figure 7. From Figure 7, we can see that the predicted value fit well with the measured data.

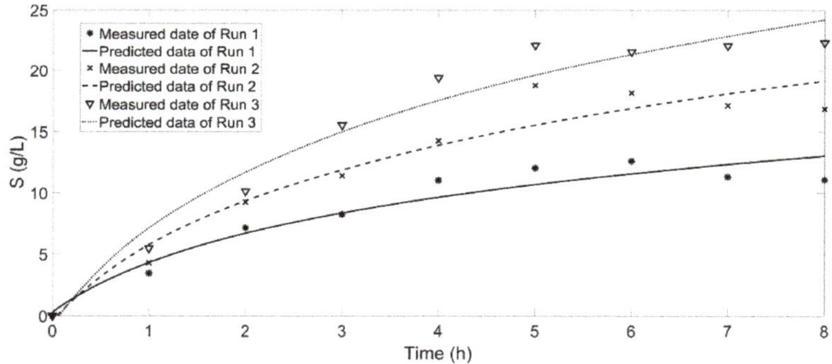

Figure 7. Comparison of predicted and measured values of TSS (S is the TSS concentration).

Error analysis was conducted to determine the difference between the predicted and measured data. The average absolute error $\overline{\Delta_a}$ and average relative error $\overline{\Delta_r}$ were calculated by:

$$\overline{\Delta_a} = \frac{1}{N} \sum_1^N abs\left((c_d)_{computed} - (c_d)_{measured} \right) \tag{32}$$

$$\overline{\Delta_r} = \left[\frac{1}{N} \sum_1^N abs\left(\frac{(c_d)_{computed} - (c_d)_{measured}}{(c_d)_{measured}} \right) \right] \times 100\% \tag{33}$$

where N is the number of measured lines in the cross section for each case, and $(c_d)_{computed}$ and $(c_d)_{measured}$ are the calculated and measured values of the DP concentration.

The DP predicted results are shown in Table 9, along with the measured data. The average relative errors $\overline{\Delta_r}$ are 10.85%, 8.41%, and 13.01% for simulations 1 to 3, respectively. The average relative errors $\overline{\Delta_r}$ are 11.67%, 8.84%, and 3.76% for simulations 4 to 6, respectively. Both the suspension model and the non-suspension model show satisfying results. The comparison of results between actual measurement and simulation verified the effectiveness of the LBM.

Table 9. Comparison between predicted value and measured value: DP concentration.

	Experiment	1 h	2 h	3 h	4 h	5 h	6 h	7 h	8 h	9 h	$\overline{\Delta_a}$	$\overline{\Delta r}$
1	Measured Value	0.048	0.057	0.056	0.080	0.109	0.101	0.102	0.105	0.104	0.0078	10.85%
	Predicted Value	0.0504	0.0632	0.0795	0.0926	0.0952	0.0978	0.0991	0.1017	0.1017		
2	Measured Value	0.034	0.078	0.110	0.143	0.140	0.149	0.152	0.159	0.163	0.0053	8.41%
	Predicted Value	0.0513	0.0833	0.1126	0.1370	0.1461	0.1526	0.1565	0.1604	0.1617		
3	Measured Value	0.034	0.092	0.096	0.174	0.188	0.194	0.201	0.211	0.207	0.0105	13.01%
	Predicted Value	0.0522	0.0983	0.1352	0.1683	0.1813	0.1904	0.1970	0.2022	0.2048		
4	Measured Value	0.031	0.068	0.087	0.093	0.082	0.106	0.115	0.123	0.123	0.0103	11.67%
	Predicted Value	0.0330	0.0695	0.0915	0.1045	0.1225	0.1245	0.1250	0.1250	0.1250		
5	Measured Value	0.039	0.080	0.098	0.112	0.098	0.135	0.137	0.140	0.136	0.0094	8.84%
	Predicted Value	0.0380	0.0795	0.1050	0.1200	0.1405	0.1430	0.1435	0.1435	0.1435		
6	Measured Value	0.030	0.085	0.110	0.128	0.134	0.141	0.139	0.146	0.146	0.0051	3.76%
	Predicted Value	0.0300	0.0830	0.1095	0.1255	0.1470	0.1495	0.1500	0.1500	0.1500		

5. Conclusions

The release characteristics of the TP and DP from sediment were experimentally investigated under different hydrodynamic conditions, with and without sediment suspension. By analyzing the measured and predicted data, the main results can be summarized as follows:

In the case of sediment suspension and no-sediment suspension, at the beginning of the experiment, TP and DP were quickly released into the overlying water, and then gradually slowed down. In this experiment, the propeller speeds corresponding to the wind speed in the field condition could improve our result in the field work. LBM was verified to be effective in simulating the process of DP release from sediment under hydrodynamic conditions within experimental conditions, and this technique could provide some theoretical references for the application of the Mochou Lake simulation.

Author Contributions: Conceptualization, Y.B. and J.G.; Methodology, Y.B.; Software, Y.B.; Validation, T.Z., Y.B. and J.G.; Formal analysis, T.Z.; Investigation, J.G.; Resources, J.G.; Data curation, J.G.; Writing—original draft preparation, Y.B.; Writing—review and editing, Y.B.

Funding: This research received no external funding.

Acknowledgments: The authors thank the anonymous reviewers, Associate Editor and Editor for their constructive comments.

Conflicts of Interest: The authors declare no conflict of interest.

Notation

b_1	first-order rate constant: absorption efficiency of sediment bed
b_2	first-order rate constant: absorption efficiency of TSS
c_d	dissolved phosphorus concentration in water
c_e	equilibrium concentration of TP in water
c_p	quantity of phosphorus adsorption in overlying water
c_s	quantity of phosphorus adsorption in sediment bed
c_{se}	sediment contamination level
D	particle diameter
D_*	dimensionless particle diameter
D_{zm}	molecular diffusion coefficient in water
D_{zs}	diffusion coefficient of nutrient in sediment
D_{zt}	turbulent diffusion coefficient
Fr_m	Froude number in the laboratory condition
Fr_p	Froude number in the field condition
f_k	particle distribution function in terms of a discrete particle along the direction k
f'_k	the value of a particle before migration along the direction k.
f_k^{eq}	equilibrium distribution function along the direction k
g	gravitational acceleration
P_s	propeller speed
r	radius of the stirring rod

S	TSS concentration in the overlying water
S_r	re-suspension flux
S_s	deposition flux
u_m	tangential speed in the laboratory experiment
u_w	wind speed positioned at 0.2 m above the water surface
u_z	flow velocity positioned at 0.5 m above the bed under field conditions
ρ_s	density of particles
ω_p	relaxation frequency in pore water
Δt	time step
v	kinematic viscosity of water
ρ	density of fluid
φ	porosity of sediment
ω	relaxation frequency

References

1. Blom, G.; Van Duin, E.H.; Aalderink, R.H.; Lijklema, L.; Toet, C. Modelling sediment transport in shallow lakes—Interactions between sediment transport and sediment composition. *Hydrobiologia* **1992**, *235*, 153–166. [CrossRef]

2. Carpenter, S.R. Eutrophication of aquatic ecosystems: Bistability and soil phosphorus. *Proc. Natl. Acad. Sci. USA* **2005**, *102*, 10002–10005. [CrossRef] [PubMed]

3. Mei, C.C.; Fan, S.J.; Jin, K.R. Resuspension and transport of fine sediments by waves. *J. Geophys. Res. Ocean.* **1997**, *102*, 15807–15821. [CrossRef]

4. Bachmann, R.W.; Hoyer, M.V.; Canfield, D.E., Jr. The potential for wave disturbance in shallow Florida lakes. *Lake Reserv. Manag.* **2000**, *16*, 281–291. [CrossRef]

5. Huang, J.; Ge, X.; Wang, D. Distribution of heavy metals in the water column, suspended particulate matters and the sediment under hydrodynamic conditions using an annular flume. *J. Environ. Sci.* **2012**, *24*, 2051–2059. [CrossRef]

6. Chao, X.; Jia, Y.; Cooper, C.M.; Shields, F.D., Jr.; Wang, S.S. Development and application of a phosphorus model for a shallow oxbow lake. *J. Environ. Eng.* **2006**, *132*, 1498–1507. [CrossRef]

7. Horppila, J.; Nurminen, L. Effects of different macrophyte growth forms on sediment and P resuspension in a shallow lake. *Hydrobiologia* **2005**, *545*, 167–175. [CrossRef]

8. Couceiro, F.; Fones, G.R.; Thompson, C.E.L.; Statham, P.J.; Sivyer, D.B.; Parker, R.; Kelly-Gerreyn, B.A.; Amos, C.L. Impact of resuspension of cohesive sediments at the oyster grounds (North Sea) on nutrient exchange across the sediment-water interface. *Biogeochemistry* **2013**, *113*, 37–52. [CrossRef]

9. Nilsson, P.; Jansson, M. Hydrodynamic control of nitrogen and phosphorus turnover in an eutrophicated estuary in the Baltic. *Water Res.* **2002**, *36*, 4616–4626. [CrossRef]

10. Le Fevre, N.M.; Lewis, G.D. The role of resuspension in enterococci distribution in water at an urban beach. *Water Sci. Technol.* **2003**, *47*, 205–210. [CrossRef]

11. Wang, J.; Pang, Y.; Li, Y.; Huang, Y.; Luo, J. Experimental study of wind-induced sediment suspension and nutrient release in Meiliang bay of lake Taihu, China. *Environ. Sci. Pollut. Res. Int.* **2015**, *22*, 10471–10479. [CrossRef] [PubMed]

12. Ishikawa, M.; Nishimura, H. Mathematical model of phosphate release rate from sediments considering the effect of dissolved oxygen in overlying water. *Water Res.* **1989**, *23*, 351–359. [CrossRef]

13. Chen, S.; Yang, B.; Zheng, C. Simulation of double diffusive convection in fluid-saturated porous media by lattice boltzmann method. *Int. Heat Mass Transf.* **2017**, *108*, 1501–1510. [CrossRef]

14. Wolf-Gladrow, D. A lattice Boltzmann equation for diffusion. *J. Stat. Phys.* **1995**, *79*, 1023–1032. [CrossRef]

15. Jiaung, W.S.; Ho, J.R.; Kuo, C.P. Lattice Boltzmann method for the heat conduction problem with phase change. *Numer. Heat Transf. Part. B Fundam.* **2001**, *39*, 167–187.

16. Li, Z.; Tang, H.; Xiao, Y.; Zhao, H.; Li, Q.; Ji, F. Factors influencing phosphorus adsorption onto sediment in a dynamic environment. *J. Hydro-Environ. Res.* **2016**, *10*, 1–11. [CrossRef]

17. Yuan, L. Potassium Persulfate Oxidation—Determination of Total Phosphorus by Mo-Sb anti-spectrophotometry method. *Environ. Monit. China* **1987**, *3*, 120–124. (In Chinese)

18. Huang, J.; Xi, B.D.; Xu, Q.J.; Wang, X.X.; Li, W.P.; He, L.S.; Liu, H.L. Experiment study of the effects of hydrodynamic disturbance on the interaction between the cyanobacterial growth and the nutrients. *J. Hydrodyn.* **2016**, *28*, 411–422. [CrossRef]

19. Higashino, M.; Stefan, H.G. Non-linear effects on solute transfer between flowing water and a sediment bed. *Water Res.* **2011**, *45*, 6074–6086. [CrossRef]

20. Wang, H.T. Dynamics of Fluid Flow and Contaminant Transport. In *Porous Media*; Higher Education Press: Beijing, China, 2008; pp. 36–37.

21. Tseng, R.L.; Wu, F.C.; Juang, R.S. Characteristics and applications of the LAGERGREN'S FIRST-order equation for adsorption kinetics. *J. Taiwan Inst. Chem. Eng.* **2010**, *41*, 661–669. [CrossRef]

22. Yuan, L.; Han, L.; Bo, W.; Chen, H.; Gao, W.; Chen, B. Simulated oil release from oil-contaminated marine sediment in the Bohai Sea, China. *Mar. Pollut. Bull.* **2017**, *118*, 79–84. [CrossRef] [PubMed]

23. Zhang, Y.; He, F.; Xia, S.; Kong, L.; Xu, D.; Wu, Z. Adsorption of sediment phosphorus by porous ceramic filter media coated with nano-titanium dioxide film. *Ecol. Eng.* **2014**, *64*, 186–192. [CrossRef]

24. Liang, D.F.; Wang, X.L.; Bockelmann-Evans, B.N.; Falconer, R.A. Study on nutrient distribution and interaction with sediments in a macro-tidal estuary. *Adv. Water Resour.* **2013**, *52*, 207–220. [CrossRef]

25. Ullman, W.J.; Aller, R.C. Diffusion coefficients in nearshore marine sediments. *Limnol. Oceanogr.* **1982**, *27*, 552–556. [CrossRef]

26. Mohamad, A.A. *Lattice Boltzmann Method: Fundamentals and Engineering Applications with Computer Codes*; Springer Science & Business Media: Berlin, Germany, 2011.

27. Inoue, T.; Nakamura, Y. Effects of hydrodynamic conditions on DO transfer at a rough sediment surface. *J. Environ. Eng.* **2010**, *137*, 28–37. [CrossRef]

28. Van Rijn, L.C. Sediment transport, part III: Bed forms and alluvial roughness. *J. Hydraul. Eng.* **1984**, *110*, 1733–1754. [CrossRef]

29. Chandler I, D. Vertical variation in diffusion coefficient within sediments. Ph.D. Thesis, University of Warwick, Coventry, UK, May 2012.

30. Zhu, H.W.; Zhong, B.C.; Wang, D.Z. Hydrodynamic effects on contaminants release due to rususpension and diffusion from sediments. *J. Hydrodyn.* **2013**, *25*, 731–736. [CrossRef]

31. Gao, G.; Falconer, R.A.; Lin, B. Numerical modelling of sediment–bacteria interaction processes in surface waters. *Water Res.* **2011**, *45*, 1951–1960. [CrossRef]

32. Song, Z.; Wu, T.; Xu, F.; Li, R. A simple formula for predicting settling velocity of sediment particles. *Water Sci. Eng.* **2008**, *1*, 37–43.

Article

The Dynamics of Water Wells Efficiency Reduction and Ageing Process Compensation

Krzysztof Polak [1],*, Kamil Górecki [2] and Karolina Kaznowska-Opala [3]

[1] Centre of Energy, Faculty of Mining and Geoengineering, AGH University of Science and Technology, 30059 Krakow, Poland

[2] WellsCtrl.Tech, 30059 Krakow, Poland; kamil.gorecki85@gmail.com

[3] Department of Surface Mining, Faculty of Mining and Geoengineering, AGH University of Science and Technology, 30059 Krakow, Poland; kazn@agh.edu.pl

* Correspondence: kpolak@agh.edu.pl

Received: 26 November 2018; Accepted: 4 January 2019; Published: 10 January 2019

Abstract: Water wells play an increasingly important role in providing water for the civilian population all over the world. Like other engineering structures, wells are subject to ageing processes resulting in degradation, which is observed as a reduction in hydraulic efficiency throughout their lifespan. To date, it has been found that the ageing process of a well is determined by a number of factors. The mathematical description of this process can be simplified. Drawing on Jacob's equation, this paper presents the course of the degradation process as a variable depending on operation time, well loss and flow rate. To apply the determined relationships in practice, simplifying assumptions were adopted, which make it possible to determine the moment of ageing compensations of the degradation processes. It was also demonstrated that the degradation process may be slowed down by the appropriate selection of initial operating parameters. The presented discussion highlights the significance of parameters α, δ and exponent β. The relation between hydraulic resistances in an aquifer and in the engineering structure is closely connected with these values. The presented arguments indicate that step drawdown tests provide the necessary information which allows tracking changes in the ageing processes occurring in the engineering structure. The analysis of the drawdown test results makes it possible to determine the moment when the necessary adjustments in the operating parameters of a water well should be performed. Eventually, it allows maintaining the high hydraulic efficiency of the intake and extending the lifespan of the well in accordance with the principle of sustainability.

Keywords: water well; hydraulic efficiency; degradation; engineering structure; well ageing; lifespan; well operation; water well management; sustainable efficiency

---

1. Introduction

Groundwater extraction is still growing all over the world. It is estimated that within half a century, i.e., between 1960 and 2010, groundwater use doubled [1]. The drilling of new groundwater wells is required. However, the number of boreholes is increasing not only due to population growth, but also due to a decrease in productivity of the existing assets. As well as other technical facilities wells are undergoing the process of degradation. As with any other technical facility, the current technical condition of a drilled well can be described with an efficiency index. Well efficiency is defined by Rorabaugh as the ratio of the theoretical drawdown computed by assuming that no turbulence is present to the drawdown in the well [2]. Walton defines the efficiency of a well as the ratio of the theoretical specific capacity to the actual specific capacity of the well [3]. Bierschenk concludes that well efficiency may be defined as the ratio of the theoretical drawdown to the measured drawdown inside the well. He also presents multiple examples of efficiency curves calculated using step-drawdown tests carried out for wells located in the Middle East [4]. Therefore, well loss determines the hydraulic efficiency of a wellbore.

Well loss can be evaluated using the step-drawdown test. In Jacob's equation, well loss is described as the non-linear (non-laminar) flow regime component CQ^2 [5]. In some cases, changes of the drawdown can be explained by the Rorabaugh formula, where the power exponent p in the expression CQ is different from 2. Mackie, as cited by Atkinson et al. [6], reviewed the results of more than 20 carefully conducted step-drawdown tests of wells completed in fractured rock aquifers and concluded that most of the responses fell into one of the three categories where specific drawdown versus discharge rate is: linear, polynomial with power exponent equal to 2, or different from 2. The results of a study by Motyka and Wilk regarding the determination of the non-linear flow zone around several dozen wells drilled in fractured rocks indicate that the radius of turbulent flow zones are usually from 0.5 to 5.0 m, although in most cases they do not exceed 1 m [7].

Atkinson et al. concludes that high pumping rate moves the turbulent flow regime into the fissured aquifer, which is the reason for power exponent changes in Rorabaugh formula [8]. Klich et al. considered that the nature of the drawdown versus well-discharge curve depends on the range of well discharge [9]. Shekhar carried out the series of filed test for unconsolidated soils. He noticed that flow into the well screen is of a turbulent nature even for low discharge. The decrease in efficiency with increase of discharge can be considered from the perspective of increase in turbulence on account of the increase in discharge. As the drawdown increases the curvature of flow path increases leading to greater head loss. Therefore, well efficiency calculated can be regarded as a reflection of head loss on account of the laminar flow from the aquifer [10].

Much of the literature on the step-drawdown test focuses on the methods of determining aquifer resistance coefficient for linear water flow B, well resistance coefficient for non-linear flow C and p (if assumed not to be 2) parameters. Several methods are compiled by Kruseman and de Ridder [11]. One of the simplest graphical approaches is attributed to Jacob [5], Bruin and Hudson [12], Bierschenk and Wilson [13]. The parameters of the step-drawdown test can also be determined numerically. Miller and Webber [14] present an iterative method for solving the equation, while Labadie and Helweg developed a computer program for step-drawdown test data interpolation with the FASTEP procedure [15]. Avci proposes a method of analysis which calculates the aquifer and well loss from best fitting the log-log relationships of the difference of specific capacity and superposition of incremental pumping to the flow rate [16]. Wen et al. built the analytical model for all the types of curves considered for water well in confined aquifer. They noticed that all the curves at the wellbore approach the same asymptotic straight line in log s–log t scales [17]. Determining hydraulic characteristics of production wells from step-drawdown test data were also proposed by Jha et al. The characteristics of production wells such as aquifer loss coefficient, well loss coefficient, well specific capacity and well efficiency were determined by both the Genetic Algorithm optimization technique and the widely used graphical method. The developed computer programs also provide information about the condition of production wells and facilitate the construction of well characteristic curves [18].

Helweg proposes using General Well Function (GWF) to interpret the step-drawdown test. It allows more reliable results to be achieved for situations when the time that wells are continuously pumped varies greatly [19]. Kawecki presents a method for calculating total well loss based on the step drawdown test. In this method, the total loss can be estimated as a function of discharge rate. However, the real range radius of the borehole and also the specific storage of the confined aquifer are changeable at different stages of the step drawdown test [20]. An analytical solution for the constant pumping test in fissured porous media is presented by de Smedt. The solution is based on the dual-porosity approach with pseudo-steady state exchange between fissures and matrix. Proposed solution provides approach to analyze pumping test in fissured porous media characterized by double-porosity effects [21].

Singh proposes variable pumping replacing the step drawdown test, which does not require steady-state conditions at each step. This method allows the simultaneous estimation of both aquifer and well loss parameters [22]. The same author proposes a method for identifying head-loss from early drawdowns, where well loss could be evaluated from a short-duration pumping test in transient conditions. However, this method requires measurements in the pumped well and observation wells [23]. In contrast to this, Avci et al. proposed an analysis technique for interpreting transient step-drawdown test data. The method is based on taking the derivative of the drawdown with respect to time for the entire pumping test period, eliminating the constant well-loss terms. The conducted tests showed that the method allows the generation of the aquifer function $B(t)$ for the entire duration of the step-drawdown test [24]. The choice of the model is the most difficult choice that the analyst of such a hydraulic test has to make since a wrong model can only lead to the wrong conclusions and failure of the borehole [25]. Although the physical model is well-known and widely used, there is a strong need to improve the techniques for estimating uncertainty associated with parameters derived from the interpretation of well tests [26].

The results of the step drawdown test could be applied to hydrogeology research. Dufresne presents the data used to develop a regional groundwater model which facilitates water planning and sustainability [27]. However, the step drawdown test is the most common tool to assess well performance and the hydraulic statement of the clogging. Many types of clogging processes are described in the book by Houben and Treskatis [28]. It involves a detailed discussion of the causes and effects of physical, chemical, electrochemical, and biochemical clogging. It also describes the methods and possibilities of rehabilitation of the well screen, gravel-pack and well-tube.

Van Beek presents the rehabilitation of mechanically clogged discharge wells. The research conducted so far has shown that the increase in flow velocity around a wellbore mobilizes fine particles, which are then transported through the porous structure. This may lead to the clogging of slots of the screen, thus reducing both porosity and permeability [29]. The mechanical clogging process in unconsolidated aquifers near the water supply wells is presented by de Zwart [30] and de Zwart et al. [31]. Blackwell et al. analyses particulate damage, which is often cited as a cause of permeability impairment around boreholes, resulting in a decline in well performance. Particulate movement and redistribution under borehole operating conditions have been assessed for a range of artificial and natural formations. Particulate damage cannot be effectively eliminated using normal development and rehabilitation techniques. This movement may occur at different stages in the life of the borehole, i.e., drilling, development and operational damage. Therefore, the operating regime has an important effect on well performance, which should be regularly monitored. If deterioration reaches approximately 10%, rehabilitation action can be initiated. Allowing deterioration to exceed 25% can greatly increase the cost of successful rehabilitation [32].

The example of chemical clogging including iron and iron-reducing bacteria is also presented by Hitchon [33]. The clogging of deep infiltration wells was also presented by de la Loma Gonzales, where several different types of chemical clogging and rehabilitation methods were compared and evaluated using a specific case of a well system [34]. The requirements for the economical construction and operation of drilled wells was also described by Treskatis et al [35]. These authors propose preventive maintenance for well rehabilitation as an essential guard against the development of non-soluble incrustations of Fe- and Mn-(hydr)oxides by back wash procedures or high-velocity, horizontal jetting techniques at an early stage of biochemical development aid to keep well efficiency at a high level. They also point out that surveillance of well efficiency and changes of drawdown can help to reduce the effort of chemicals for well rehabilitation. Van Beek concludes that well bore clogging may be prevented by regular intermittent abstraction. During rest, as there is no incoming flow, the particle accumulations will disintegrate. However, there is no clear picture of this disintegration process, an objective criterion for the rest time is lacking. It might be possible to shorten the rest period by reversing the flow direction with the abstraction flow velocity at least. Well bore clogging may also be counteracted by proper well development. Currently the criterion for a developed well is the absence of sand in the abstracted water. Actually, well development should be maximal until there is no further increase in specific capacity [36].

Houben et al. conclude that dramatic increases in head losses occur when clogging has reduced the effective porosity of the gravel pack by ~65%, the open area of the screen by $\geq$98%, and the casing diameter by ~50%. However, the clogging of the gravel pack is the most important object influencing well ageing. Moreover, obstructions in the gravel pack are much more difficult to remove. So, regular monitoring of well performance is needed, since processes in the gravel pack are difficult to track directly [37]. Therefore, low well efficiency contributes to the increased operating costs of wells. Helweg and Bengstone considered the economic approach the most profitable pumping rate in the whole life-span of the well, including of substitute well drilling [38]. Cost optimization throughout the life-span of a well is also the subject of a paper by Hurynowicz and Syczewa, who conclude that maintaining relatively low operating costs of a well depends on its appropriate design and operating conditions over several dozen years of its operation [39].

To sum up the above, well ageing causes well loss which is time-dependent. However, well loss is a naturally occurring phenomenon that depends on the hydraulic features of the aquifer, well construction, hydrogeochemistry, borehole drilling, well development and operation. The literature review presented above demonstrates that step-drawdown tests are typically used to estimate the hydrogeological parameters, well loss and well efficiency. Step-drawdown test results are also used to assess the current state and the effects of rehabilitation operations.

2. Materials and Methods

Water supply wells should be characterized by stable operation over a long period of time. They should therefore be operated with the highest possible hydraulic efficiency. After by van Tonder et al. [25], the rate at which a borehole can be pumped without lowering the water level below a set level is called "sustainable yield". Piscoppo and Summa discussed an experiment to estimate the pumping rate to ensure that the sustainable yield concept is realized [40]. At the beginning a step drawdown test was carried out, then the efficiency of the well was determined. Based on the results of the tests performed, the constant-head for the functioning of the well was defined. It was calculated that for a drawdown of around 12 m, the well efficiency is greater than 75%. The well was equipped with suitable submersible pump unit, then it was monitored on-line by one year. The pumping rate was defined as the discharge rate that will not cause the water level in the well to drop below a prescribed limit. In conclusions, authors emphasize that in some cases the constant-head pumping can be an alternative method of well management. Analysis of the recorded data changes shown that the discharge rate of the well trend with time was similar to that of the springs' hydrograms. Moreover, the water volume extracted did not exceed the recharge [40]. Subsequently, constant head model was compared with a constant flow rate model by Baiocchi et al. Based on numerical modelling, the authors concluded that constant flow rate model could be particularly useful when the problem is one of determining the sustainable yield of a single well from aquifers with low hydraulic diffusivity and when an extensive monitoring of the aquifer is not economically viable [41]. Unfortunately, the researchers did not evaluate whether both models had an impact on the well efficiency at the end of the research period.

In contrast to examples mentioned above, the authors of this paper have developed a concept of well operation while maintaining its constant efficiency, which could be also called "sustainable efficiency". This concept required the adoption of certain initial assumptions and was then determined theoretically using formulas evaluating the hydraulic efficiency of the well, taking into account the elements of the optimization calculus and also the ageing function of technical objects. Literature on hydrogeological issues states that the efficiency of a well can be described using an expression involving a rational function in the form proposed by Bierschenk [2]:

$$\eta = \frac{s_1}{s} = \frac{s_1}{s_1+s_2} = \frac{BQ}{BQ+CQ^2} = \frac{1}{1+\frac{C}{B}Q}$$
$$\eta = \frac{1}{1+\alpha Q}, for\ \alpha = \frac{C}{B} \tag{1}$$

where:

η hydraulic efficiency [-];
s_1 aquifer loss [L];
s drawdown observed in the pumping well [L];
s_2 well loss in the well-screen adjacent zone [L];
Q volumetric flow rate [L^3/T];
B aquifer resistance coefficient for laminar water flow [T/L^2];
C well resistance coefficient for turbulent flow [T^2/L^5];
α parameter describing the prevalence of turbulent flow [T/L^3].

This concept is presented in a schematic Figure 1.

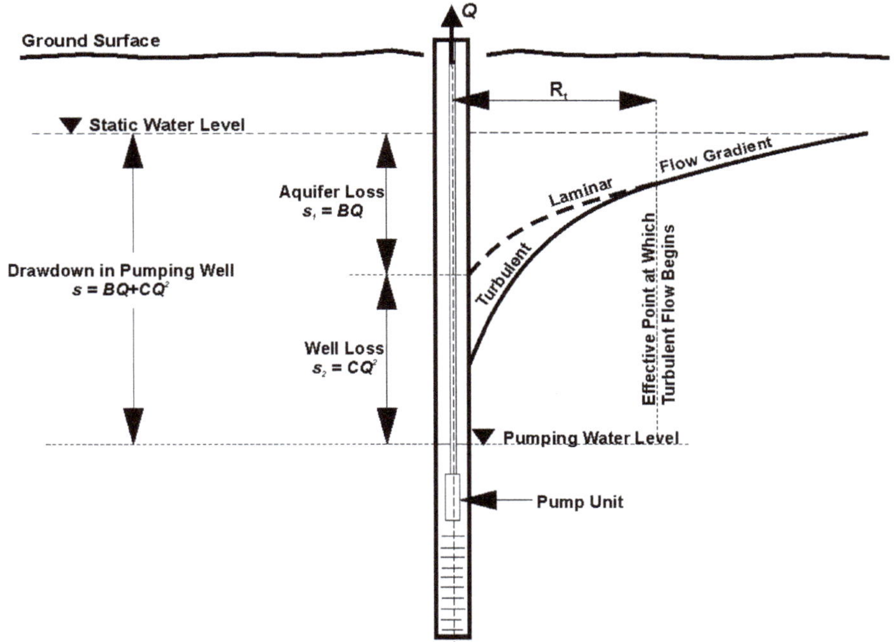

Figure 1. Head losses in pumped well [4]; where: R_t—distance from center of well to effective point formation where transition from laminar to turbulent flow takes place.

At the beginning of the consideration presented below, it was assumed that the state of knowledge is sufficient to be used to select the optimum operating parameters of new wells and also to perform ongoing adjustments to counteract the intensive ageing of operated wells. So far, this subject matter has not been discussed in detail in the literature. For this reason, the authors present the results of their own research and considerations regarding the dynamics of the ageing process and the possibilities of reducing its rate. Below the article presents a process of degradation of a water well as an engineering structure. As previously mentioned, well efficiency function is time dependent. It means that a degradation process can be described as efficiency function in time. Therefore, well ageing is derivative from well efficiency in time.

For this purpose, the initial assumptions relating to the ageing process of the well could be defined by taking into account the state of knowledge presented in previous section:

Assumption 1. *Aquifer resistances are slow-variable. This assumption is valid for water supply wells where hydrogeological conditions do not change over time. In relation to drainage wells, the condition is usually not applicable due to the lowering of the water level and the relatively short life-span of the wells.*

Assumption 2. *The value of well loss (s_2) should not significantly exceed the value of aquifer loss (s_1). This assumption is usually relevant to new wells where the well-loss is insignificant in comparison to total drawdown. However, such criteria is also applicable to other wells in good hydraulic condition.*

Assumption 1 assumes that aquifer resistances are slow variables. This means that $\dot{B}(t) \approx 0$ in comparison to the rates of changes in the well resistance parameters (C) or its discharge (Q)—$\dot{B}(t) \ll \dot{C}(t)$ (or $\dot{Q}(t)$) (accurate to the consistency factor of the dimension of a physical quantity). Therefore, if the said coefficients depend on time in a way that: $C = f(t)$, $B = const(t)$, $\alpha = \frac{C}{B} = g(t)$ and $Q = h(t)$, then well efficiency will also be a function of time $\eta = \eta(t)$. The well efficiency function can be expanded into a Taylor series around moment $t = t_0$ as:

$$\eta(t) = \eta(t_0) + \sum_{n=1}^{k} \frac{1}{n!} \frac{d^{(n)}\eta}{dt^{(n)}} \bigg|_{t=t_0} (t - t_0)^n \tag{2}$$

or by writing the above equation as follows:

$$\eta(t) = \eta(t_0) + \frac{1}{1!}\dot{\eta}(t_0)\cdot(t - t_0) + \frac{1}{2!}\ddot{\eta}(t_0)\cdot(t - t_0)^2 + \frac{1}{3!}\dddot{\eta}(t_0)\cdot(t - t_0)^3 + \ldots \tag{3}$$

Increases in the relevant variables may be used here—the increase in the independent variable (time) $\Delta t = t - t_0$ and in the dependent variable (well efficiency) $\Delta\eta = \eta - \eta_0$. Then, the following relationship is present:

$$\Delta\eta(t) = \dot{\eta}(t_0)\cdot\Delta t + \frac{1}{2}\ddot{\eta}(t_0)\cdot\Delta t^2 + \frac{1}{6}\dddot{\eta}(t_0)\cdot\Delta t^3 + \ldots \tag{4}$$

The time derivative of well efficiency is:

$$\dot{\eta}(t) = \frac{d}{dt}\left(\frac{1}{1+\alpha Q}\right) = \frac{-1}{(1+\alpha Q)^2}(\dot{\alpha}Q + \alpha\dot{Q})\dot{\eta}(t) = -(\dot{\alpha}Q + \alpha\dot{Q})\cdot\eta^2 \tag{5}$$

The second-order time derivative can be described by the expression:

$$\ddot{\eta}(t) = \tfrac{d}{dt}(\dot{\eta}) = (-1)\frac{(\dot{\alpha}Q+\alpha\dot{Q})'\cdot(1+\alpha Q)^2-[(1+Q)^2]'\cdot(\dot{\alpha}Q+\alpha\dot{Q})}{[(1+\alpha Q)^2]^2}$$

$$\ddot{\eta}(t) = (-1)\frac{(\ddot{\alpha}Q+\dot{\alpha}\dot{Q}+\dot{\alpha}\dot{Q}+\alpha\ddot{Q})\cdot(1+\alpha Q)^2-2(1+\alpha Q)\cdot(\dot{\alpha}Q+\alpha\dot{Q})\cdot(\dot{\alpha}Q+\alpha\dot{Q})}{(1+\alpha Q)^4}$$

$$\ddot{\eta}(t) = (-1)\frac{(\ddot{\alpha}Q+2\dot{\alpha}\dot{Q}+\alpha\ddot{Q})\cdot(1+\alpha Q)^2-2(1+\alpha Q)\cdot(\dot{\alpha}Q+\alpha\dot{Q})^2}{(1+\alpha Q)^4} \tag{6}$$

$$\ddot{\eta}(t) = (-1)\frac{[(\ddot{\alpha}Q+2\dot{\alpha}\dot{Q}+\alpha\ddot{Q})\cdot(1+\alpha Q)-2(\dot{\alpha}Q+\alpha\dot{Q})^2](1+\alpha Q)}{(1+\alpha Q)^4}$$

$$\ddot{\eta}(t) = \frac{2(\dot{\alpha}Q+\alpha\dot{Q})^2-(\ddot{\alpha}Q+2\dot{\alpha}\dot{Q}+\alpha\ddot{Q})\cdot(1+\alpha Q)}{(1+\alpha Q)^3}$$

The above can be expressed in a simplified manner as:

$$\ddot{\eta}(t) = \tfrac{2}{\eta}(\dot{\eta})^2 - \eta^2(\ddot{\alpha}Q + 2\dot{\alpha}\dot{Q} + \alpha\ddot{Q})$$
$$\ddot{\eta}(t) = \tfrac{2}{\eta}(\dot{\eta})^2 - \eta^2\tfrac{d^2}{dt^2}(\alpha Q) \tag{7}$$

In Assumption 2 the value of well loss (s_2) should not significantly exceed the value of aquifer loss (s_1). This means that:

$$\frac{s_2}{s_1} \ll 1$$
$$\frac{s_2}{s_1} = \frac{CQ^2}{BQ} = \frac{C}{B}Q = \alpha Q \ll 1 \tag{8}$$

Therefore, the rate of change is small in comparison with other elements in the expression for the second-order time derivative of well efficiency:

$$\frac{d^2}{dt^2}\left(\frac{s_2}{s_1}\right) = \frac{d^2}{dt^2}(\alpha Q) \approx 0$$
$$\ddot{\eta}(t) = \frac{2}{\eta}(\dot{\eta})^2 - \eta^2 \frac{d^2}{dt^2}(\alpha Q) \approx \frac{2}{\eta}(\dot{\eta})^2 \tag{9}$$
$$\ddot{\eta}(t) \approx \frac{2}{\eta}(\dot{\eta})^2$$

Including in the expression the first-order derivative in the form: $\dot{\eta}(t) = -(\dot{\alpha}Q + \alpha\dot{Q})\cdot\eta^2$ provides the clear form:

$$\ddot{\eta}(t) = \frac{2}{\eta}\left[(-1)(\dot{\alpha}Q + \alpha\dot{Q})\cdot\eta^2\right]^2 \tag{10}$$
$$\ddot{\eta}(t) = 2(\dot{\alpha}Q + \alpha\dot{Q})^2\cdot\eta^3$$

The above considerations can be simplified by applying approximations in the solution, in which higher-order terms are negligible.

3. Results

The above equations describe the function of well degradation as a derivative of efficiency in time. However, it can be used to determinate the rate of ageing as well as to counteract this process. Issues can be considered in a simplified form as first-order approximation (kinetic), or in expansions of second-order approximations (dynamic). The higher-order terms, starting from the third order may be considered negligible. Changes in the function of the state of the well are presented in two independent variants, applying the principle of caeteris paribus. The first variant describes the ageing of the object and the decrease in hydraulic efficiency. The second variant describes the recovery of the lost efficiency of the well by changing the operational parameters of the well.

3.1. First-Order Approximation

In this case, higher-order terms are negligible, starting from the second order. In such a case the expression expanding the Taylor series for a function to a power series is identical to the expression for the differential of the function (adjacent to point (t_0; $\eta(t_0)$)):

$$\Delta\eta(t) \approx \dot{\eta}(t_0)\cdot\Delta t \tag{11}$$

which, after inserting the time derivative value for moment t_0 equals:

$$\Delta\eta = -(\dot{\alpha}Q + \alpha\dot{Q})\cdot\eta_0^2\cdot\Delta t \tag{12}$$

To ensure clarity, subscripts are omitted for derivatives of functions Q and α at the moment of time t_0. However, this fact should be taken into account when performing numerical calculations.

Time ranges can be matched in such a way that only one of the parameters, α or Q, will change distinctively over time, which would allow it to provide a significant contribution to the expression for changes in well efficiency throughout its lifespan. This will be discussed below on two extreme cases, which are viable on an engineering scale.

Case 1–Well Ageing

In this case it was assumed that the flow rate is invariable or slow-variable, while well resistances in relation to aquifer resistances increase considerably. This causes the deepening of the dynamic water table while maintaining a constant flow rate. This leads to a considerable reduction in the hydraulic efficiency of the well. Therefore, the turbulent flow of the reservoir fluids migrating in the near well zone acquires a high significance with time. This is connected with well ageing, which is a result of sediment accumulation in the functional part of the screen, clogging of the porous or fissure area of the medium directly adjacent to the screen and even the appearance of particulates inside the screen of the pumping well.

In other words: $\dot{Q} \approx 0$ *and* $\dot{\alpha} > 0$. Therefore:

$$\Delta\eta_{age} = -(\dot{\alpha}Q + \alpha\dot{Q})\cdot\eta_0^2\cdot\Delta t_{age} \Delta\eta_{age} = -\dot{\alpha}Q\cdot\eta_0^2\cdot\Delta t_{age} \tag{13}$$

As can be easily determined, the inequality $\Delta\eta_{age} < 0$ over time will always be satisfied.

Case 2—Compensation of Ageing

In this case the flow rate decreases with time and the well resistances do not increase significantly. This leads to the compensation of ageing in the operated pumping well by reducing the flow rate. For $\dot{Q} < 0$ *and* $\dot{\alpha} \approx 0$ the following can be specified:

$$\Delta\eta_{cmp} = -(\dot{\alpha}Q + \alpha\dot{Q})\cdot\eta_1^2\cdot\Delta t_{cmp} \Delta\eta_{cmp} = -\alpha\dot{Q}\cdot\eta_1^2\cdot\Delta t_{cmp} \tag{14}$$

Obviously, in the above case the inequality $\Delta\eta_{cmp} > 0$ over time will always be satisfied.

It should be pointed out here that a well with an initial efficiency of η_0 after the expiry of the period of ageing Δt_{age} reaches a state with the hydraulic efficiency of η_1. This state represents the initial state for the later compensation of the well loss value. After the compensation time Δt_{cmp} the well reaches an efficiency of η. The number of iterations depends on the manner of transition between the extreme states of the well desired by the user.

3.2. Second-Order Approximation

In this case, it was assumed that higher-order terms, starting from the third order, are negligible. Therefore, the power series expansion of the well efficiency function is as follows:

$$\Delta\eta(t) \approx \dot{\eta}(t_0)\cdot\Delta t + \frac{1}{2}\ddot{\eta}(t_0)\cdot\Delta t^2 \tag{15}$$

while ensuring that the time derivatives at the moment of time $t = t_0$ are equal to:

$$\begin{aligned} \dot{\eta}(t_0) &= -(\dot{\alpha}Q + \alpha\dot{Q})\cdot\eta_0^2 \\ \ddot{\eta}(t) &\approx \frac{2}{\eta_0}[\dot{\eta}(t_0)]^2 = 2(\dot{\alpha}Q + \alpha\dot{Q})^2\cdot\eta_0^3 \end{aligned} \tag{16}$$

Therefore, the relationship for the well efficiency change will be equal to:

$$\begin{aligned} \Delta\eta(t) &\approx -(\dot{\alpha}Q + \alpha\dot{Q})\cdot\eta_0^2\cdot\Delta t + \frac{1}{2}2(\dot{\alpha}Q + \alpha\dot{Q})^2\cdot\eta_0^3\cdot\Delta t^2 \\ \Delta\eta(t) &\approx -(\dot{\alpha}Q + \alpha\dot{Q})\cdot\eta_0^2\cdot\Delta t + (\dot{\alpha}Q + \alpha\dot{Q})^2\cdot\eta_0^3\cdot\Delta t^2 \\ \Delta\eta(t) &\approx (\dot{\alpha}Q + \alpha\dot{Q})\cdot\eta_0^2\cdot\Delta t[(\dot{\alpha}Q + \alpha\dot{Q})\cdot\eta_0\cdot\Delta t - 1] \end{aligned} \tag{17}$$

The first element of the difference provided in square brackets could be significantly smaller than one. This is possible because the time derivative of the product $\dot{\alpha}Q + \alpha\dot{Q} = \frac{d}{dt}(\alpha Q)$ for a short time (where $\Delta t \to 0$) is low or slow-variable. Therefore, for $\frac{d}{dt}(\alpha Q) \approx 0$ or $\alpha Q \approx const$ and $\Delta t \to 0$ an accurate expression for well efficiency change can be obtained at first-order approximation.

Case 1—Well Ageing

Well ageing occurs with an invariable flow rate and an increase in the drawdown observed in the pumping well through a significant increase in well resistances and insignificant increase in aquifer resistances.

For $\dot{Q} \approx 0$ *and* $\dot{\alpha} > 0$ the simplified form of relationship (17) applies:

$$\Delta\eta_{age} = -\dot{\alpha}Q\cdot\eta_0^2\cdot\Delta t_{age}[1 - \dot{\alpha}Q\cdot\eta_0\cdot\Delta t_{age}] \tag{18}$$

The above equation will be satisfied for the loss of hydraulic efficiency of a well ($\Delta\eta_{age} < 0$) if the passage of time is expressed in a limited range, given as the strict inequality $\Delta t_{age} < (\dot{\alpha}Q\cdot\eta_0)^{-1}$.

Case 2—Well Ageing Compensation

Flow rate decreases in an appropriately short time, while well resistances do not significantly increase over this time. For $\dot{Q} < 0$ *and* $\dot{\alpha} \approx 0$ it will be:

$$\Delta\eta_{cmp} = -\alpha\dot{Q}\cdot\eta_1^2\cdot\Delta t_{cmp}[1 - \alpha\dot{Q}\cdot\eta_1\cdot\Delta t_{cmp}] \tag{19}$$

As can be easily noticed, expression (19) describes inequality $\Delta\eta_{cmp} > 0$ with time and decreasing flow rate in the well over time. In other words, it is possible to recover at least a part of the hydraulic efficiency lost. As before, after time Δt_{age}, well efficiency decreased from η_0 to η_1, and it constitutes the initial value for the compensation of the well loss through a reduction in the flow rate of the borehole at time Δt_{cmp}. The number of iterations depends on the manner of transition between the extreme states of the well desired by the user.

4. Discussion

The archived results can be used in practice to compensate for well ageing. Taking into account that equations are difficult to use in practice, they can be expressed in a more practical way. Particular emphasis is put on the first order-approximation. For this purpose, characteristic indicators for full compensation of ageing will be presented below. In addition, expressions defining the necessary intervals to be taken in order to carry out corrective action are presented. The usage of those indicators for the selected example will be presented in the following section.

4.1. First Order Approximation

4.1.1. Compensation of Well Ageing

It follows from case 1 that efficiency η_1 is:

$$\eta_1 = \eta_0 - \dot{\alpha}Q\cdot\eta_0^2\cdot\Delta t_{age} \tag{20}$$

while obviously always $\Delta\eta_{age} = \eta_1 - \eta_0 < 0$. Case 2 indicates that the final efficiency η value is:

$$\eta = \eta_1 - \alpha\dot{Q}\cdot\eta_1^2\cdot\Delta t_{cmp} \tag{21}$$

while this time always $\Delta\eta_{cmp} = \eta - \eta_1 > 0$.

The insertion of the expression for efficiency η_1 results in a change in efficiency as a result of compensation as:

$$\begin{aligned}
\Delta\eta_{cmp} &= -\alpha\dot{Q}\cdot\eta_1^2\cdot\Delta t_{cmp} \\
\Delta\eta_{cmp} &= -\alpha\dot{Q}\cdot(\eta_0 - \dot{\alpha}Q\cdot\eta_0^2\cdot\Delta t_{age})^2\cdot\Delta t_{cmp} \\
\Delta\eta_{cmp} &= -\alpha\dot{Q}\cdot\eta_0^2\cdot\Delta t_{cmp}(1 - \dot{\alpha}Q\cdot\eta_0\cdot\Delta t_{age})^2
\end{aligned} \tag{22}$$

4.1.2. Full Compensation of Well Ageing

It is readily visible that the total change of well efficiency equals:

$$\Delta\eta = \eta - \eta_0 = (\eta - \eta_1) + (\eta_1 - \eta_0)\Delta\eta = \Delta\eta_{cmp} + \Delta\eta_{age} \tag{23}$$

the sign of the total change (difference) in efficiency is determined by the size of the individual elements.

Assuming that time moves towards moment t_0, i.e., $t \to t_0$, so the time range becomes infinitesimal ($\Delta t_i \to 0$) then the decrease in well efficiency as a result of ageing also becomes infinitesimal ($\Delta\eta_{age} \to 0$ or $\eta_1 \to \eta_0$). Furthermore, an increase in well efficiency as a result of compensating the well loss will also be infinitesimal ($\Delta\eta_{cmp} \to 0$ or $\eta \to \eta_1$). As for the rate of each of the above changes, it will be given using the "0/0" indeterminate form. However, it follows from the principle of transitivity of implication that $\eta \to \eta_0$ also occurs, therefore in general the transition of the function to the limit $\Delta\eta \to 0$ where $\Delta t \to 0$ must occur. This means that the equality of function limits will be satisfied.

$$\lim_{\Delta t \to 0} \Delta\eta = \lim_{\Delta t \to 0} \Delta\eta_{age} = \lim_{\Delta t \to 0} \Delta\eta_{cmp} = 0 \tag{24}$$

Equations (23) and (24) provide knowledge about the possibility of the following relationship:

$$\Delta\eta = \Delta\eta_{age} + \Delta\eta_{cmp} = 0 \tag{25}$$

It is always satisfied not only for initial moments of time, but also independently of time, which occurs with a full compensation of the decrease in well efficiency.

The insertion of expressions (13) and (14) into (25) results in:

$$\dot{\alpha} Q \cdot \eta_0^2 \cdot \Delta t_{age} + \alpha \dot{Q} \cdot \eta_1^2 \cdot \Delta t_{cmp} = 0 \tag{26}$$

Applying Leibniz's rules to the differential calculus for function composition (the chain rule), it can be specified that if the proposition $\frac{df}{dx} = \frac{\frac{df}{dt}}{\frac{dx}{dt}} \Rightarrow \frac{df}{dx}\frac{dx}{dt} = \frac{df}{dt}$ is true, then the time from derivatives in expression (26) can be easily erased to obtain a first-order homogenous differential equation depending on the variable parameters α and Q:

$$\frac{d\alpha}{dQ} \cdot \eta_0^2 \cdot \Delta t_{age} + \frac{\alpha}{Q} \cdot \eta_1^2 \cdot \Delta t_{cmp} = 0 \tag{27}$$

The separation of variables makes it possible to express the formula in the following form:

$$\begin{aligned}
-\frac{d\alpha}{\alpha} &= \frac{dQ}{Q} \cdot \left(\frac{\eta_1}{\eta_0}\right)^2 \left(\frac{\Delta t_{cmp}}{\Delta t_{age}}\right) \\
-\frac{d\alpha}{\alpha} &= \beta \frac{dQ}{Q}
\end{aligned} \tag{28}$$

The integration of relationship (28) with the constant of integration written as $\ln\gamma_0 = const$ and assuming for further simplicity constancy of coefficient β will result:

$$\begin{aligned}
-\int \frac{d\alpha}{\alpha} &= \int \beta \frac{dQ}{Q} = \beta \int \frac{dQ}{Q} \\
-\ln\alpha &= \beta \ln Q + const \\
\ln\alpha^{-1} &= \ln Q^\beta + \ln\gamma_0 = \ln\left(\gamma_0 Q^\beta\right)
\end{aligned} \tag{29}$$

After introducing coefficient $\delta = \gamma_0^{-1}$ the above expression can be written in an antilogarithm (exponential) formula:

$$\begin{aligned}
\alpha^{-1} &= \gamma_0 Q^\beta \\
\gamma_0^{-1} &= \delta = \alpha Q^\beta
\end{aligned} \tag{30}$$

where exponent β of the power is equal to:

$$\beta = \left(\frac{\eta_1}{\eta_0}\right)^2 \left(\frac{\Delta t_{cmp}}{\Delta t_{age}}\right) \tag{31}$$

Exponent β can assume different values, but some characteristic values can be listed. The following is a discussion of cases where: $\beta = 0$ and $\beta = 1$.

When $\beta = 0$, it should be expected that there is a strict inequality for the hydraulic efficiencies of the well at specific moments of time, i.e., $\eta_1 < \eta_0$. Therefore, their quotient is $\frac{\eta_1}{\eta_0} < 1$ and the square of quotient is $\left(\frac{\eta_1}{\eta_0}\right)^2 \ll 1$. Furthermore, it can be assumed that the time available for compensating well loss is shorter than the well ageing time. It is therefore realistic to assume that the relationship $\Delta t_{cmp} < \Delta t_{age}$ will be true each time. This means that the exponent in Equation (31) will be equal to $\beta \ll 1$. In the context of mathematical analysis, it proceeds in its limit to the value of $\beta \to 0$. Hence, expression (30) must assume the form:

$$\lim_{\beta \to 0} \delta = \lim_{\beta \to 0}\left(\alpha Q^\beta\right) = \lim_{\beta \to 0}\left(\alpha Q^0\right) = \lim_{\beta \to 0} \alpha = \alpha$$
$$\lim_{\beta \to 0} \delta = \alpha = \tfrac{C}{B} = \delta_0 \tag{32}$$

Therefore, when the user of a water supply well is determined to achieve full compensation of the well loss in a wellbore used over a span of many years, then a certain stable state between the resistances of the aquifer and the resistances of the well should be established for the water flow in the aquifer. Indeed, as provided by Equation (32):

$$C = \alpha \cdot B \quad or \quad C = \delta_0 \cdot B \quad when \quad \beta \approx 0 \tag{33}$$

The delta (δ_0) parameter does depend on the property of the area directly adjacent to the functional part of the well screen and the permeability of the rock-soil medium conducting the fluid, as well as the type of the flowing reservoir medium.

Taking into account Equations (1) and (33), the hydraulic efficiency of the well can be equal to:

$$\eta = \frac{1}{1 + \alpha Q} = \frac{1}{1 + \delta_0 Q} \quad when \quad \beta \approx 0 \tag{34}$$

In other words, the ageing and compensation time ranges assumed for the well are similar in value ($\Delta t_{age} \approx \Delta t_{cmp}$) and the well efficiencies compared to the limits of the structure's ageing period range are almost identical ($\eta_0 \approx \eta_1$), then the exponent from Equation (31) equals $\beta \approx 1$. Obviously, this case will occur for initial moments—in situations where $\Delta t_{age} \to 0$ and $\Delta t_{cmp} \to 0$, i.e., where $\eta_1 \to \eta_0$ (or $\Delta \eta_{age} \to 0$). Therefore, the limit:

$$\lim_{\beta \to 1} \delta = \lim_{\beta \to 1}\left(\alpha Q^\beta\right) = \lim_{\beta \to 1}\left(\alpha Q^1\right) = \lim_{\beta \to 1} \alpha Q = \alpha Q$$
$$\lim_{\beta \to 1} \delta = \alpha \cdot Q = \tfrac{C}{B} \cdot Q = \delta_1 \tag{35}$$

Then:

$$\delta_1 = \alpha \cdot Q = \frac{C}{B} \cdot Q = \frac{CQ^2}{BQ} \quad or \quad \delta_1 = \frac{s_2}{s_1} \quad when \quad \beta \approx 1 \tag{36}$$

Parameter δ_1 for the initial moment under the conditions described above defines the mutual relationships between well loss and aquifer depression (or generally for a pool of any reservoir medium).

The hydraulic efficiency of a water supply well in accordance with the initial Jacob's formula, Equation (1), is then equal to:

$$\eta = \frac{s_1}{s_1 + s_2} = \frac{s_1}{s_1 + \delta_1 \cdot s_1} = \frac{1}{1 + \delta_1} \quad when \quad \beta \approx 1 \tag{37}$$

A table listing was prepared to summarize the discussion on the full compensation of well loss in the operated pumping well (Table 1).

Table 1. A listing of the calculation results for the full compensation of ageing-related well efficiency decrease, relationship of the type $\delta = \alpha \cdot Q^\beta$ for $\Delta\eta = \Delta\eta_{age} + \Delta\eta_{cmp} = 0$.

β	δ_β	η	Note
0	$Q_0^{-1} = \alpha = C/B$	$1/(1 + \delta_0 \cdot Q)$	Full compensation—long period of well operation
1	$s_2/s_1 = \alpha \cdot Q$	$1/(1 + \delta_1)$	Full compensation—initial moments of well operation

4.1.3. Compensation Time

From Equations (30) and (31) the time required for the full compensation of hydraulic efficiency of the water supply well can be calculated. The following expression can be written:

$$\beta = \frac{\ln \frac{\delta}{\alpha}}{\ln Q} = \left(\frac{\eta_1}{\eta_0}\right)^2 \left(\frac{\Delta t_{cmp}}{\Delta t_{age}}\right) \tag{38}$$

Therefore, compensation time can be calculated using the formula:

$$\Delta t_{cmp} = \frac{\ln \frac{\delta}{\alpha}}{\ln Q} \cdot \left(\frac{\eta_0}{\eta_1}\right)^2 \cdot \Delta t_{age} \tag{39}$$

4.2. Second Order Approximation

Omitting the second element of the factor provided in square brackets in Equations (18) and (19) makes it possible to obtain a first-order approximation in such special cases. Writing down further equations would blur clarity due to complicated notation. At this stage Equations (18) and (19) provide information which allows taking action aimed at improving the hydraulic efficiency of the well.

The above discussion clearly demonstrates that periodical step drawdown tests play a fundamental role in the assessment of hydraulic state and also the potential actions aimed at extending the period of intake operation at a high level of hydraulic efficiency. However, the increase in well loss could be observed during regular or occasional inspections of operating parameters, i.e., drawdown and flow rate. In practice, compensation of the loss of hydraulic efficiency is achieved by methods such as valve throttling, switching the type series of pump units or, alternatively, regulating the rotational speed of the pump impeller.

Under conditions of a full compensation of wellbore ageing, parameter δ_0 is numerically related to the flow of the reservoir medium for which the flow rate of a water supply well $Q_0 = (C/B)^{-1}$ results in the hydraulic efficiency of the well equal to 0.5 (50%), i.e., where well loss (s_2) is equal to aquifer loss (s_1). An increase in the value of parameter δ_0 means that the well reaches the same level of hydraulic efficiency with a lower volumetric flow rate. This is combined with a progressive increase in the resistance coefficient value of the well in turbulent flow at the well-screen near zone in relation to the resistance coefficient of laminar flow in the aquifer. In other words, parameter δ_0 may constitute a reliable indicator of well condition.

For a sufficiently long period of operation of the well the exponent is $\beta = 0$. Then for factor $\delta = \delta_0 = \alpha = C/B$ the time required to compensate well loss $\Delta t_{cmp} = 0$, i.e., in the conditions of well operation, is $\Delta t_{cmp} << \Delta t_{age}$. This means that for the long and intensive operation of water supply wells, flow rate reductions and compensation of hydraulic efficiency loss must not be delayed. For exponent $\beta = 1$, i.e., at the initial stage of the lifespan of a water supply well, factor $\delta = \delta_1 = s_2/s_1 = \alpha Q$ ensures compensation time at a level equal to the time of ageing, i.e., operation of the structure—$\Delta t_{cmp} \approx \Delta t_{age}$, provided that the efficiencies assume similar values ($\eta_0 \approx \eta_1$). In other words, the time available for taking corrective actions for the well becomes extended and is similar in its order of magnitude to the time of operation of the wellbore.

Furthermore, the higher the initial efficiency of the well (in relation to that expected after the period of operation of the well), the later it will be possible to begin compensating the loss of hydraulic efficiency by regulation through discharge, e.g., in an active or passive manner. On the other hand, excessively intensive pumping from the well will result in reducing the time available for compensating the effect of well loss. The well loss should increase with the clogging of the well-screen zone, manifested in the increase in the resistance coefficient value C, i.e., with the turbulent flow in the aquifer. Ultimately, this manifests itself in a decrease in the hydraulic efficiency curve of the well. Moreover, greater curvature of the graph is observed.

4.3. Case Study

Figures 2 and 3 present an example of using the solution presented above. Initially, Well 1 (Figure 2) operated at the flow rate of 12.5×10^{-3} m^3/s, however parameter $\alpha^{-1} = B/C = Q_0 = 37.4 \times 10^{-3}$ m^3/s. After 10 years of uninterrupted operation there was an unforeseen failure connected with the exposure of the pumping unit. A pumping test demonstrated a well loss of up to 80% of total drawdown. To protect the new pump unit, the pumping rate was limited to 5.6×10^{-3} m^3/s, which corresponded to the hydraulic efficiency of the well of 30%. The conducted rehabilitation procedures failed to bring the expected results. The tests demonstrated that the final low efficiency of the well was caused by the relatively low initial well efficiency and the lack of efficiency compensation during operation. It has led to gravel-pack clogging and particulate damage. Finally, the value Q_0 decreased to 2.4×10^{-3} m^3/s, which corresponds to an efficiency of 50%.

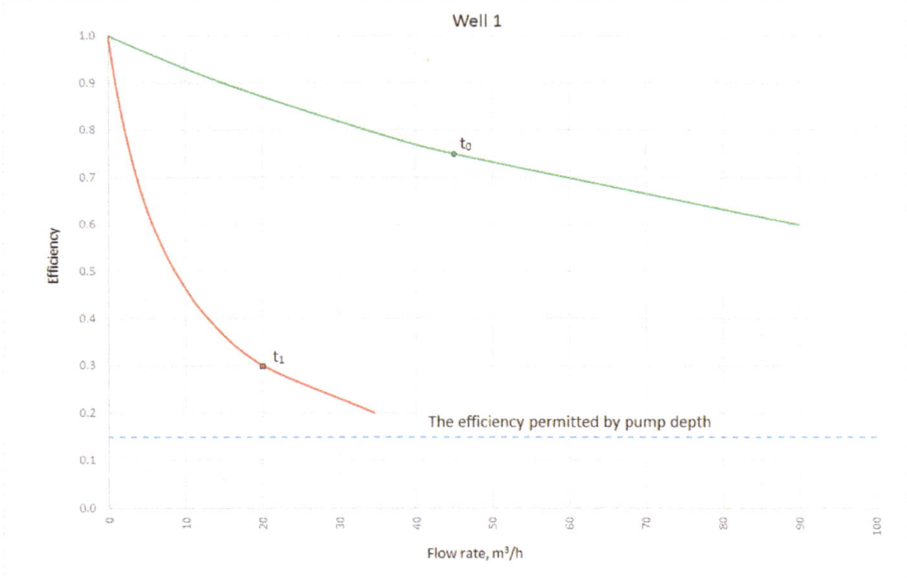

Figure 2. The hydraulic efficiency of the well without flow rate compensation: t_0—at the beginning of operation time; t_1—at the end of operation time.

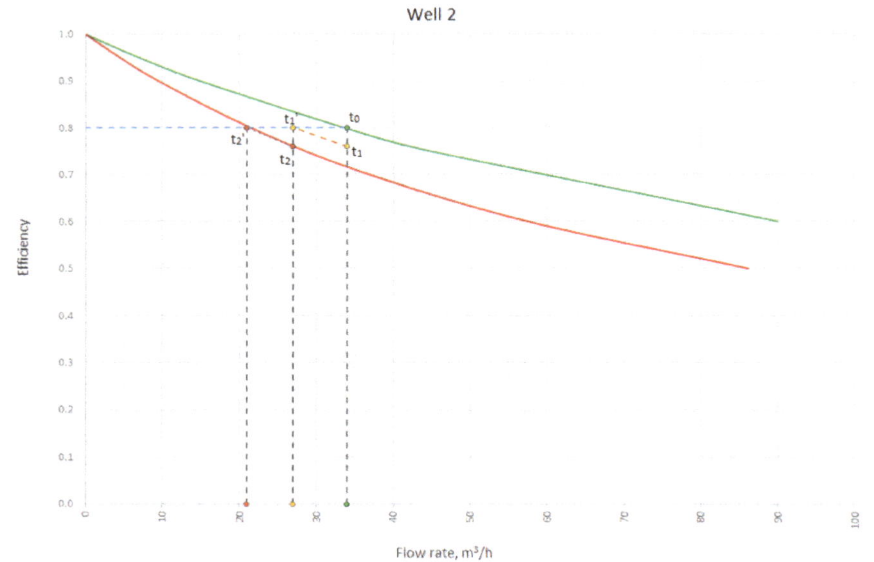

Figure 3. The hydraulic efficiency of the well with flow rate compensation: t_0—the working point at the beginning of operation; t_1—the working point before first flow rate compensation; $t_{1'}$—the working point after first flow rate compensation; t_2—the working point before second flow rate compensation; $t_{2'}$—the working point after second flow rate compensation; η_0—the origin efficiency curve; η_2—the current efficiency curve.

To reduce operating costs, the company conducting groundwater extraction constructed a replacement well. The methods presented in this article were used during its operation (Figure 3). Initially, the maintenance of an efficiency of 80% was established, which corresponded to the flow rate of 9.4×10^{-3} m^3/s, however parameter $Q_0 = 37.1 \times 10^{-3}$ m^3/s. After a further eight years the reduced flow rate was 5.8×10^{-3} m^3/s with a constant efficiency. In this time, the value Q_0 decreased to 23.9×10^{-3} m^3/s, which corresponds to an efficiency of 50%.

It should be noted that flow rate was significantly reduced for both wells. Currently, Well 1 is used occasionally due to high operating costs, while the costs of water extraction from Well 2 are approximately 65% lower. In addition, Well 2 operates at an efficiency which guarantees its reliable operation in the long term. The efficiency values, which are marked green and red in Figures 2 and 3, were confirmed by step drawdown tests. In practice, efficiency compensation in the second case was achieved by valve throttling at the beginning of operation and replacing the pump unit type series at appropriate time [16].

5. Conclusions

The solution presented in this paper involves ageing compensation and maintaining as high a hydraulic efficiency as possible. It concerns water supply wells in which well-loss has a minor share in the drawdown generated by water pumping. So, the presented methodology is valid to both new and old wells in good hydraulic condition, where hydrogeological conditions do not generally change over time. The fundamental meaning in presented method has parameter α. This parameter is a relation C to B, and may constitute a reliable well condition index which is relevant to 0.5 (50%) efficiency of the well.

Unlike "sustainable yield", concept of efficiency, compensation is not constant-rate or constant-head method. The "sustainable efficiency" concept requires regulation of both flow rate and drawdown. In relation to aquifer recharge fluctuations, operational parameters could be temporarily increased when water larger resources are available. However, in general, operating parameters of the well are reduced over time. This approach allows for the maintenance of a constant efficiency of the well, as long as the water output is sufficient to cover the water demand. Then, rehabilitation of the well is needed. Taking into consideration compensation time, it should be expressed that regulation time is dependent on current efficiency of the well. When the initial efficiency is relatively high, the compensation can be postponed or delayed. Alternatively, when the initial efficiency is low the time to reducing operational parameters is shortened.

The methodology was applied in a groundwater extraction company for new Well 2, drilled as a substitute of a Well 1, where efficiency was dramatically reduced over 10 years of constant water extraction. The periodical efficiency reduction of Well 2 during similar times of operation forced both the pumping rate and drawdown reduction. Currently, it has resulted in a relatively low cost of water extraction. Furthermore, the relation B to C still remains on a relatively high-level value. This indicates that the well-loss is quite small and the particulate damage in the near-well zone was minimized. The upcoming rehabilitation of Well 2 will probably be much more effective in this case. It can also be noticed that theoretical considerations presented in this paper were verified by step-drawdown tests at the beginning and also at the end of the research period. So, the method can be applied as a practical solution in well management to prevent inefficiencies.

Author Contributions: K.P. and K.K.-O. conceived of the presented idea. K.G. developed the theory and performed the computations. K.P. verified the analytical methods, performed the measurements and field test data calculations. K.K.-O. contributed to the interpretation of the results. All authors discussed the results and contributed to the final manuscript.

Funding: This research received no external funding. The APC was funded by AGH University of Science and Technology, Krakow 30-059, Poland.

Acknowledgments: The authors are grateful for the constructive comments made by anonymous reviewers which helped to improve the manuscript.

Conflicts of Interest: The authors declare no conflict of interest.

References

1. Wada, Y.; van Beek, L.P.H.; van Kempen, C.M.; Reckman, J.W.T.M.; Vasak, S.; Bierkens, M.F.P. Global depletion of groundwater resources. *Geophys. Res. Lett.* **2010**, *37*, L20402. [CrossRef]
2. Rorabaugh, M.J. Graphical and Theoretical Analysis of Step Drawdown Test of Artesian Well. *Proc. Am. Soc. Civ. Eng.* **1953**, *79*, 1–23.
3. Walton, W.C. Ground-Water Hydraulics as an Aid to Geologic Interpretation. *Ohio J. Sci.* **1955**, *55*, 13–20.
4. Bierschenk, W.H. *Determining Well Efficiency by Multiple Step-Drawdown Tests*; International Association of Scientific Hydrology Publication: London, UK, 1964; pp. 493–507.
5. Jacob, C.E. Drawdown test to determine effective radius of artesian well. *Trans. Am. Soc. Civ. Eng.* **1947**, *112*, 1047–1064.
6. Atkinson, L.C.; Gale, J.E.; Dudgeon, C.R. New Insight to the Step-Drawdown Test in Fractured-Rocks Aquifers. *Appl. Hydrogeol.* **1994**, *2*, 9–18. [CrossRef]
7. Motyka, J.; Wilk, Z. Calculated radius of nonlinear zone around fissured-karstified rocks tapping well. *Przegląd Geologiczny* **1984**, *32*, 90–94. (In Polish)
8. Atkinson, L.C.; Kemping, P.G.; Wright, J.C.; Liu, H. The Challenges of Dewatering at the Victor Diamond Mine in Northern Ontario, Canada. *Mine Water Environ.* **2010**, *29*, 99–101. [CrossRef]
9. Klich, J.; Polak, K.; Sobczyński, E. Description of the method to assessment of water supply and drainage wells. In Proceedings of the III International Scientific and Technical Conference on Water Supply for Cities and Villages, Poznań, Poland, 1–3 June 1998; pp. 161–176. (In Polish)
10. Shekhar, S. An approach to interpretation of step drawdown tests. *Hydrogeol. J.* **2006**, *14*, 1018–1027. [CrossRef]

11. Kruseman, G.P.; de Ridder, N.A. *Analysis and Evaluation of Pumping Test Data*, 2nd ed.; International Institute for Land Reclamation and Improvement: Wageningen, The Netherlands, 1990; pp. 199–215, ISBN 13 978-9070754204.

12. Bruin, J.; Hudson, H.E., Jr. *Selected Methods for Pumping Test Analysis*, 3rd ed.; Report of Investigation; Illinois State Water Survey: Champaign, IL, USA, 1961; pp. 6–37.

13. Bierschenk, W.H.; Wilson, G.R. *The Exploitation and Development of Ground Water Resources in Iran*; International Association of Scientific Hydrology Publication: London, UK, 1961; pp. 338–342.

14. Miller, C.T.; Weber, W.J. Rapid solution of the nonlinear step-drawdown equation. *Ground Water* **1983**, *21*, 584–588. [CrossRef]

15. Labadie, J.W.; Helweg, O.J. Step-drawdown test analysis by computer. *Ground Water* **1975**, *13*, 46–52. [CrossRef]

16. Avci, C.B. Parameter Estimation for Step-Drawdown Tests. *Ground Water* **1992**, *30*, 338–342. [CrossRef]

17. Wen, Z.; Huang, G.; Zhan, H.; Li, J. Two-region non-Darcian flow toward a well in a confined aquifer. *Adv. Water Resour.* **2008**, *31*, 818–827. [CrossRef]

18. Jha, M.K.; Manda, G.; Samuel, M.P. Determining Hydraulic Characteristics of Production Wells using Genetic Algorithm. *Water Resour. Manag.* **2004**, *18*, 353–377. [CrossRef]

19. Helweg, O.J. General Solution to the Step-Drawdown Test. *Ground Water* **1994**, *32*, 363–366. [CrossRef]

20. Kawecki, M.W. Meaningful Interpretation of Step-Drawdown Tests. *Ground Water* **1995**, *33*, 23–32. [CrossRef]

21. De Smedt, F. Analytical Solution for Constant-Rate Pumping Test in Fissured Porous Media with Double-Porosity Behaviour. *Transp. Porous Media* **2011**, *88*, 479–489. [CrossRef]

22. Singh, S.K. Well Loss Estimation: Variable Pumping Replacing Step Drawdown Test. *J. Hydraul. Eng.* **2002**, *128*, 343–348. [CrossRef]

23. Singh, S.K. Identifying Head Loss from Early Drawdowns. *J. Irrig. Drain. Eng.* **2008**, *134*, 107–110. [CrossRef]

24. Avci, C.B.; Ciftci, E.; Sahin, A.U. Identification of aquifer and well parameters from step-drawdown tests. *Hydrogeol. J.* **2010**, *18*, 1591–1601. [CrossRef]

25. Van Tonder, G.J.; Botha, J.F.; Chiang, W.-H.; Kunstmann, H.; Xu, Y. Estimation of the sustainable yields of boreholes in fractured rock formations. *J. Hydrogeol.* **2001**, *241*, 70–90. [CrossRef]

26. Renard, P. The future of hydraulic tests. *Hydrogeol. J.* **2005**, *13*, 259–262. [CrossRef]

27. Dufresne, D.P. Developing Better Regional Groundwater Flow Models with Effective Use of Step-Drawdown Test Results. *Fla. Water Resour. J.* **2011**, *63*, 36–40.

28. Houben, G.; Treskatis, C. *Water Well Rehabilitation and Reconstruction*, 1st ed.; McGraw-Hill Education: New York, NY, USA, 2007; p. 15, ISBN 13 978-0071486514.

29. Van Beek, C.G.E.M. Rehabilitation of clogged discharge wells in the Netherlands. *Q. J. Eng. Geol. Hydrogeol.* **1989**, *22*, 75–80. [CrossRef]

30. De Zwart, A.H. *Investigation of Clogging Processes in Unconsolidated Aquifers near Water Supply Wells*; Ponaen & Looyen BV: Delft, The Netherlands, 2007; pp. 2–6.

31. De Zwart, B.-R.; van Beek, K.; Houben, G.; Treskatis, C. Mechanische Partikelfiltration als Ursache der Brunnenalterung. *Fachmagazin Für Brunnen Und Leitungsbau* **2006**, *57*, 42–49.

32. Blackwell, I.M.; Howsam, P.; Walker, M.J. Borehole performance in alluvial aquifers: Particulate damage. *Q. J. Eng. Geol. Hydrogeol.* **1995**, *28*, 151–162. [CrossRef]

33. Hitchon, B. "Rust" contamination of formation waters producing wells. *Appl. Geochem.* **2000**, *15*, 1527–1533. [CrossRef]

34. De la Loma Gonzalez, B. Clogging of Deep Well Infiltration Recharge Systems in the Netherlands. In *Clogging Issues Associated with Managed Aquifer Recharge Methods*; Martin, R., Ed.; IAH Commission on Managing Aquifer Recharge: Australia, 2013; pp. 163–173. Available online: https://recharge.iah.org/files/2015/03/Clogging_Monograph.pdf (accessed on 10 January 2019).

35. Treskatis, C.; Volgnandt, P.; Wessollek, H.; Puronpää-Schäfer, P.; Gerbl-Rieger, S.; Blank, K.-H. Anforderungsprofile an den wirtschaftlichen Bau und Betrieb von Bohrbrunnen. *Grundwasser* **1998**, *3*, 117–128. [CrossRef]

36. Van Beek, C.G.E.M. Cause and Prevention of Clogging of Wells Abstracting Groundwater from Unconsolidated Aquifers. Ph.D. Thesis, FSC, Bonn, Germany, 2010.

37. Houben, G.; Wachenhausen, J.; Guevara, C. Effects of ageing on the hydraulics of water wells and the influence of non-Darcy flow. *Hydrogeol. J.* **2018**, *26*, 1285–1294. [CrossRef]

38. Helweg, O.J.; Bengtson, M. Water Well Efficiency. In Proceedings of the Water Resource Engineering and Water Resources Planning and Management, Minneapolis, MN, USA, 30 July–2 August 2000; pp. 1–5. [CrossRef]

39. Hurynovich, A.; Syczewa, E. The methodology in estimation of cost of life cycle of deep wells of underground waters intakes. *Ekonomia i Środowisko* **2013**, *4*, 67–76. (In Polish)

40. Piscopo, V.; Summa, G. Experiment of pumping at constant-head: An alternative possibility to the sustainable yield of a well. *Hydrogeol. J.* **2007**, *15*, 679–687. [CrossRef]

41. Baiocchi, A.; Lotti, F.; Piacentini, S.M.; Piscopo, V. Comparison of pumping at constant head and at a constant rate for determining the sustainable yield of a well. *Environ. Earth Sci.* **2013**, *72*, 989–996. [CrossRef]

Article

Quantitative Assessment of the Influences of Three Gorges Dam on the Water Level of Poyang Lake, China

Dan Wang [1,2], Shuanghu Zhang [2,*], Guoli Wang [1], Qiaoqian Han [3], Guoxian Huang [4], Hao Wang [2], Yin Liu [2] and Yanping Zhang [2]

[1] School of Hydraulic Engineering, Dalian University of Technology, Dalian 116024, China
[2] State Key Laboratory of Simulation and Regulation of Water Cycle in River Basin, Beijing 100038, China
[3] China Water International Engineering Consulting Co., Ltd., Beijing 100053, China
[4] Chinese Research Academy of Environmental Sciences, Beijing 100012, China
*   Correspondence: sxslzsh@163.com; Tel.: +86-010-6878-5508

Received: 4 June 2019; Accepted: 18 July 2019; Published: 22 July 2019

Abstract: Lakes are important for global ecological balance and provide rich biological and social resources. However, lake systems are sensitive to climate change and anthropogenic activities. Poyang Lake is an important wetland in the middle reach of the Yangtze River, China and has a complicated interaction with the Yangtze River. In recent years, the water level of Poyang Lake was altered dramatically, in particular showing a significant downward trend after the operation of the Three Gorges Dam (TGD) in 2003, thus seriously affecting the lake wetland ecosystem. The operation of the TGD changed both the hydrological regime and the deeper channel in the middle reach of the Yangtze River, and affected the river–lake system between the Yangtze River and Poyang Lake. This study analyzed the change in the water level of Poyang Lake and quantified the contributions of the TGD operation, from the perspectives of water storage and erosion of the deeper channel in the middle reach of the Yangtze River, through hydrodynamic model simulation. The erosion of the deeper channel indicated a significant decrease in annual water level. However, due to the water storage of the TGD in September and October, the discharge in the Yangtze River sharply decreased and the water level of Poyang Lake was largely affected. Especially in late September, early October, and mid-October, the contributions of water storage of the TGD to the decline in the water level of Poyang Lake respectively reached 68.85%, 59.04%, and 54.88%, indicating that the water storage of the TGD was the main factor in the decrease in water level. The erosion of the deeper channel accelerated the decline of the water level of Poyang Lake and led to about 10% to 20% of the decline of water level in September and October. Due to the combined operation of the TGD and more reservoirs under construction in the upper TGD, the long-term and irreversible influence of the TGD on Poyang Lake should be further explored in the future.

Keywords: water level; Three Gorges Dam; hydrodynamic model; river–lake system; Poyang Lake

---

1. Introduction

Lakes provide valuable ecosystem services for riparian communities and play an important role in sustaining ecological security and sustainable development [1,2]. Water levels and surface areas of lakes are sensitive to climatic factors and anthropogenic activity and greatly influence the species distribution and functions in lake ecosystems [3,4]. Many lakes are connected with rivers to form complicated river–lake systems [5]. Anthropogenic activities in rivers and lakes, such as dam construction, reservoir operation, farming, and landscape modification, would influence the water levels of lakes and river–lake systems. The variation in water level is the direct response to the hydrological regime change and anthropogenic activities in lakes, and influences lake productivity, stability, species diversity, and succession of wetland vegetation communities [6,7].

Poyang Lake in the middle reach of the Yangtze River is the largest freshwater lake in China and one of the world's most important wintering grounds of migratory birds in the center of East Asian–Australasia flyway. It was added to the List of Wetlands of International Importance (the Ramsar List) in 1992 and also designated as one of the world's important ecological zones by the World Wide Fund for Nature. There is a complicated river–lake system between Poyang Lake and the Yangtze River, and the hydrological regimes of Poyang Lake catchment and the Yangtze River contribute to the water level of Poyang Lake differently in over the year [8].

Since the twenty-first century, especially after the operation of Three Gorges Dam in 2003, the water level of Poyang Lake was extremely low during the dry season and lake and wetland areas varied significantly [3,9–11]. This phenomenon has led to the seasonal bottomlands appearing in advance, reducing the suitable habitats of mainly migratory birds. In addition, the living spaces of fishes and migration of Yangtze finless porpoises were seriously influenced. The operation of the Three Gorges Dam (TGD) has greatly changed the runoff regime in the middle and lower reaches of the Yangtze River Basin. In particular, water storage of the TGD in September every year leads to the lower water level and earlier dry season of the Yangtze River [12,13]. Therefore, the supporting force of the Yangtze River on the water level of Poyang Lake is decreased, leading to the declining water level of Poyang Lake, the earlier dry season, and the altered seasonal inundation pattern [14–16]. Seasonal bottomlands appeared one month in advance, thus decreasing community species diversity and seriously damaging the wetland ecosystem and habitats of migratory birds [17,18].

The variation in water level of Poyang Lake is related to climatic factors and anthropogenic activities, such as hydrological regimes of lake catchment and the Yangtze River, dam construction and reservoir operation in the lake and Yangtze River, and geographical changes caused by river channel erosion and sand extraction [7,10]. In recent years, the influences of the TGD and climatic change on the variation of water level of Poyang Lake have been extensively explored. The dominant role of the TGD or climate change in the variation of water level of Poyang Lake is controversial [19–21]. Mei et al. [10] considered Poyang Lake as a frustum, quantified the influences of precipitation variation and anthropogenic activities in Poyang Lake catchment on the water level of Poyang Lake, and then got the contribution of the TGD. Mei et al. [10] indicated that the operation of the TGD was the main cause for the lowered water level of Poyang Lake in October. However, the variation in water level of Poyang Lake is a consequence of water storage of the TGD and the erosion of the deeper channel in the middle reach of the Yangtze River [22], Mei et al. [10] did not differentiate the influences of the two factors.

This study aims to identify and quantify the influences of the TGD on the water level of Poyang Lake. Firstly, the variations in the water level of Poyang Lake, and the starting time and duration of the dry season after the operation of the TGD were discussed. Secondly, the contributions of water storage of the TGD and geographical change to the variations in water level throughout the year and the dry season were quantified with a hydrodynamic model. This study is unique in that it quantifies the influences of TGD on water level variation of Poyang Lake for the first time, analyzes the discharge at Hukou and provides the basis for understanding the reason for the water level variation of Poyang Lake.

2. Study Area and Data

2.1. Study Area

Poyang Lake is in the north of Jiangxi Province, China (115°49′–116°46′ E, 28°24′–29°46′ N) (Figure 1). There is another lake, Dongting Lake, the second largest freshwater lake in China, located between the TGD and Poyang Lake. The basin area of Poyang Lake is 1.62×10^5 km^2. Its water source mainly from the Xiu River, Gan River, Fu River, Xin River, and Rao River and then flows into the main steam of the Yangtze River at Hukou from south to north.

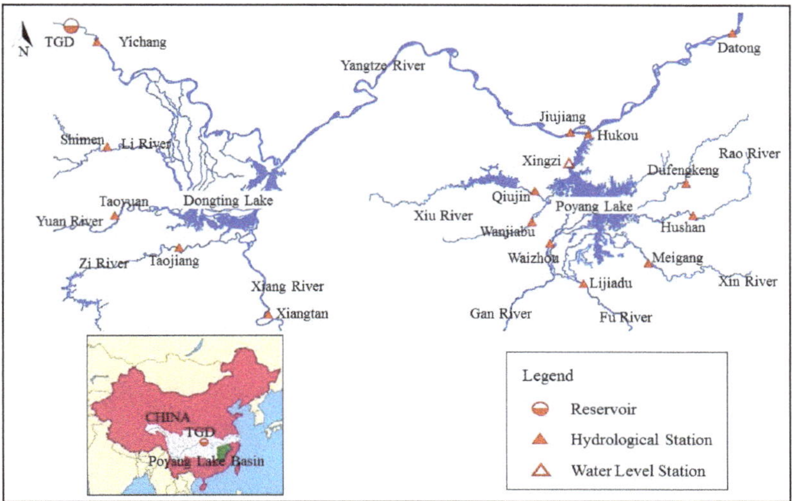

Figure 1. Locations of Poyang Lake and sketch map of hydrodynamic model domain. TGD: Three Gorges Dam.

Poyang Lake is an important flood storage and detention basin and water source of the middle and lower reaches of the Yangtze River Basin. The water level of Poyang Lake shows strong seasonal variations and has a complicated relationship with the Yangtze River. During the flood season of Poyang Lake from April to July, the water level rises due to the increasing inflow from the lake catchment. The flood season of the Yangtze River is later than that of Poyang Lake. Due to the high water level of the Yangtze River from July to September, the water level of Poyang Lake is also high. When the water level of Poyang Lake is high, the water surface is wide. After October, the discharge and water level of the Yangtze River gradually decrease and the water level of Poyang Lake decreases. Due to the smaller inflow of the Poyang Lake catchment, Poyang Lake enters the dry season with a lower water level and then the water–land transition zone and independent seasonal bottomlands appear, the lake area is decreased, and separated river channels are formed. The water–land transition zone is constantly moving with the decrease in the water level and forms new food-rich habitats for migratory birds [23]. The large water level range, unique water situation and environmental conditions have created the high biodiversity and significantly affected wetland ecosystems [24].

2.2. Data

In previous studies [25,26], the influences of the TGD on the Yangtze River and lakes were usually explored in two stages: the period before the operation of the TGD from 1991 to 2002 and the period after the operation of the TGD in 2003. Thus, the observed daily water level data of Xingzi Station, a representative water level station in Poyang Lake, from 1991 to 2014 were used to analyze the variations of water level, daily discharge and water level of each hydrological station in the Yangtze River, Poyang Lake, and Dongting Lake catchments from 1991 to 2014 and daily inflow and outflow data of the TGD from 2003 to 2014 were used to simulate and analyze the influences of the TGD on the variation of the water level of Poyang Lake in this study. Geographical data in 2003 of the main stream, lakes, and main tributaries, which were interpolated with linear interpolation or curvilinear grid-based methods, were used to represent the pre-TGD geography. The newly measured geographic data in 2013 was used to indicate the changed geography after the TGD operation. The missing geographical data of partial areas were calculated by the linear interpolation of irregular triangular mesh.

3. Methodology

In this study, a non-parametric method, Mann–Kendall (MK) [27,28] test was used to detect the trends of water level variation and influence of the TGD on Poyang Lake. A newly developed one-dimensional hydrodynamic model was used to interpret the reasons for water level variation of Poyang Lake.

3.1. MK Test

MK test has been widely used in hydrology and meteorology to examine the trends of time series. In MK test, the sample data are not required to follow a specific distribution or be disturbed by a small number of outliers. For the sample data $(x_1, x_2, ..., x_n)$, the statistical S can be calculated with Equation (1).

$$S = \sum_{i=1}^{n-1} \sum_{j=i+1}^{n} sgn(x_j - x_i).$$

(1)

The $sgn(\ )$ in this equation is a sign function which can be calculated with Equation (2).

$$sgn(x_j - x_i) = \begin{cases} 1 & x_j - x_i > 0 \\ 0 & x_j - x_i = 0 \\ -1 & x_j - x_i < 0 \end{cases}.$$

(2)

When the number of sample data $n \geq 10$, the statistical S is approximate to the normal distribution with a mean of zero and variance shown in Equation (3)

$$Var(S) = \frac{n(n-1)(2n+5) - \sum_{i=1}^{m} t_i(t_i - 1)(2t_i + 5)}{18},$$

(3)

where m is the number of groups of t_i tied observations, with the same value.

Then the standard normal distributed statistical Z can be calculated by Equation (4). In the two-sided trend test at a specified confidence level (α), if $|Z| \geq Z_{\alpha/2}$ ($Z_{\alpha/2} = 1.96$ at 0.05 confidence level), there is a significant upward trend in the sample data when $Z > 0$, and a significant downward trend when $Z < 0$.

$$Z = \begin{cases} \frac{S-1}{\sqrt{Var(S)}} & S > 0 \\ S & S = 0 \\ \frac{S+1}{\sqrt{Var(S)}} & S > 0 \end{cases}.$$

(4)

3.2. Hydrodynamic Model

The reasons for water level variation in lakes and river–lake systems are complicated. Usually, the reasons can be interpreted by hydrodynamic model simulation [29–31]. A developed one-dimensional hydrodynamic model, which covered 4772.8 km long channels (Figure 1), was used in this study. The model involves the main steam of the Yangtze River from Yichang to Datong, Poyang Lake, Dongting Lake, and numerous tributaries to the Yangtze River. The control boundaries of the hydrodynamic model include discharge boundaries (Yichang Hydrological Station and hydrological stations at each tributary of Poyang Lake and Dongting Lake) and water level boundaries (Datong Station). Saint-Venant equations are used in this hydrodynamic model to describe the governing equations (Equations (5) and (6)) and the water surface elevation and discharge are discretized into cross-sections through an implicit four-point finite-difference scheme and denoted as Z_j and Q_j, where j is the number of cross-sections:

$$\frac{\partial A}{\partial t} + \frac{\partial Q}{\partial x} = q,$$

(5)

$$\frac{\partial Q}{\partial t} + \frac{\partial}{\partial x}\left(\frac{Q^2}{A}\right) + gA\left(\frac{\partial Z}{\partial x} + s_f + s_e\right) + L = 0, \tag{6}$$

where A is the wetted cross-sectional area; t is time; Q is the flow discharge; X is the curvilinear discharge of the river channel; q is the lateral discharge per unit channel length; Z is the elevation of water surface; S_f is friction slope; S_e is local bed slope; L is the momentum of lateral discharge and can be expressed as Equation (7) when $q > 0$ or Equation (8) when $q < 0$; u_b is the magnitude of lateral velocity along the main streamline:

$$L = q(u_b - Q/A), \tag{7}$$

$$L = -qQ/A. \tag{8}$$

Hydraulic conditions at river junctions are governed by mass and energy conservation equations:

$$\sum\nolimits_{k=1}^{m} \Delta Q_l k = -A_l \frac{\Delta Z_l}{t}, \tag{9}$$

where m is the total number of sub-channels linked to junction l; k is the number of the channel linked to junction l; A_l is the storage area of junction l; Z_l is the water surface elevation at l cross-sections.

Then the model is solved by Gauss–Jordan elimination method and the three-level solution method is used to reduce the required computational time and storage. More details can be found in the report by Huang et al. [32].

3.3. Analysis of the Influence of the TGD

A hydrodynamic model can provide the water level and discharge results under different hydrological conditions and distinguish and quantify the influences of different factors, such as water storage of the TGD and geographical change of the Yangtze River and Poyang Lake Basin. In this paper, the variation of water level before and after the TGH operation is denoted as ΔZ; the percentage of water level variation is denoted as R, R_s and Rz represent the percentage of the influences of water storage of the TGD and geographical change on water level variation of Poyang Lake. Three scenarios were set to quantify the influences of the TGD as follows.

Scenario 1. Geographical data in 2003 and measured discharge data from 2003 to 2014 were used in simulation, and the calculated water level at Xingzi Station is denoted as Z_1.

Scenario 2. Natural discharge of the Yangtze River was calculated according to the inflow and outflow of the TGD from 2003 to 2014. With geographical data in 2003, the water level at Xingzi Station was calculated and denoted as Z_2. The difference between Z_1 and Z_2 can be used to indicate the influence of water storage of the TGD without considering the influences of geographical change.

Scenario 3. Geographical data in 2013 and measured discharge data from 2003 to 2014 were used in the simulation. The calculated water level at Xingzi Station is denoted as Z_3. The difference between Z_1 and Z_3 can reflect the influences of geographical change on water level.

Therefore, the contributions of water storage of the TGD and geographical change to water level variation can be calculated as follows:

$$R_s = \frac{Z_1 - Z_2}{\Delta Z} \times 100\%, \tag{10}$$

$$R_Z = \frac{Z_3 - Z_1}{\Delta Z} \times 100\%. \tag{11}$$

4. Results

4.1. Variations of Water Level and the Dry Season

Since the operation of the TGD in 2003, the water level of Poyang Lake was significantly changed, displaying the decreased annual average water level, the increased declining rate of water level from

September to December, and longer and earlier dry season. The annual water level and inner-annual 10 day mean water level at Xingzi Station from 1991 to 2014 are shown in Figure 2. The mean annual average water level from 2003 to 2014 was 1.32 m lower than that from 1991 to 2002. The statistical Z was −2.877, indicating a significant downward trend of water level. In addition, the inner-annual 10 day average water level was lowered and the declining percentage of 10 day average water level reached 10% in 21periods. Especially in late October and early November, the declining percentage of 10 day average water level reached 18.03% and 17.78%, respectively. Since October, the declining rate of water level had significantly increased, and the water level decreased by 3.36 m until early November. The decrease of the water level was increased by 56.96% compared to the value (2.14 m) in the period from 1991 to 2002. The water level had risen slightly in mid-November, displaying a different variation trend from that in the period from 1992 to 2002.

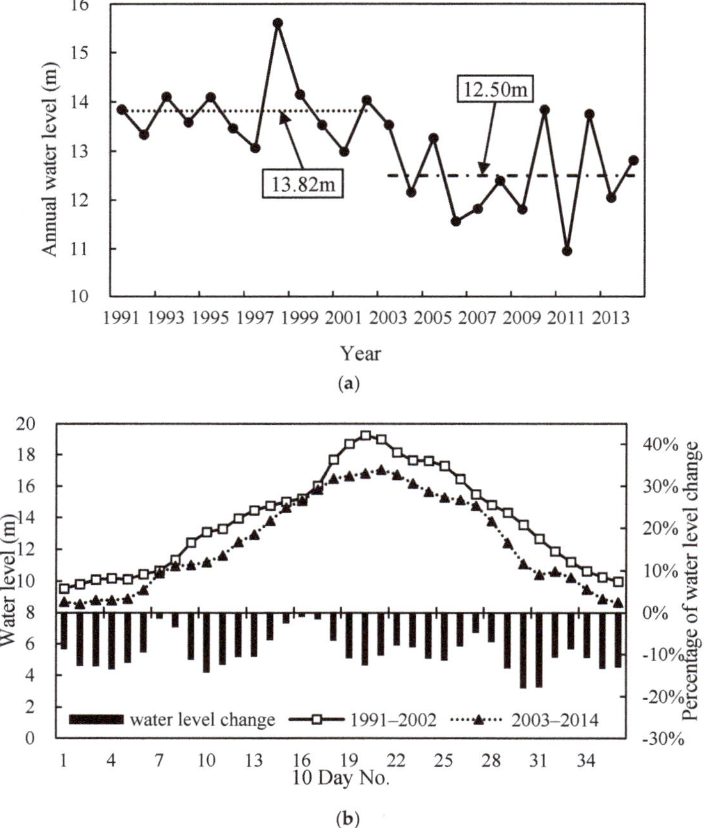

Figure 2. Water level at Xingzi Station: (**a**) annual water level from 1959 to 2014, (**b**) inner-annual 10 day mean water level during different periods.

When the water level at Xingzi Station was lower than 12 m, the surface area of Poyang Lake decreased sharply and impeded the local water supply. Therefore, water level decline lower than 12 m was usually regarded as a symbol of the dry season in the management of Poyang Lake. The duration and the first day of the dry season of Poyang Lake are shown in Figure 3. Duration of the dry season of Poyang Lake had increased significantly from 2003 to 2014 and the mean period was 57 days longer than that in the years from 1959 to 2002. The first day of the dry season was mostly in November or

December before 2003 and the mean starting date of the dry season was in mid-November. Dry season was 32 days in advance on average since 2003, especially in 2006 and 2011. Both the mean period and first day of the dry season had indicated a significant upward trend, and the statistical Z values were 3.200 and 2.555, respectively.

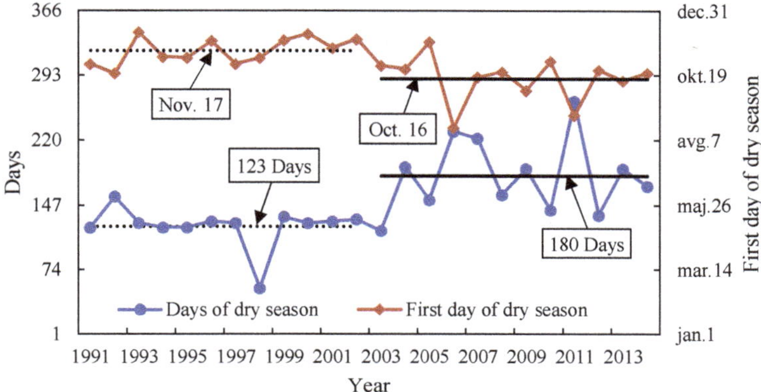

Figure 3. Duration and starting date of the dry season of Poyang Lake.

4.2. Quantitative Analysis of the Influences of Different Factors on Water Level Variation

Based on the daily results of the hydrodynamic model simulation, the difference between Scenario 1 and Scenario 2 reflected the influence of water storage of the TGD on water level without considering the influences of geographical change, and the influence of geographical change on water level could be calculated based on the difference between Scenario 1 and Scenario 3.

Figure 4 shows the annual influences of water storage of the TGD and geographical change. With the operation of the TGD, the annual influence of water storage on the water level of Poyang Lake indicated a slight upward trend, and the corresponding statistical Z was 1.713. Especially after 2008, when the TGD was fully put into operation, the upward trend was relatively stable while the geographic change significantly lowered the annual water level of Poyang Lake, and the corresponding statistical Z was 3.270.

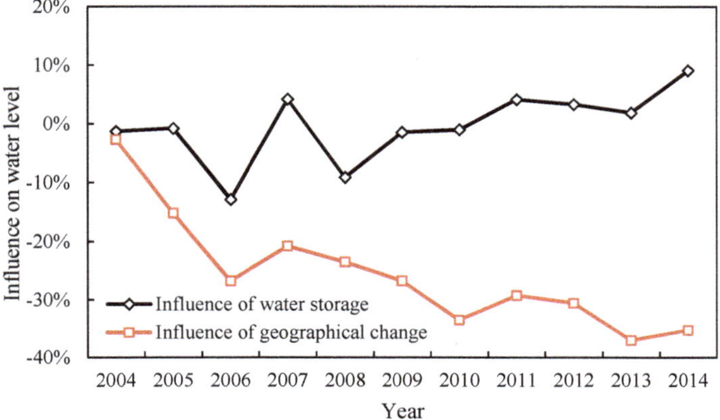

Figure 4. Annual influences of water storage of the TGD and geographical change on water level.

For the inner-annual influence, 10 day mean water level at Xingzi Station was calculated (Figure 5), and Table 1 shows the data during the period with significant influences. In general, the water level of Poyang Lake was decreased throughout the year under the influence of geographical change. Water level decrease was the most serious in early April and reached 0.37 m, whereas the water level in mid-August was just decreased by 0.10 m. The annual average decrease in the water level of Poyang Lake caused by geographical change was 0.26 m. As for the influence of water storage of the TGD on water level, the water level of Poyang Lake mainly reflected the significant decrease after the flood season. In the period from September to November, the water level of Poyang Lake was 0.51 m lower than that under natural conditions. The decrease in the water level was the most serious in October (0.95 m) and reached 0.34 m in September and 0.22 m in November. From mid-July to early August, the water level also decreased slightly and the average decrease was 0.12 m. From early December to early June of the next year, the water level was slightly increased by 0.21 m due to the water drainage of the TGD.

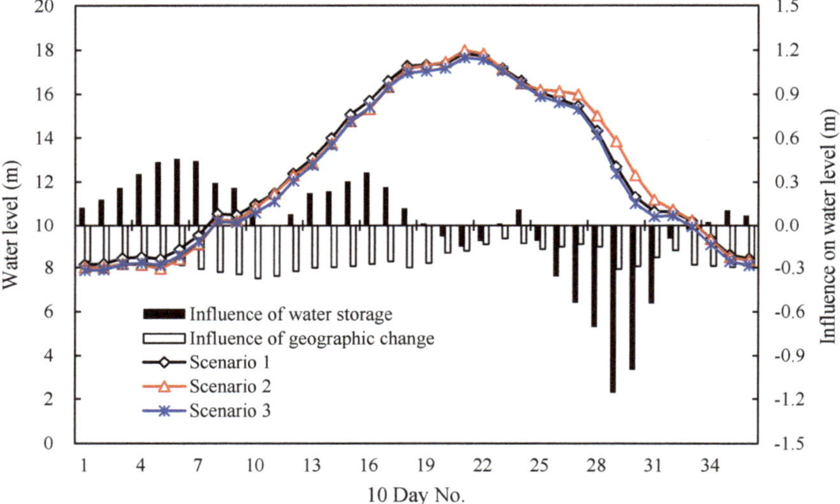

Figure 5. 10 day mean water level at Xingzi Station calculated based on hydrodynamic model simulation.

Table 1. 10 day mean water level at Xingzi Station from the model during the time of significant influences.

10 Day No.	Scenario 1 (m)	Scenario 2 (m)	Scenario 3 (m)	Influence of Water Storage (m)	Influence of Geographical Change (m)
10	10.97	10.85	10.60	0.13	−0.37
20	17.38	17.46	17.18	−0.08	−0.20
22	17.72	17.84	17.59	−0.11	−0.14
23	17.16	17.15	17.07	0.01	−0.10
25	16.08	16.19	15.91	−0.11	−0.17
26	15.78	16.14	15.62	−0.36	−0.16
27	15.45	15.99	15.32	−0.54	−0.14
28	14.32	15.03	14.17	−0.70	−0.16
29	12.70	13.86	12.39	−1.16	−0.31
30	11.31	12.31	11.02	−1.00	−0.29
31	10.65	11.20	10.42	−0.54	−0.23
32	10.65	10.74	10.46	−0.10	−0.18
33	10.22	10.25	9.94	−0.03	−0.28

The TGD mainly stores water in September and October every year and promotes the decrease in the runoff and water level at the Yangtze River, thus aggravating the declining trend of water level of Poyang Lake during the dry period. The decrease in water level and the advance of the dry period seriously affect the ecological system of Poyang Lake and more attention on this issue is needed. Therefore, this study mainly analyzed the contribution of water storage and geographical change to the water level of Poyang Lake in September and October (Table 2). The influence of water storage of the TGD on water level was mainly reflected in the period from late September to mid-October. Water storage of the TGD was the dominant factor in the decrease in water level in this period and showed a maximum contribution of 68.85% in late September. The contribution of geographical change to the decline in the water level of Poyang Lake was lower than 20%.

Table 2. Quantitative analysis of the factors of water level decrease in September and October.

Month	Water Level (m)			Influence of Water Storage (R_s)		Influence of Geographic Change (R_z)	
	1991–2002	2003–2014	ΔZ	(m)	(%)	(m)	(%)
Sep. 1–Sep. 10	17.28	15.31	−1.97	−0.11	5.75	−0.17	8.74
Sep. 11–Sep. 20	16.46	14.99	−1.46	−0.36	24.58	−0.16	10.65
Sep. 21–Sep. 30	15.39	14.60	−0.79	−0.54	68.85	−0.14	17.46
Oct. 1–Oct. 10	14.81	13.62	−1.19	−0.70	59.04	−0.16	13.05
Oct. 11–Oct. 20	14.31	12.20	−2.11	−1.16	54.88	−0.31	14.66
Oct. 21–Oct. 31	13.55	10.93	−2.62	−1.00	38.17	−0.29	11.11
Sep.	16.37	14.97	−1.41	−0.34	24.03	−0.15	11.02
Oct.	14.22	12.25	−1.97	−0.95	48.33	−0.25	12.76

5. Discussion

5.1. Influences of the TGD on Poyang Lake

The water storage of the TGD from 2003 to 2014 is shown in Figure 6. The TGD mainly stored water in September and October. Water storage of the TGD in September and October was respectively 48×10^8 m^3 and 61×10^8 m^3 and accounted for 8.0% and 15.9% of the runoff, respectively. As a result, the discharge and water level of the Yangtze River was significantly reduced. Based on the flow capacity of the exit section at Poyang Lake and the relationship between the water level of Poyang Lake and the discharge of the Yangtze River, Fang et al. [33] indicated that the change of 1 m^3/s in the discharge of Jiujiang Station corresponded to the change of 0.89–0.99 m^3/s in the discharge of Hukou Station. The decrease in runoff and water level at the Yangtze River caused by water storage of the TGD in September and October accelerated the outflow at Hukou Station and the decrease in water level of Poyang Lake.

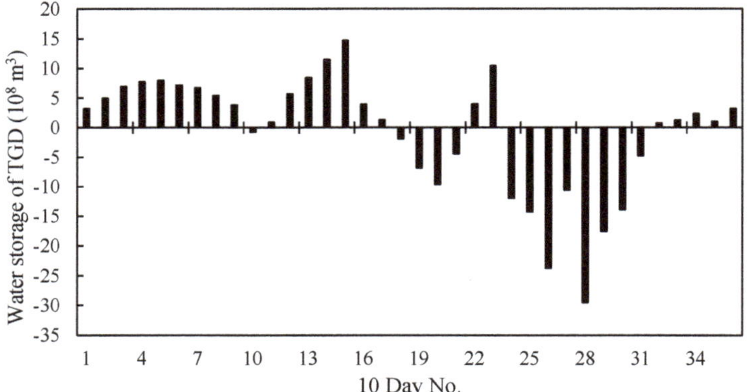

Figure 6. Annual average water storage of TGD (Notes: positive values represent increased drainage, negative values represent water storage).

After the operation of the TGD, the middle reach of Yangtze River was mainly subjected to deeper channel erosion, which resulted in the decrease in water level under the same discharge. The stage-discharge curve of Jiujiang Station in the study period is shown in Figure 7. The obvious erosion of the deeper channel changed the river–lake system and increased the hydraulic gradient between Poyang Lake and the Yangtze River. As a result, the outflow at Hukou increased. The water level at Xingzi Station is highly correlated with that at Jiujiang Station. The coefficient of determination was 0.9824 from 1991 to 2002 and 0.9911 from 2003 to 2014 (Figure 8). The water level at Xinzi Station, which had the same water level as that of the Yangtze River, was decreased.

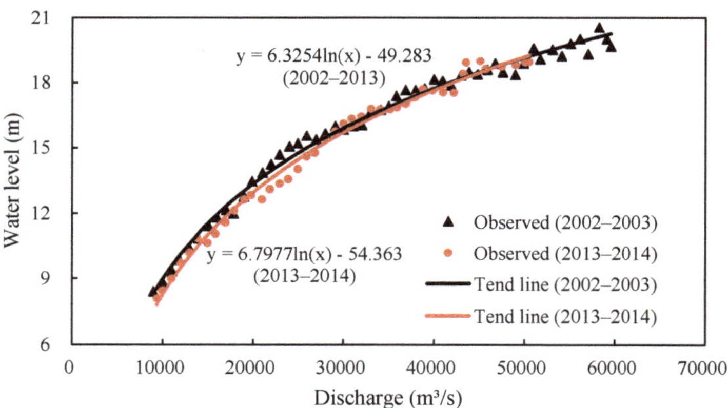

Figure 7. Water level–discharge curve of Jiujiang Station.

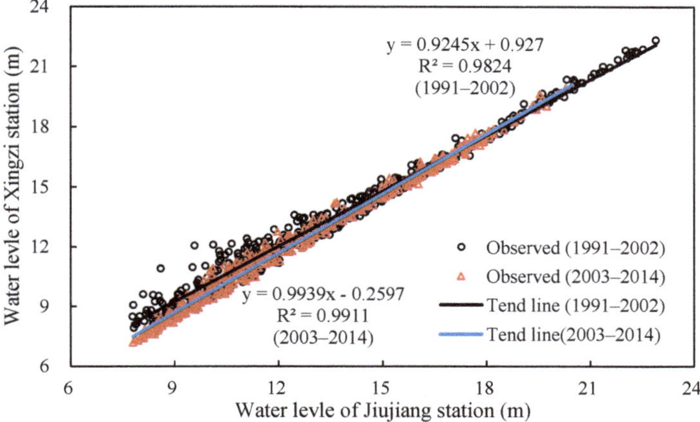

Figure 8. Relationship between the 10 day water level measured at Xiangzi Station and Jiujiang Station.

Based on the simulation of three scenarios, Figure 9 shows the influences of water storage of the TGD and geographical change on the outflow at Hukou Station. The geographical change mainly increased the outflow of Hukou Station by 8.55×10^8 m^3 from March to June during the flood season of Poyang Lake. The TGD started to store water in late June and water was stored mainly in September and October. Water storage of the TGD mainly increased the outflow at Hukou in July and September. The outflow from late June to early October was increased by 19.86×10^8 m^3. Water storage of the TGD accelerated the declining rate of water level in the dry season of Poyang Lake and the water quantity in Poyang Lake decreased significantly in advance. Due to the significant decrease in the water quantity of Poyang Lake, the outflow of Hukou decreased gradually from mid-October.

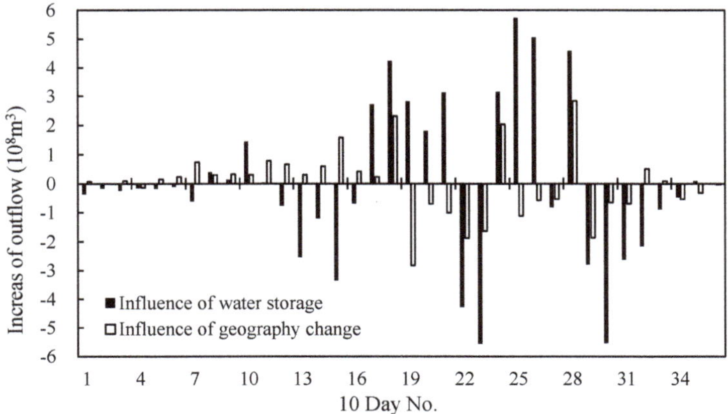

Figure 9. Influences of TGD water storage and geographical change on the outflow of Hukou.

5.2. Influences of Water Level Variation

Due to the unique geographical features, Poyang Lake forms numerous lakes in low water conditions and capacious water surface in high water conditions. Figure 10 shows the relationship between the water level and water surface area of Poyang Lake. When the water level at Xingzi Station was below 15.5 m, every 0.5 m of the decrease in water level for led to the significant reduction in the water surface area, especially when the water level at Xingzi Station was between 11 m and 14 m.

The mean water level of Poyang Lake in September and October from 2003 to 2014 is 13.74 m. Water storage of the TGD in September and October could cause the serious and earlier decrease in water level and water surface area. With the decline in water level, seasonal bottomlands were exposed in advance and the land–water transitional zone moved downward. Furthermore, the distribution of the habitats of migratory birds and other ecological systems were affected.

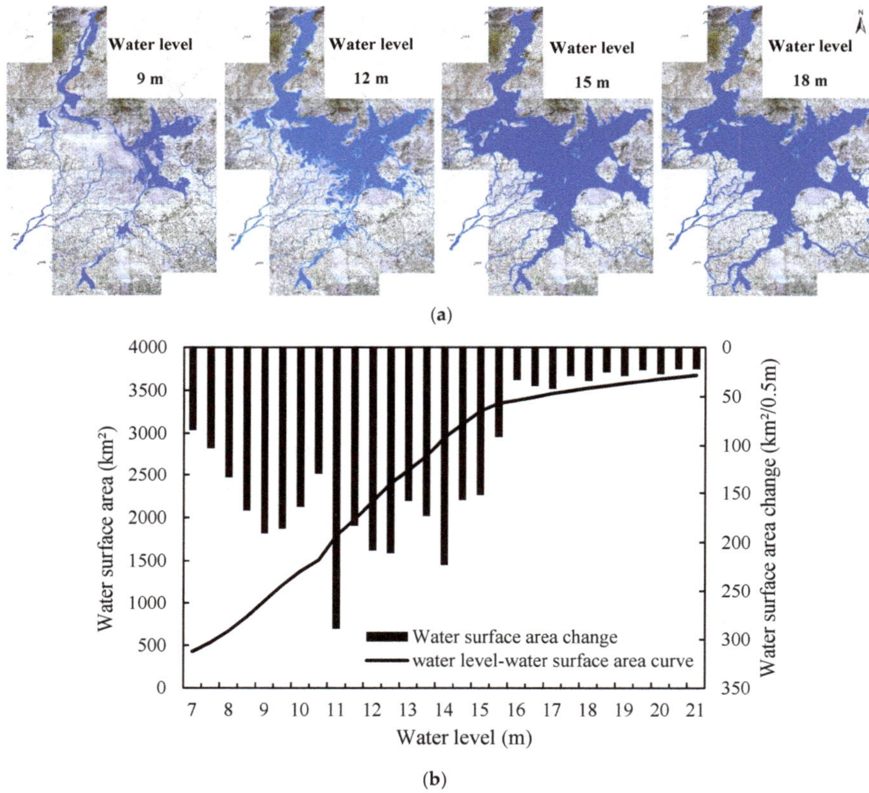

Figure 10. Relationship between water level and water surface area of Poyang Lake: (**a**) water surface area of Poyang Lake under different water levels, (**b**) water level–water surface area curve and the change rate of water surface area.

6. Conclusions

Since 2003, the water level of Poyang Lake has decreased, displaying a significant downward trend. The annual average water level decreased by 1.32 m and an especially sharp decline occurred in the period from September to November. The duration of the dry season was longer and the dry season occurred about one month in advance. The aggravation phenomenon of the dry season of Poyang Lake was significant and the habitats of migratory birds and wetland ecosystems were largely affected.

The operation of the TGD changed the hydrological regime and deeper channel geography in the middle reach of the Yangtze River. This study quantified the influences of the TGD in detail. The geographical change significantly lowered the annual water level of Poyang Lake. After flood season, due to water storage of the TGD in September and October, the sharp decrease in the discharge and water level in the Yangtze River accelerated the outflow at Hukou and enhanced the decrease in water level and water surface area of Poyang Lake. The water storage of the TGD was the main factor in the decrease in water level during this period and the average decreases in water levels in September and October were respectively 0.34 m and 0.95 m. Especially in late September, early October, and mid-October, the contributions of water storage of the TGD to the decline in the water level of Poyang Lake respectively reached 68.85%, 59.04%, and 54.88%. The change in deeper channel geography accelerated the decrease in water level and the contributions of geographical change to the decrease in water level in September and October reached 11.02% and 12.76%, respectively. This study deeply explored the influences of the TGD on the variation in the water level of Poyang Lake. The increasing number of reservoirs in operation in the upper stream will generate adverse and irreversible influences on the water level and ecological environment of Poyang Lake, which should be widely concerned and well protected.

Author Contributions: S.Z. and G.W. developed the original ideas. D.W., Q.H. and G.H. developed the original ideas and completed the study of this paper. Y.L. and Y.Z. processed the raw data. D.W. drafted the manuscript and it was revised substantially by all authors.

Funding: This study was funded by the National Key Basic Research Program of China (Grant No: 2015CB452701) and the CAS "Light of West China" Program (Grant No: Y8R2230230).

Conflicts of Interest: The authors declare no conflict of interest.

References

1. Wantzen, K.M.; Junk, W.J.; Rothhaupt, K.O. Ecological effects of water-level fluctuations in lakes: An urgent issue. *Hydrobiologia* **2008**, *613*, 1–4. [CrossRef]
2. Williamson, C.E.; Saros, J.; Vincent, W.F.; Smol, J.P. Lakes and reservoirs as sentinels, integrators, and regulators of climate change. *Limnol. Oceanogr.* **2009**, *54*, 2273–2282. [CrossRef]
3. Li, M.F.; Zhang, Q.; Li, Y.L.; Yao, J.; Tan, Z.Q. Inter-annual variations of Poyang Lake area during dry seasons: Characteristics and implications. *Hydrol. Res.* **2016**, *47*, 40–50. [CrossRef]
4. Reid, J.R.W.; Colloff, M.J.; Arthur, A.D.; McGinness, H.M. Influence of catchment condition and water resource development on waterbird assemblages in the Murray-Darling Basin, Australia. *Biol. Conserv.* **2013**, *165*, 25–34. [CrossRef]
5. Yao, J.; Zhang, Q.; Li, Y.L.; Li, M.F. Hydrological evidence and cause of seasonal low water levels in a large river-lake system: Poyang Lake, China. *Hydrol. Res.* **2016**, *47*, 24–39. [CrossRef]
6. You, H.L.; Xu, L.G.; Liu, G.L.; Wang, X.L.; Wu, Y.M.; Jiang, J.H. Effects of Inter-Annual Water Level Fluctuations on Vegetation Evolution in Typical Wetlands of Poyang Lake, China. *Wetlands* **2015**, *35*, 931–943. [CrossRef]
7. Yuan, Y.J.; Zeng, G.M.; Liang, J.; Huang, L.; Hua, S.S.; Li, F.; Zhu, Y.; Wu, H.P. Variation of water level in Dongting Lake over a 50-year period: Implications for the impacts of anthropogenic and climatic factors. *J. Hydrol.* **2015**, *525*, 450–456. [CrossRef]
8. Shao, J.; Wang, J.; Lv, S.Y.; Bing, J.P. Spatial and temporal variability of seasonal precipitation in Poyang Lake basin and possible links with climate indices. *Hydrol. Res.* **2016**, *47*, 51–68. [CrossRef]
9. Liu, C.L.; Tan, Y.J.; Lin, L.S.; Tao, H.N.; Tan, H.R. The wetland water level process and habitat of migratory birds in Lake Poyang. *J. Lake Sci.* **2011**, *23*, 129–135. (In Chinese)
10. Feng, L.; Hu, C.M.; Chen, X.L.; Zhao, X. Dramatic inundation changes of China's two largest freshwater lakes linked to the Three Gorges Dam. *Environ. Sci. Technol.* **2013**, *47*, 9628–9634. [CrossRef]
11. Mei, X.F.; Dai, Z.J.; Du, Z.; Chen, J.Y. Linkage between Three Gorges Dam impacts and the dramatic recessions in China's largest freshwater lake, Poyang Lake. *Sci. Rep.* **2015**, *5*, 18197. [CrossRef] [PubMed]
12. Xie, D.M.; Zheng, P.; Deng, H.B.; Zhao, J.Z.; Fan, Z.W.; Fang, Y. Landscape responses to changes in water levels at Poyang Lake wetlands. *Acta Ecol. Sin.* **2011**, *31*, 1269–1276. (In Chinese)
13. Zhang, Q.; Li, L.; Wang, Y.G.; Werner, A.D.; Xin, P.; Jiang, T.; Barry, D.A. Has the Three-Gorges Dam made the Poyang Lake wetlands wetter and drier. *Geophys. Res. Lett.* **2012**, *39*, L20402.1–L20402.7. [CrossRef]
14. Lai, X.; Liang, Q.H.; Huang, Q.; Jiang, J.H.; Lu, X.X. Numerical evaluation of flow regime changes induced by the Three Gorges Dam in the Middle Yangtze. *Hydrol. Res.* **2016**, *47*, 149–160. [CrossRef]
15. Guo, H.; Hu, Q.; Zhang, Q.; Feng, S. Effects of the Three Gorges Dam on Yangtze River flow and river interaction with Poyang Lake, China: 2003–2008. *J. Hydrol.* **2012**, *416*, 19–27. [CrossRef]

16. Lai, X.J.; Liang, Q.H.; Jiang, J.H.; Huang, Q. Impoundment Effects of the Three-Gorges-Dam on Flow Regimes in Two China's Largest Freshwater Lakes. *Water Resour. Manag.* **2014**, *28*, 5111–5124. [CrossRef]

17. Wu, G.P.; Liu, Y.B. Assessment of the Hydro-Ecological Impacts of the Three Gorges Dam on China's Largest Freshwater Lake. *Remote Sens.* **2017**, *9*, 1069. [CrossRef]

18. Mei, X.F.; Dai, Z.J.; Fagherazzi, S.; Chen, J.Y. Dramatic variations in emergent wetland area in China's largest freshwater lake, Poyang Lake. *Adv. Water Resour.* **2016**, *96*, 1–10. [CrossRef]

19. Hu, Z.P.; Ge, G.; Liu, C.L.; Chen, F.S.; Li, S. Structure of Poyang Lake Wetland Plants Ecosystem and Influence of Lake Water Level for the Structure. *Resour. Environ. Yangtze Basin* **2010**, *19*, 597–605. (In Chinese)

20. Hu, Q.; Feng, S.; Guo, H.; Chen, G.Y.; Jiang, T. Interactions of the Yangtze River flow and hydrologic processes of the Poyang Lake, China. *J. Hydrol.* **2007**, *347*, 90–100. [CrossRef]

21. Ye, X.C.; Zhang, Q.; Liu, J.; Li, X.H.; Xu, C.Y. Distinguishing the relative impacts of climate change and human activities on variation of streamflow in the Poyang Lake catchment, China. *J. Hydrol.* **2013**, *494*, 83–95. [CrossRef]

22. Zhang, Q.; Ye, X.C.; Werner, A.D.; Li, Y.L.; Yao, J.; Li, X.H.; Xu, C.Y. An investigation of enhanced recessions in Poyang Lake: Comparison of Yangtze River and local catchment impacts. *J. Hydrol.* **2014**, *517*, 425–434. [CrossRef]

23. Zhang, W.; Yuan, J.; Han, J.Q.; Huang, C.T.; Li, M. Impact of the Three Gorges Dam on sediment deposition and erosion in the middle Yangtze River: A case study of the Shashi Reach. *Hydrol. Res.* **2016**, *47*, 175–186. [CrossRef]

24. Zhou, W.B.; Wan, J.B.; Jiang, J.H. *Changes in Water Level of Poyang Lake and Influence of Ecosystem*; Science Press: Beijing, China, 2011.

25. Fang, C.M.; Hu, C.H.; Chen, X.J. Impacts of Three Georges Reservoir's operation on outflow of the three outlets of Jingjiang River and Dongting Lake. *J. Hydraul. Eng.* **2014**, *45*, 36–41. (In Chinese)

26. Zhang, R.; Zhang, S.H.; Xu, W.; Wang, B.D.; Wang, H. Flow regime of the three outlets on the south bank of Jingjiang River, China: An impact assessment of the Three Gorges Reservoir for 2003–2010. *Stoch. Environ. Res. Risk Assess.* **2015**, *29*, 2047–2060. [CrossRef]

27. Mann, H.B. Nonparametric tests against trend. *Econometrica* **1945**, *13*, 245–259. [CrossRef]

28. Kendall, M.G. *Rank Correlation Methods*; Griffin: London, UK, 1975.

29. Lai, X.J.; Jiang, J.H.; Liang, Q.H.; Huang, Q. Large-scale hydrodynamic modeling of the middle Yangtze River Basin with complex river-lake interactions. *J. Hydrol.* **2013**, *492*, 228–243. [CrossRef]

30. Hu, D.C.; Zhong, D.; Zhu, Y.H.; Wang, G.Q. Prediction-correction method for parallelizing implicit 2D hydrodynamic models II: Application. *J. Hydraul. Eng.* **2015**, *141*, 1–10. [CrossRef]

31. Pu, J.H.; Huang, Y.F.; Shao, S.D.; Hussain, K. Three-Gorges Dam Fine Sediment Pollutant Transport: Turbulence SPH Model Simulation of Multi-Fluid Flows. *J. Appl. Fluid Mech.* **2016**, *9*, 1–10. [CrossRef]

32. Huang, G.X.; Zhou, J.J.; Lin, B.L.; Xu, X.F.; Zhang, S.H. Modelling flow in the middle and lower Yangtze River, China. *Water Manag.* **2016**, *170*, 1–12. [CrossRef]

33. Fang, C.M.; Cao, W.H.; Mao, J.X.; Li, H. Relationship between Poyang Lake and Yangtze River and influence of Three Georges Reservoir. *J. Hydraul. Eng.* **2012**, *39*, 175–181. (In Chinese)

Communication

Experimental Investigation of Erosion Characteristics of Fine-Grained Cohesive Sediments

Bommanna Gounder Krishnappan [1,*], Mike Stone [2], Steven J. Granger [3], Hari Ram Upadhayay [3], Qiang Tang [3], Yusheng Zhang [3] and Adrian L. Collins [3]

[1] National Water Research Institute, Burlington, ON L7R 4A6, Canada
[2] Department of Geography and Environmental Management, University of Waterloo, Waterloo, ON N2L 3G1, Canada; mstone@uwaterloo.ca
[3] Sustainable Agriculture Sciences Department, Rothamsted Research, North Wyke, Okehampton, Devon EX20 2SB, UK; steve.granger@rothamsted.ac.uk (S.J.G.); hari.upadhayay@rothamsted.ac.uk (H.R.U.); qiangtang@imde.ac.cn (Q.T.); yusheng.zhang@rothamsted.ac.uk (Y.Z.); adrian.collins@rothamsted.ac.uk (A.L.C.)
* Correspondence: krishnappan@sympatico.ca

Received: 28 April 2020; Accepted: 22 May 2020; Published: 25 May 2020

Abstract: In this short communication, the erosion process of the fine, cohesive sediment collected from the upper River Taw in South West England was studied in a rotating annular flume located in the National Water Research Institute in Burlington, Ontario, Canada. This study is part of a research project that is underway to model the transport of fine sediment and the associated nutrients in that river system. The erosion experimental data show that the critical shear stress for erosion of the upper River Taw sediment is about 0.09 Pa and it did not depend on the age of sediment deposit. The eroded sediment was transported in a flocculated form and the agent of flocculation for the upper River Taw sediment may be due to the presence of fibrils from microorganisms and organic material in the system. The experimental data were analysed using a curve fitting approach of Krone and a mathematical model of cohesive sediment transport in rotating circular flumes developed by Krishnappan. The modelled and measured data were in good agreement. An evaluation of the physical significance of Krone's fitting coefficients is presented. Variability of the fitting coefficients as a function of bed shear stress and age of sediment deposit indicate the key role these two factors play in the erosion process of fluvial cohesive sediment.

Keywords: erosion; cohesive sediments; rotating circular flume; mathematical modelling; fitting coefficients; sediment deposition; flocculation; bed shear stress; consolidation

---

1. Introduction

Fine-grained cohesive sediment plays an important role in the transportation of pollutants and it is a key driver of water quality degradation in rivers. Rigorous quantification of cohesive sediment transport processes is fundamental for predicting sediment and associated contaminant transport in aquatic systems (Horowitz and Elrick [1]; Luoma and Rainbow [2]). The development of reliable numerical models to simulate cohesive sediment transport dynamics requires an accurate description of fundamental sediment transport processes such as erosion, deposition, and transport of solids in suspension (Grabowski et al. [3]). Factors affecting the erosion characteristics of cohesive sediments include the rate of bed shear (Amos et al. [4]), the degree of consolidation/age of deposit (Lick and McNeil [5]), bio-stabilisation effects by microorganisms (Friend et al. [6]) and the initial conditions that created the deposit (Lau et al. [7]). In a re-examination of the sediment erosion data from flume studies conducted by Roberts et al. [8] and Zreik et al. [9], Krone [10] demonstrated the importance of the sediment bed matrix structure on erosion characteristics of fine sediment deposits. Factors controlling the deposition characteristics of cohesive sediment include the phyco-chemical properties of sediment water mixture, the concentration of organic matter including microorganisms in the water column and the turbulence characteristics of the flow field which in turn, can cause the suspended sediment particles to flocculate and change their porosity, density and settling characteristics. At the present state of knowledge, numerical models of cohesive sediment transport rely mainly on laboratory experiments using specialised flumes such as a rotating annular flume for the determination of transport parameters that include the critical shear stresses for erosion and deposition, erosion and deposition rates and properties of sediment flocs for site-specific sediments.

A research project is underway in the upper River Taw in South West England to model the transport of fine-grained cohesive sediments. The upper River Taw has been instrumented as a landscape scale observatory for exploring the interactions between climate, land use, farm management and water quality in conjunction with a larger strategic programme exploring pathways for improving the sustainability of agriculture. The upper River Taw drains both organic-rich peaty and podzolic upland moorland soils near its source and clayey gley and brown earth soils in the more intensive agricultural areas (beef and sheep grazing, cereals, maize, oil seed rape) on the more intensively farmed lowland adjacent to the moor. Sediment stress in the study area has been highlighted as a critical factor impacting on lithophilous fish species dependent upon clean bed gravels for their early life stages including the incubation of progeny in redds cut into bed gravels.

As part of the research project on modelling cohesive sediment transport, a survey of 18 cross-sections located ~0.5 km apart was conducted in the upper River Taw (Figure 1) during the summer low flows in 2018. The bank-full width of the study river is ~10 m and the average depth is ~1 m. The riverbed is armoured with large stones including pebbles and cobbles. Bedrock outcrops are also observed. Although the matrix bed material in the study reach is very coarse, fine-grained sediment is present in the river channel bed because of the entrapment process, which retains fine material in the lee of large rocks or directly within the gravel bed matrix as interstitial fines. Fine-grained sediment deposits are also present in recirculation eddy zones along the edges of the river channel where the bed shear stresses of the flow field are low. Representative samples of fine-grained sediment from the study reach were collected using a conventional bed sediment remobilisation technique (Lambert and Walling [11]; Duerdoth et al. [12]; Naden et al. [13]) at all 18 cross sections and 18 one L bottles of sediment slurry were then shipped to Canada for testing in a rotating annular flume located in the National Water Research Institute in Burlington, Ontario, Canada.

In this short communication, preliminary results from this research project on cohesive sediment transport are presented. Specifically, a component of the cohesive sediment transport, namely, the erosion process, was investigated by carrying out erosion experiments using a rotating annular flume. A novel approach was used to analyse the erosion data. The approach is based on a methodology proposed by Krone [10] and a mathematical model of cohesive sediment transport in rotating annular flumes developed by Krishnappan [14].

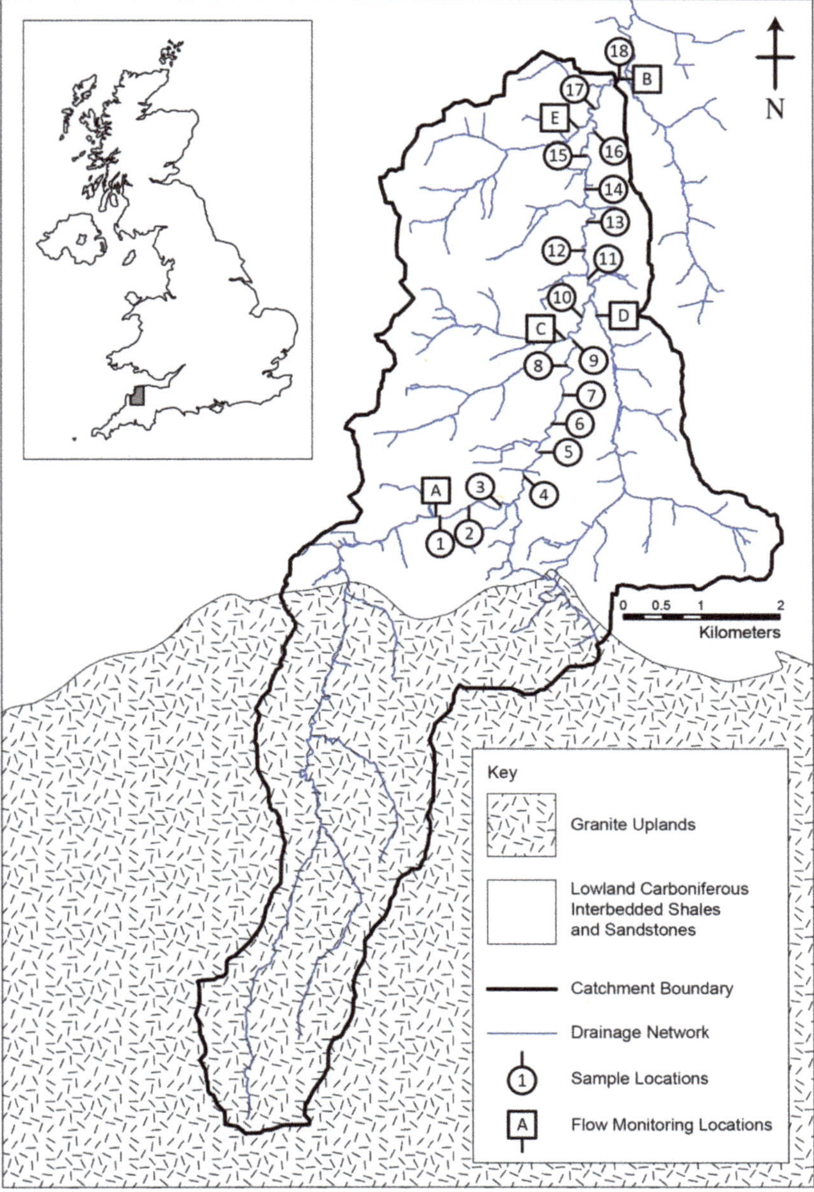

Figure 1. Map of the upper River Taw study catchment, showing location in south west England, channel bed cohesive sediment sampling locations and flow gauging stations.

2. Material and Methods

2.1. Rotating Annular Flume

The rotating circular flume used in this study was 5.0 m in mean diameter, 30 cm wide and 30 cm deep and it rests on a rotating platform which was 7 m in diameter. An annular lid fitted inside the flume with close tolerance (about 1 mm gap all around) and it rotated in the opposite direction to the flume's rotation. The annular lid maintained contact with the water surface within the flume during experiments. The generated flow fields in such assemblies were nearly two dimensional with near constant bed shear stress across the channel and with minimum secondary circulation in the transverse direction (Petersen and Krishnappan [15]). The flow field in this rotating annular flume assembly was computed using the 3D hydrodynamic flow model, PHOENICS (Rolston and Spalding [16]). The bed shear stress computed by the model was verified using direct measurements of bed shear stress using a preston tube (Krishnappan and Engel [17]). The main advantage of rotating flumes over linear flumes is that detrimental effects of the pump and the pipe system on the floc structure of the cohesive sediment were avoided, thereby permitting reliable studies on floc behaviour to be conducted. A complete description of the flume can be found in Krishnappan [18].

The sediment samples shipped from the UK were stored in a cold room until the start of the experiments. A large composite sediment sample was prepared by combining all 18 bottles of samples and wet sieved using a 200 mesh sieve before being added to the flume. The erosion experiments were conducted by using the standard procedure which involves operating the flume at high speed initially to suspend the sediment and then lowering the flume speed gradually to allow the suspended sediment to deposit and form a uniform bed. The bed was then allowed to age for three different time periods (herein referred to as age of deposit), namely, 22 h, 38 h and 160 h. During an erosion test, bed shear stresses were increased over time in steps as a stair-case function. The shear stress steps used were 0.06, 0.09, 0.12, 0.17, 0.27 and 0.33 Pa, with each maintained for a period of one hour. In each shear stress step, sediment samples were collected every 10 min to determine the concentration of the eroded sediment as a function of time. When the concentration reached a steady-state value, the flume speed was increased to the next level. Whenever there was sufficient sediment suspended in the water column, sediment samples were collected for size analysis using a LISST (Laser In-situ Scattering and Transmissometry 100X; Sequoia Scientific, Bellevue, WA, USA) particle size analyser and an image analysis system. This procedure was repeated until the maximum permissible flume speed was reached.

2.2. Methodology of Krone

Krone [10] developed his methodology when he analysed erosion data reported by Zreik et al. [9] who studied the erosion behaviour of cohesive sediments from Boston Harbour using a rotating circular flume. Measured sediment concentration data from Zreik et al. [9] showed that the concentration of eroded sediment increased rapidly at the beginning of each applied shear stress step change but then levelled off with time before the next incremental step in shear stress.

Krone [10] fitted the following equation to describe the variation in sediment concentration during a shear stress step:

$$C = \frac{\frac{1}{c_1}t}{\frac{c_0}{c_1}t + 1} \tag{1}$$

where C = incremental concentration of eroded sediment during a shear stress step; t = time after the shear stress change; c_0 and c_1 = constants. This equation fitted the experimental data of Zreik et al. [9] very well and, importantly, the constants c_0 and c_1 have physical meaning. For example, when t tends to infinity, the ratio $1/c_0$ takes on the value of the concentration near the end of the shear stress step in question. When t = 0, the ratio $1/c_1$ assumes the value of the erosion rate, i.e., $dC/dt = 1/c_1$ at $t = 0$. The constants c_0 and c_1 were evaluated by knowing sediment concentrations at two different times (near the beginning and near the end of the shear stress step) using the following relationships:

$$c_1 = \left(\frac{1}{C_T} - \frac{1}{C_L}\right)T \qquad \text{and } c_0 = \frac{1}{C_L} \tag{2}$$

where C_T = concentration near the beginning of the shear stress step, C_L = limit concentration at the end of the shear stress step and T is the elapsed time from the beginning of the shear stress step till the time when C_T is specified. Knowing c_0 and c_1, Equation (1) can be used to calculate the concentration of the eroded sediment for the entire duration of each shear stress step. In addition, Equation (1) is used to derive an erosional rate function that was used in the model of cohesive sediment transport developed by Krishnappan [2]. The erosion rate of sediment can be expressed as follows:

$$E = h\frac{dC}{dt} \tag{3}$$

where E is the erosion rate in gm/m²s, h is the depth of water in the flume in metres and dC/dt is the concentration gradient which can be evaluated from Equation (1). Substituting the expression of dC/dt in Equation (3), the erosion rate function can be derived as:

$$E = h\frac{(1/c_1)}{\left(\frac{c_0}{c_1}t + 1\right)^2} \tag{4}$$

2.3. Mathematical Model of Cohesive Sediment Transport Developed by Krishnappan

A mathematical model of cohesive sediment transport in the rotating circular flume (called the FLUME model) was used to simulate the erosion process of the sampled upper Taw River sediment. The FLUME model incorporates the erosion rate function of Krone [10] i.e., Equation (4). A full description of the model can be found in Krishnappan [14]. Here, a summary of the salient features of the model is outlined for the sake of completeness. The FLUME model treats the motion of sediment in the rotating flume in two stages: a transport/settling stage and a flocculation stage. Some salient features of these two fundamental stages of sediment transport are briefly discussed below:

2.3.1. Settling Stage

The governing equation for the settling stage was obtained from mass balance considerations. Assuming that the flow in the rotating flume is uniform in the longitudinal and tangential directions, the one-dimensional mass balance equation as shown below was adopted:

$$\frac{\partial C_k}{\partial t} + w_k\frac{\partial C_k}{\partial z} = \frac{\partial}{\partial z}\left(\Gamma\frac{\partial C_k}{\partial z}\right) + S \tag{5}$$

where, C_k is the concentration of sediment in size fraction k, w_k is the settling velocity of that fraction and Γ is the dispersion coefficient in the vertical direction. The symbol t represents the time axis and z represents the coordinate axis in the vertical direction. S is the source/sink term. The boundary conditions used for the settling stage are:

At the free surface:

$$-w_kC_k - \Gamma\frac{\partial C_k}{\partial z} = 0 \tag{6}$$

At the bed:

$$-w_kC_k - \Gamma\frac{\partial C_k}{\partial z} = q_e + q_d + q_{entrap} \tag{7}$$

The free surface boundary condition expresses the balance between the settling flux and the dispersive flux so that there is no external input of sediment at this boundary. At the bed surface, it is assumed that the settling and dispersive fluxes are balanced by the net amount of sediment exchanged at the sediment-water interface. The sediment exchange at the sediment-water interface can occur when: (1) sediment is eroded from the bed and entrained into the water column (q_e) (2) sediment settles to the bed and stays on the bed as the deposited sediment (q_d) (3) sediment settles to the bed and a portion of the deposited sediment can ingress into the interstitial pores of the gravel bed and become unavailable for further erosion and entrainment (q_{entrap}).

The first two components of the sediment exchange at the sediment-water interface have been studied extensively in the literature, including by Partheniades [19], Mehta and Partheniades [20], Parchure [21], Lick [22], Krone [10], Krishnappan [23], Krone [24], Mehta and Partheniades [20], and Lick [22]. In this study, the approaches proposed by Krone [10] and Krone [24] for erosion and deposition respectively have been adopted. Accordingly, the erosion flux was calculated as:

$$q_e = E \tag{8}$$

where the erosion rate, E is given by Equation (4) and the deposition flux was calculated as:

$$q_d = Pw_kC_{kb} \tag{9}$$

where P is a probability parameter which gives a measure of the probability that a sediment particle settling to the bed, stays at the bed. C_{kb} is the near-bed concentration of the sediment fraction k. Krone [24] proposed a relationship for P as:

$$P = \left(1 - \frac{\tau}{\tau_{crd}}\right) \tag{10}$$

where τ is the bed shear stress and τ_{crd} is the critical shear stress for deposition, which is defined as the bed shear stress above which none of the initially suspended sediment would deposit.

The third component, namely, the entrapment component, is assumed to be directly proportional to the settling flux near the bed and is expressed as:

$$q_{entrap} = \alpha w_k C_{kb} \tag{11}$$

where α is the proportionality constant and is termed the entrapment coefficient. The entrapment coefficient is expected to be a function of porosity of the gravel substrate, the thickness of the gravel bed layer and the permeability of the substrate. Further research is needed to quantify this term as a function of the bulk properties governing the entrapment process.

2.3.2. Flocculation Stage

The flocculation stage was modelled using the coagulation equation (Fuchs [25]) shown below:

$$\frac{\partial N(i,j)}{\partial t} = -\beta N(i,t) \sum_{i-1}^{\infty} K(i,j)N(j,t) + \frac{1}{2}\beta \sum_{i-1}^{\infty} K(i-j,j)N(i-j,j)N(j,t) \tag{12}$$

where $N(j,t)$ and $N(j,t)$ are number concentrations of particle size classes i and j, respectively. $K(i,j)$ is the collision frequency function, which is a measure of the probability that a particle of size i collides with a particle of size j in unit time and β is the cohesion factor, which defines the probability that a pair of collided particles will coalesce and form a new floc. The β term accounts for the effects of cohesion properties such as the electrochemical properties of the sediment-water mixture and the effects of polymers secreted by microorganisms. In this study, is treated as an empirical parameter and was determined through the calibration of the model using experimental data from the flume.

The first term on the right side of Equation (12) gives the reduction of particles in size class i because of flocculation of these particles in size class i with all other size classes. The second term gives the generation of new particles in the size class i because of flocculation of particles in smaller size classes. In this process, the mass of particles is assumed to be conserved. Equation (12) was simplified by considering the particle sizes in discrete size classes where the continuous particle size space is considered in discrete size ranges. Each range is considered as a bin containing particles of a certain size range. For example, r_i is the geometric mean radius of particles in bin 1. The particle ranges were selected in such a way that the mean volume of particles in bin i is twice that of the preceding bin.

Under this scheme, when particles of bin i flocculate with particles in bin j ($j < i$), the newly formed particles will fit into bins i and $i + 1$. The proportion of particles going into these bins can be calculated by considering the mass balance of particles before and after flocculation. With this simplification, the coagulation equation can be expressed in discrete form as follows:

$$\frac{\Delta N_i}{\Delta t} = -\sum \beta K(r_i, r_j)\, N_i N_j + \sum f_{i,j}\beta K(r_i, r_j) N_i N_j + \sum (1 - f_{i,j})\beta K(r_i, r_j) N_{i-1} N_j \tag{13}$$

where $f_{i,j}$ is the allocation function given by:

$$f_{i,j} = \left(\rho_{s,j} V_i + \rho_{s,j} V_j - \rho_{s,j+1} V_{i+1}\right) / \left(\rho_{s,i} V_i - \rho_{s,j+1} V_{i+1}\right) \tag{14}$$

where, ρ_s denotes the density of the flocs and V is the volume of the flocs. The density of the flocs is dependent on floc size and, here, an empirical relationship proposed by Lau and Krishnappan [26] was adopted in this study. The form of the relationship is as follows:

$$\rho_s - \rho = \left(\rho_p - \rho\right)\exp(-bD^c) \tag{15}$$

where ρ_s, ρ and ρ_p are densities of flocs, water and parent material forming the flocs, respectively. D is the diameter of the floc and b and c are empirical coefficients to be determined through model calibration. The settling velocity of flocs, needed for the settling stage of the sediment particle motion, was calculated using the above density relation and Stoke's Law. The resulting expression is as follows:

$$w_k = \left(\frac{1.65}{18}\right)\left(\frac{gD_k^2}{v}\right)\exp(-bD^c) \tag{16}$$

where g stands for acceleration due to gravity and v stands for the kinematic viscosity of water.

The collision frequency function $K(i, j)$ takes different functional forms depending on the collision mechanisms that bring the particles to close proximity. The mechanisms considered in this study include (1) Brownian motion (K_b); (2) turbulent fluid shear (K_{sh}); (3) inertia of particles in turbulent flow (K_I), and; (4) differential settling of flocs (K_{ds}). An effective collision frequency function, K_{ef}, was calculated in terms of the individual collision functions as follows:

$$K_{ef} = K_b + \sqrt{\left(K_{sh}^2 + K_I^2 + K_{ds}^2\right)} \tag{17}$$

The collision frequency functions for the individual collision mechanisms can be found in Valioulis and List [27].

The break-up of flocs due to turbulent fluctuations of the flow was modelled using a scheme proposed by Tambo and Watanabe [28]. According to this scheme, a "collision-agglomeration" function was introduced as a multiplier to the collision frequency function K_{ef}. This produced an effective collision frequency that resulted in an optimum floc size distribution for a given level of turbulence. The function proposed by Tambo and Watanabe [28] is as follows:

$$\alpha_R = \alpha_0\left(1 - \frac{R}{R_m + 1}\right)^n \tag{18}$$

where R is the number of primary particles contained in a floc and R_m is the number of particles contained in the biggest floc for the given level of turbulence. The parameters α_0 and n assume values of 1/3 and 6, respectively. The flow field and the turbulence characteristics in the rotating circular flume needed for the FLUME model were predicted using PHOENICS (Rosten and Spalding [16]).

The FLUME model was used in this study to simulate the erosion process, even though the model is capable of predicting the total transport of cohesive sediment in rotating circular flumes including the deposition and the entrapment processes. The FLUME model, therefore, can be used to determine all parameters governing the transport of cohesive sediments by carrying out experiments using a rotating circular flume and then use these parameters in models that can predict the transport of cohesive sediment under field conditions.

3. Results and Discussion

The results of erosion tests from all three runs carried out for the present study are presented in Figure 2. The erosion data indicate that the sediment deposit was completely stable until the shear stress reached a value of 0.09 Pa, which can be considered as the critical shear stress for the erosion of the surficial layer of the sediment deposit. The data presented also show that six out of the seven shear stress steps at each age of deposit produced erosion of the riverbed sediment. At each of the six shear stress steps when erosion of the sediment bed occurred, the eroded sediment concentration increased suddenly after the shear stress change, suggesting that the sediment erosion follows the pattern of "bulk" erosion. As the sediment erosion continues, the erosion rate decreases due to the increasing strength of the bed with bed depth. Erosion during the later stages of the shear stress step is defined as the "surface" erosion when the removal of the material from the bed is particle by particle and the concentration in the water column approaches a steady state value. Additionally, the influence of the age of deposit is evident for the lower shear stress steps, since the concentration of the eroded sediment decreased as the age of deposit increased (Figure 2). For higher shear stress steps (i.e., for 0.27 and 0.33 Pa), the influence of age of deposit is not as pronounced as for the lower shear stress steps.

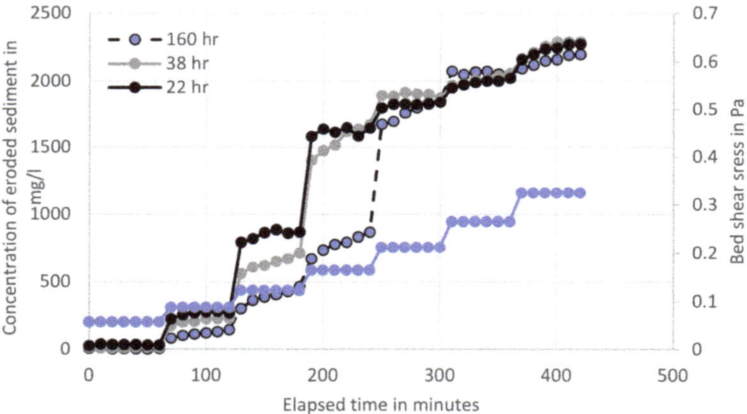

Figure 2. Results from the erosion tests.

The size distribution of the eroded sediment measured using a LISST and the image analysis system indicate that the sediment is flocculated. Figure 3 shows the size distribution data measured using LISST for the shear stress step of 0.33 Pa and where the age of deposit is 160 h.

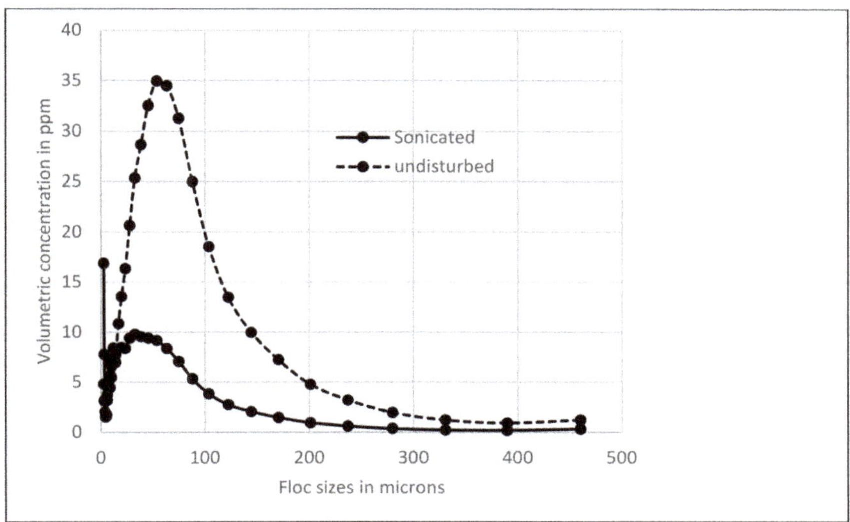

Figure 3. Size distribution of eroded sediment at the shear stress step of 0.33 Pa and age of deposit equal to 160 h.

The undisturbed and sonicated size distribution of eroded sediment for one set of experimental conditions (0.33Pa and deposit age of 160 h) are presented in Figure 3. The dotted line represents the size distribution of the eroded sediment as determined by the LISST for the undisturbed sample, whereas the solid line represents the distribution for the sonicated sample. Sonication breaks up all the flocs that are present in the sample. From these two distributions, we can conclude that the eroded sediment in the water column is in a flocculated state and the sediment exhibits cohesive tendencies. The change in the predominant size of the flocculated sediment from ~50 μm to ~25 μm after sonication highlights how flocculation can influence particle morphology and, by extension, volumetric concentration (Figure 3). A particle size of 63 microns is considered in the literature as the division between cohesive and non-cohesive sediments.

A photomicrograph of the eroded sediment sample collected for the image analysis is shown in Figure 4.

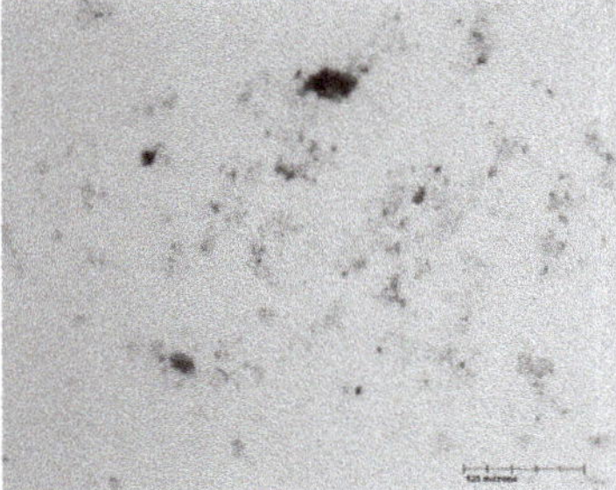

Figure 4. A photomicrograph of the eroded sediment for shear stress step of 0.33 Pa and age of deposit equal to 160 h.

This figure provides visual evidence of flocculation of the particles and also shows that mico-flocs are the building blocks of larger flocs as previously reported by Stone et al. [29]. The particles are interconnected through loose fibril material that might have been secreted by the microorganisms or the organic material that is present in the system. Therefore, flocculation is enhanced by microorganisms and the organic material and this process will have to be accurately represented in models to simulate sediment transport in the upper River Taw River.

Krone's curve fitting approach was used for the data shown in Figure 2. For each shear stress step where the sediment erosion was present, the constants c_0 and c_1 were calculated using Equation (2) with T = 10 min. The values of these constants, termed herein as "fitting coefficients" are listed in Table 1.

Table 1. Fitting coefficients c_0 and c_1

Shear Stress Steps	Age of Deposit 22 h		Age of Deposit 38 h		Age of Deposit 160 h	
	c_0 (m^3/gm)	c_1 (m^{3}sec/gm)	c_0 (m^3/gm)	c_1 (m^{3}sec/gm)	c_0 (m^3/gm)	c_1 (m^{3}sec/gm)
0.09 Pa	4.2×10^{-3}	0.57	4.6×10^{-3}	0.88	7.1×10^{-3}	3.65
0.12 Pa	1.7×10^{-3}	0.15	2.1×10^{-3}	0.57	3.2×10^{-3}	1.95
0.17 Pa	1.3×10^{-3}	0.07	1.0×10^{-3}	0.24	2.5×10^{-3}	1.43
0.21 Pa	5.1×10^{-3}	0.87	4.4×10^{-3}	0.18	1.0×10^{-3}	0.14
0.27 Pa	5.5×10^{-3}	2.22	5.4×10^{-3}	3.08	4.6×10^{-3}	0.35
0.33 Pa	4.0×10^{-3}	2.03	4.4×10^{-3}	3.05	6.6×10^{-3}	8.57

With values of c_0 and c_1, Equation (1) was applied to all the shear stress steps and the concentration variation as a function of time was calculated. A comparison of the fitted concentration variation with the measured data is presented in Figure 5 for a shear stress step of 0.17 Pa and consolidation time of 38 h as an example. The modelled and measured data are in agreement for all six shear stress steps in all three sediment consolidation time tests.

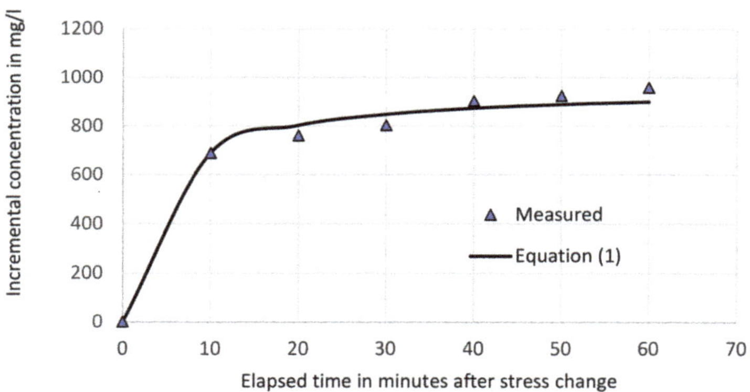

Figure 5. Comparison between measured and fitted sediment concentrations using Equation (1) for shear stress step: 0.17 Pa; Age of deposit: 38 h ($r^2 = 0.993$).

The FLUME model was applied to the present erosion experiments to simulate the concentration of the eroded sediment in the water column for all three experiments. The model was applied to each shear stress step using the appropriate parameters, c_0 and c_1, corresponding to that particular shear stress step and the age of the deposit. The deposition and entrapment fluxes were suppressed for the present simulations. The simulated concentration of the eroded sediment is compared to the measured concentration in Figure 6 and a favourable agreement between the two was observed. The erosion rate function that was derived from the fitting equation (Equation (1)) of Krone [10] works well with the model able to produce sediment concentrations that agree well with the measured data.

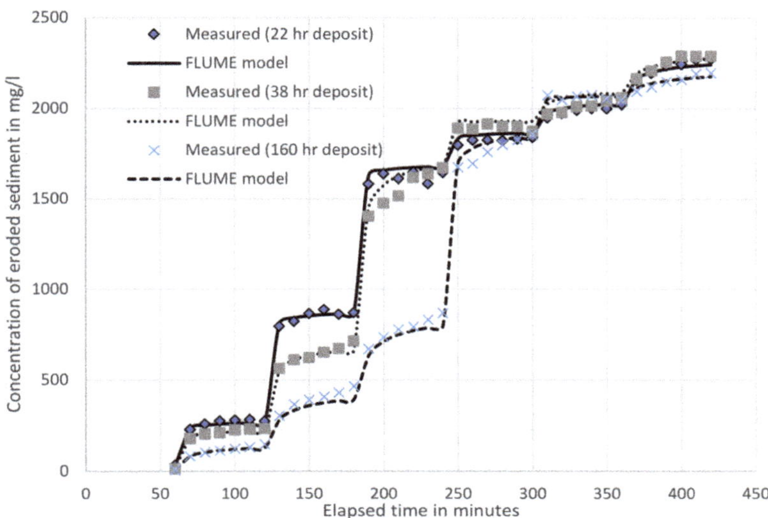

Figure 6. Comparison of measured and modelled sediment concentrations predicted by the FLUME model.

The fitting coefficients, c_0 and c_1, have different values for each shear stress step and age of deposit (Table 1). The variability of these coefficients with each shear stress and age of deposit is shown in Figure 7 which shows that these coefficients vary as a function of both shear stress and age of the deposit.

The coefficients exhibit a complex behaviour with respect to bed shear stress. Both coefficients decrease initially as the shear stress increases, reach minimum values and then increase as the shear stress is increased further. Notably, this behaviour was observed for all three ages of deposit tested. The minimum conditions for the coefficients imply maximum erosion rate and the maximum amount of sediment eroded (implying "bulk" erosion). The minimum condition for the coefficients shifts to the right as the age of deposit increases (Figure 7). Notably, when the age of deposit is 22 h, minimum conditions for both coefficients are at around 0.15 Pa, whereas for the age of deposit of 160 h, the minimum conditions are shifted to 0.21 Pa for c_0 and to 0.25 Pa for c_1. Accordingly, this suggests that the bed shear strength has increased due to the age of the deposit and higher shear stresses are needed to cause "bulk" erosion. The variability of the fitting coefficients demonstrates the importance of the roles that the bed shear stress and the age of deposit play in determining the erosion behaviour of the upper River Taw sediment. This finding will be useful in developing a modelling framework for predicting cohesive sediment transport in the upper River Taw.

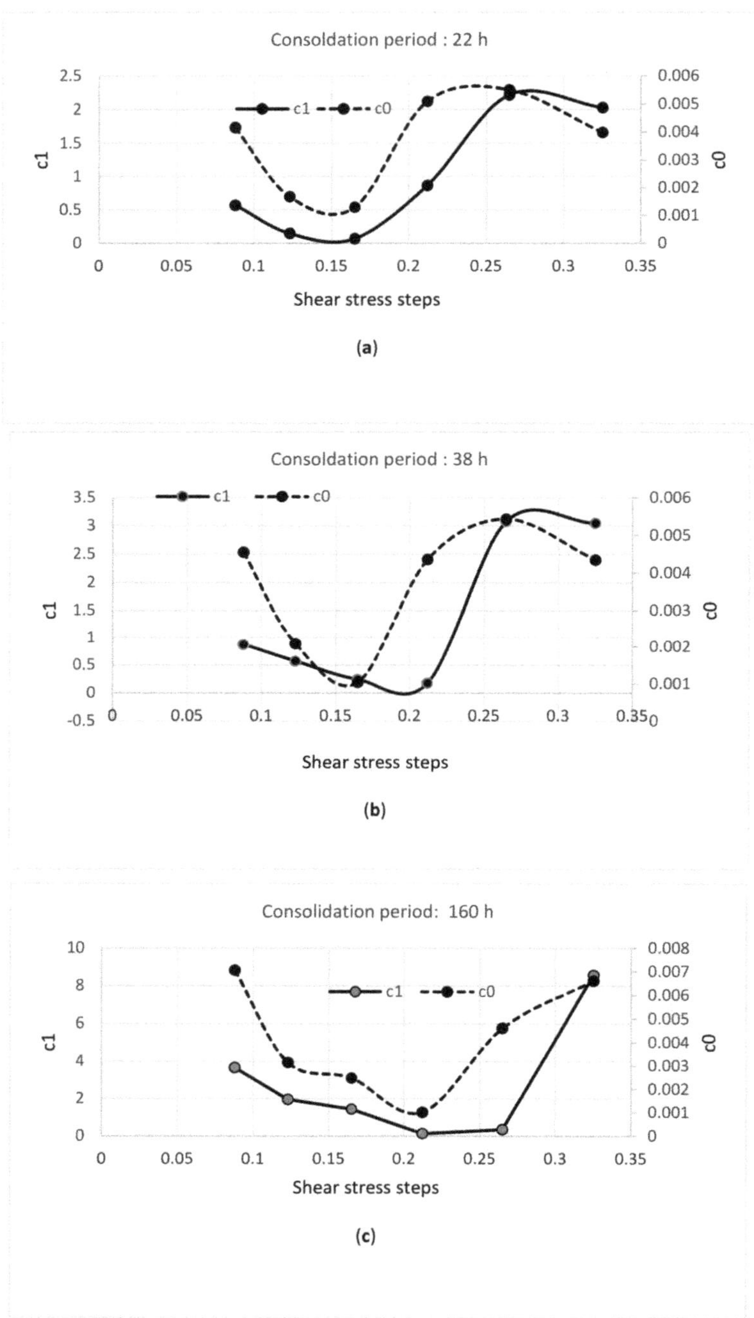

Figure 7. The variability of the fitting constants as a function of shear stress and age of deposit (**a**): 22 h; (**b**): 38 h and (**c**): 160 h.

4. Conclusions

A better understanding of cohesive sediment transport processes is required to develop models that can simulate scenarios that address both scientific and policy questions. In this short communication, the erosion process of the fine, cohesive sediment collected from the upper River Taw in South West England was studied by conducting erosion experiments using a rotating annular flume located in the National Water Research Institute in Burlington, ON, Canada. The erosion experimental data show that the critical shear stress for erosion of the upper River Taw sediment is about 0.09 Pa and it did not depend on the age of the deposit. The size distribution measurements for the eroded sediment indicates that the eroded sediment was transported in a flocculated form and hence the upper River Taw sediment exhibits cohesive properties. Photomicrographs of the eroded sediment samples obtained by image analysis suggest that the agent of flocculation for the sampled sediment may be due to the presence of fibrils from microorganisms and organic material present in the River Taw system. The experimental data from the erosion experiments were analysed using an approach proposed by Krone [10] and a mathematical model of cohesive sediment transport in rotating annular flumes developed by Krishnappan [14]. The approach of Krone [10] was applied to the experimental data and the fitting coefficients were established as a function of bed shear stress and age of sediment deposit. Using the approach of Krone [1], an erosion rate function was calculated and was used in the FLUME model of Krishnappan [14]. A comparison of the model predictions and the experimental data confirmed that the fine sediment transport model is capable of simulating accurately the erosion experiments in a rotating circular flume. The variability of the fitting coefficients, as a function of bed shear stress and the age of the deposit, was also examined in the present study. Future work will incorporate the experimental results reported herein into a catchment scale cohesive sediment transport modelling framework for exploring the implications of future climate and land management scenarios on sediment transport and behaviour.

Author Contributions: Conceptualization, B.G.K., M.S. and A.L.C.; methodology, B.G.K.; Software, B.G.K.; Validation, S.J.G., H.R.U., Q.T. and Y.Z.; formal analysis, B.G.K.; investigation, S.J.G., H.R.U., Q.T. and Y.Z.; resources, S.J.G.; data curation, S.J.G., H.R.U., Q.T. and Y.Z.; writing-original draft preparation, B.G.K.; review and editing, M.S. and A.L.C.; visualization, S.J.G.; supervision, M.S. and A.L.C.; project administration, M.S. and A.L.C.; funding acquisition, A.L.C. and M.S. All authors have read and agreed to the published version of the manuscript.

Funding: Rothamsted Research received strategic funding from UKRI-BBSRC (UK Research and Innovation—Biotechnology and Biological Sciences Research Council) and the contribution of institute staff to this paper was funded by the Soil Nutrition institute strategic programme project 3 grant (BBS/E/C/000I0330). Funding for the flume experiments was from an NSERC Discovery Grant RGPIN-2020-06963 awarded to M. Stone.

Acknowledgments: The authors acknowledge the cooperation of Ian Droppo of Environment and Climate Change Canada and the use of the Rotating Circular Flume for their study. The flume experiments were carried out by Robert Stephens, a retired Research Technician of Environment and Climate Change Canada.

Conflicts of Interest: The authors declare no conflict of interest.

References

1. Horowitz, A.J.; Elrick, K.A. The relation of stream sediment surface area, grain size and composition to trace element chemistry. *Appl. Geochem.* **1987**, *2*, 437–451. [CrossRef]
2. Luoma, S.N.; Rainbow, P.S. *Metal Contamination in Aquatic Environments: Science and Lateral Management*; Cambridge University Press: Cambridge, UK, 2008; 573p.
3. Grabowski, R.C.; Droppo, I.C.; Wharton, G. Erodibility of cohesive sediment: The importance of sediment properties. *Earth Sci. Rev.* **2011**, *105*, 101–120. [CrossRef]
4. Amos, C.L.; Li, M.Z.; Sutherland, T.F. The contribution of ballistic momentum flux to the erosion of cohesive beds by flowing water. *J. Coast. Res.* **1998**, *14*, 564–569.
5. Lick, W.; McNeil, J. Effects of sediment bulk properties on erosion rates. *Sci. Total Environ.* **2001**, *266*, 41–48. [CrossRef]

6. Friend, P.L.; Ciavola, P.; Cappucci, S.; Santos, R. Bio-dependent bed parameters as a proxy tool for sediment stability in mixed habitat intertidal areas. *Cont. Shelf Res.* **2003**, *23*, 1899–1917. [CrossRef]

7. Lau, Y.L.; Droppo, I.G.; Krishnappan, B.G. Sequential erosion/deposition experiments demonstrating the effects of depositional history on sediment erosion. *Water Res.* **2001**, *35*, 2767–2773. [CrossRef]

8. Roberts, J.; Jepsen, R.; Gotthard, D.; Lick, W. Effects of particle size and bulk density on erosion of quartz particles. *J. Hydraul. Eng. ASCE* **1998**, *124*, 1261–1267. [CrossRef]

9. Zreik, D.A.; Krishnappan, B.G.; Germaine, J.T.; Masden, O.S.; Ladd, C.C. Erosional and mechanical strengths of deposited cohesive sediments. *J. Hydraul. Eng. ASCE* **1998**, *124*, 1076–1085. [CrossRef]

10. Krone, R.B. Effects of bed structure on erosion of cohesive sediments. *J. Hydraul. Eng. ASCE* **1999**, *125*, 1297–1301. [CrossRef]

11. Lambert, C.P.; Walling, D.E. Measurement of channel storage of suspended sediment in a gravel-bed river. *Catena* **1988**, *15*, 65–80. [CrossRef]

12. Duerdoth, C.P.; Arnold, A.; Murphy, J.F.; Naden, P.S.; Scarlett, P.; Collins, A.L.; Sear, D.A.; Jones, J.I. Assessment of a rapid method for quantitative reach-scale estimates of deposited fine sediment in rivers. *Geomorphology* **2015**, *230*, 37–50. [CrossRef]

13. Naden, P.S.; Murphy, J.S.; Old, G.H.; Newman, J.; Scarlett, P.; Harman, M.; Duerdoth, C.P.; Hawczak, A.; Pretty, J.L.; Arnold, A.; et al. Understanding the controls on deposited fine sediment in the streams of agricultural catchments. *Sci. Total Environ.* **2016**, *347*, 366–381. [CrossRef]

14. Krishnappan, B.G. Recent Advances in basic and applied research in cohesive sediment transport in aquatic systems. *Can. J. Civ. Eng.* **2007**, *34*, 731–743. [CrossRef]

15. Petersen, O.; Krishnappan, B.G. Measurement and analysis of flow characteristics in a circular flume. *J. Hydraul. Res. IAHR* **1994**, *32*, 483–494. [CrossRef]

16. Rosten, H.I.; Spalding, D.B. *The PHOENICS Reference Manual*; TR/200; CHAM Ltd.: Wimbledon, London, UK, 1984.

17. Krishnappan, B.G.; Engel, P. Distribution of bed shear stress in Rotating Circular Flume. *J. Hydraul. Eng.* **2004**, *130*, 324–331. [CrossRef]

18. Krishnappan, B.G. Rotating Circular Flume. *J. Hydraul. Eng. ASCE* **1993**, *119*, 758–767. [CrossRef]

19. Partheniades, E. A Study of Erosion and Deposition of Cohesive Soils in Salt Water. Ph.D. Thesis, University of California, Berkeley, CA, USA, 1962.

20. Mehta, A.J.; Partheniades, E. An investigation of the depositional properties of flocculated fine sediments. *J. Hydraul. Res. IAHR* **1975**, *13*, 361–381. [CrossRef]

21. Parchure, T.M. Erosional Behaviour of Deposited Cohesive Sediments. Ph.D. Thesis, University of Florida, Gainesville, FL, USA, 1984.

22. Lick, W. Entrainment, deposition and transport of fine grained sediments in Lakes. *Hydrobiologia* **1982**, *91*, 31–40. [CrossRef]

23. Krishnappan, B.G. Erosion behaviour of fine sediment deposits. *Can. J. Civ. Eng.* **2004**, *31*, 759–766. [CrossRef]

24. Krone, R.B. *Flume Studies of the Transport of Sediments in Estuarial Shoaling Processes*; Report; Hydr. Engr. Lab. and Sanitary Engr. Lab. University of California: Berkeley, CA, USA, 1962.

25. Fuchs, N.A. *The Mechanics of Aerosols*; Pergamon: New York, NY, USA, 1964; 408p.

26. Lau, Y.L.; Krishnappan, B.G. *Measurement of Size Distribution of Settling Flocs*; Report No. 97–223; National Water Research Institute; Environment Canada; CCIW: Burlington, ON, Canada, 1997.

27. Valioulis, I.A.; List, E.J. Numerical simulation of a sedimentation basin. 1 Model development. *Environ. Sci. Technol.* **1984**, *18*, 242–247. [CrossRef]

28. Tambo, N.; Watanabe, Y. Physical aspects of flocculation processes—I Fundamental Treatise. *Water Res.* **1979**, *13*, 429–439. [CrossRef]

29. Stone, M.; Krishnappan, B.G.; Emelko, M. The effect of bed age and shear stress on the particle morphology of eroded cohesive sediment in an annular flume. *Water Res.* **2008**, *42*, 4179–4187. [CrossRef]

Article

An Integrated Hydrological-CFD Model for Estimating Bacterial Levels in Stormwater Ponds

Farzam Allafchi [1], Caterina Valeo [1,*], Jianxun He [2] and Norman F. Neumann [3]

[1] Department of Mechanical Engineering, University of Victoria, Victoria, BC V8P 5C2, Canada; fallafchi@uvic.ca

[2] Department of Civil Engineering, University of Calgary, Calgary, Alberta, AB T2N 1N4, Canada; jianhe@ucalgary.ca

[3] School of Public Health, University of Alberta, Edmonton, Alberta, AB T6G 2R3, Canada; nfneuman@ualberta.ca

* Correspondence: valeo@uvic.ca; Tel.: +1-250-721-8623

Received: 1 May 2019; Accepted: 10 May 2019; Published: 15 May 2019

Abstract: A hydrological model was integrated with a computational fluid dynamics (CFD) model to determine bacteria levels distributed throughout the Inverness stormwater pond in Calgary, Alberta. The Soil Conservation Service (SCS) curve number model was used as the basis of the hydrological model to generate flow rates from the watershed draining into the pond. These flow rates were then used as input for the CFD model simulations that solved the Reynolds-Averaged Navier-Stokes (RANS) equations with k-ε turbulence model. *E. coli*, the most commonly used fecal indicator bacteria for water quality research, was represented in the model by passive scalars with different decay rates for free bacteria and attached bacteria. Results show good agreement with measured data in each stage of the simulations. The middle of the west wing of the pond was found to be the best spot for extracting water for reuse because it had the lowest level of bacteria both during and after storm events. In addition, only one of the four sediment forebays was found efficient in trapping bacteria.

Keywords: stormwater reuse; SCS curve number; CFD; fecal indicator bacteria; *E. coli*

1. Introduction

In recent decades stormwater has been considered as an alternative water source for reuse, specifically for applications that need less than pristine water quality. Reusing stormwater is more critical in water-scarce regions and regions where rainfall patterns and rainfall frequencies are changing [1]. Several regions are trying to reuse stormwater as a sustainable method of water resources management; thus, prompting research into the feasibility of reusing stormwater from a water quality perspective [2].

Although stormwater ponds are built with the primary objective of reducing runoff quantities in order to protect urban areas against flooding, they also improve the quality of stormwater as well [1]. The stormwater quality within a pond varies both spatially and temporally and is not only a function of the quality of the influent but also a function of local hydrological conditions and on the pond's design [3,4]. A water quality study of a very large stormwater pond in Calgary, Alberta, Canada, showed that the quality of stormwater in the pond observed over a three year period did meet the irrigation water quality requirements but only under certain circumstances [2]. Thus, stormwater recycling with pond water often requires continuous or intermittent water quality monitoring of the pond water in order to remain compliant with local regulations. Highly distributed water quality sampling in stormwater ponds is often impractical due to the sizes of these ponds and the cost. In addition, most ponds are not designed with reuse in mind and the extraction point is often located in an ad hoc fashion and possibly in a region of the pond which has higher pollution levels relative to the rest of the pond because of local hydrodynamic conditions. If retrofitting a pond with the intent to recycle the water, the municipality could undertake a water quality sampling program that collected samples distributed throughout the pond over a period of time in order to identify the optimum location for extracting the "cleanest" water in the pond (assuming there is no treatment of this water). A more cost-effective alternative is to develop a physical model to estimate the bacteria level in the pond that incorporates the factors leading to bacterial contamination of stormwater in retention ponds.

Water quality is impaired when mobilized sources of contaminants are transported away from their original location by runoff and discharged in aquatic environments [5,6]. Fine particles are either detached from the soil or washed off an impervious surface [7]. Determining levels of pathogenic microorganisms is expensive and difficult due to their large diversity; thus, microbiological water quality monitoring procedures often use fecal indicator bacteria (FIB). FIB are present in feces of human and warm-blooded animals in large numbers and can be easily detected [8]. Total coliforms (TC) and fecal coliforms (FC) were considered the main group of FIB during the twentieth century. However, nowadays, *Escherichia coli* (*E. coli*) and intestinal enterococci (IE) are enumerated and considered as FIB because some water-related epidemiological studies have shown that they are better bacterial indicators for predicting sanitary risk [9,10].

The aim of microbiological water quality monitoring of stormwater ponds is primarily to assess the level of fecal contamination in an aquatic system. This contamination varies spatially and temporally throughout the pond and, therefore, very large data sets must be collected to enable an adequate understanding of contamination levels and the processes leading to the degree of pollution in the system. Collecting such a large data set is often intractable, and thus models are often the only feasible approach for gaining greater insight into aquatic ecosystem processes. Two main types of models have been proposed to estimate bacterial levels in aquatic environments: data-driven models, which are also considered black box models, and process-based models [11].

Data-driven models use statistical methods or computational intelligence and machine learning to relate the involved parameters to the state variables (input, internal and output variables) with only a limited number of assumptions about the physical behavior of the system. In contrast, process-based methods predict the FIB concentration based on the mathematical description of sources, sinks, and internal processes influencing FIB levels. The fundamental principle underlying process-based models is conservational equations. The most important factors influencing microorganism fate in the aquatic environment are (i) environmental parameters, and (ii) whether or not they are attached to particles [11]. Few estimates of the percentage of particles with attached bacteria in any collected stormwater sample can be found in the literature. One study found up to 10–28% and 22–30% of fecal coliform and *E. coli*, respectively, were associated with suspended solids in stormflow samples at the mouth of a canal discharging water to a lake with brackish water [12]. Another study found that 30–55% of both fecal coliform and *E. coli* were attached to sediment particles in stormwater samples [13].

The fate and transport of bacteria depend heavily on the attachment to suspended solids. Attachment to sediment protects bacteria from some processes that may accelerate death and decay, such as sunlight and predation. Thus, any modelling should use different decay rates for bacteria attached to sediment vs. free-floating bacteria. For example, the decay rate of free-floating *E. coli* was considered to be twice that of *E. coli* attached to sediment in a study in the Scheldt drainage network in Belgium [14]. In another study in the Blackstone River watershed in Massachusetts, the ratio of the decay rate of free-floating bacteria to attached bacteria was considered to be 4 [15]. Some studies even assumed that attached pathogens did not decay at all [11].

The fate of bacteria is most commonly modeled by first-order kinetic decay, and this first-order model was used to study the efficiency of waste stabilization ponds. Shilton [16] used a pulse tracer study incorporated with computational fluid dynamics (CFD) to find the retention time of a waste stabilization pond in order to use the first-order kinetic decay based on the retention time [16]. In another study on an anaerobic lagoon, decay was calculated based on the retention time resulting from CFD simulations [17]. In both the study of the waste stabilization pond [16] and the anaerobic lagoon, the objective was only to study the inlet and outlet bacteria concentrations, and they neglected to consider variation within the body. In addition, in both of the studies, the simulation was assumed to be steady-state. Therefore, the retention time, which did not change with time could be an indicator of bacteria concentration. Two other studies on waste stabilization ponds used tracers integrated with CFD simulations to study different configurations of baffles [18,19]. They implemented first-order kinetic decay by a source term in the transport equation of the tracer. All of the abovementioned studies simulated the flow as steady-state and neglected wind. The fluid flow and bacteria fate and transport in stormwater ponds are intrinsically unsteady. Also, the wind cannot be neglected. In addition, the main goal of the present study is to study bacteria concentration within stormwater ponds for the purposes of determining the optimum point at which water may be withdrawn for reuse—the optimum point is that location with the cleanest water and having the lowest level of bacteria concentrations.

In the present study, a comprehensive model was developed that integrates a hydrological model for a catchment draining to a stormwater pond, with a computational fluid dynamics (CFD) model simulating the pond's hydrodynamics. The results of the hydrological model were used as inputs to the CFD simulation. The model results are validated against data collected at the pond. The overall goal of this paper was to enhance knowledge of bacteria fate and transport in stormwater ponds. Furthermore, the developed modeling approach leading to this enhanced understanding may ultimately be used as a tool to evaluate the capacity for a stormwater pond as a candidate for reuse and/or the need for the modification/retrofit. In addition, it may provide designers and planners with guidance to define standards for stormwater reuse.

2. Materials

2.1. Study Area

The City of Calgary is a semi-arid city in Alberta, Canada, with many stormwater ponds that are being considered as candidate sources of stormwater reuse for irrigating parkland and other public lands during the irrigation season [20]. The Inverness pond is one of the largest stormwater ponds in Calgary and is located in the southeast quadrant of the city. It contains approximately 235,000 m^3 of water at the permanent water level. Seven inlets convey stormwater runoff from 415 ha of catchment area into the pond. Figure 1 shows the pond and the locations of the inlets and outlets. All of the inlets are submerged except for I1. Outlet O1 at the west wing of the pond is the main outlet of the pond and is a 1.5 m diameter concrete pipe. I4 inlet, located at the south wing, conveys water from the largest subbasin of the catchment, which is 257.97 ha. The second largest subbasin corresponds to I3 inlet with an area of 89.52 ha. In addition, there are four sediment forebays located in front of I2, I3, I4 and I5 inlets. Table 1 shows the subbasin area and design parameters corresponding to each inlet.

Table 1. Inlet and outlet parameters of the Inverness stormwater pond.

Inlet	I1	I2	I3	I4	I5	I6	I7	O1	O2
Subbasin area (ha)	2.74	4.68	89.52	257.97	15.3	13.04	18.48	outlet	outlet
Residential (%)	0	27	6.48	10.04	62	83	69	-	-
Commercial (%)	0	53	0.36	0	0	7	21	-	-
Industrial (%)	0	0	0	0	0	0	0	-	-
Parks and Institution (%)	0	20	1.8	1.54	38	10	8	-	-
Major Transport Infrastructure (%)	0	0	0.36	23.97	0	0	2	-	-
Newly graded (%)	100	0	40	36.22	0	0	0	-	-
Farm (%)	0	0	52	28.22	0	0	0	-	-
Sediment Forebay	no	yes	yes	yes	yes	no	no	no	no
Invert elevation (m) from PWL [1]	-	2.44	2.12	2.83	2.11	2.48	1.86	2.80	2.80

[1] Permanent water level.

Figure 1. Arial schematic of Inverness pond showing the location of inlets and outlets (base map from Google Earth (50°54′39.4″ N 113°57′46.2″ W at white cross-hairs at center), arrows and texts added by author).

2.2. Data Collection

Over a three-year data collection campaign, several data were collected within and around the pond. Water quality and flow rate data collected during several rain events in the 2007 irrigation season were taken at a skimming manhole located just upstream of the I5 inlet. The water quality data included FIB level (*E. coli*, FC, and total coliform), total suspended solid (TSS), turbidity, pH, and dissolved oxygen (DO) [20]. Rainfall data in 5-min intervals were acquired from the City of Calgary (rain gauge #26 located 1 km north of the study site in McKenzie Towne) and wind data were extracted from the Calgary International Airport. Five rain events documented in 2007 were used in this research. Three of these were lower intensity events leading to moderate flow rates (on 28 May, 26 August and 20 September 2007), while the other two (6 June and 12 September 2007) were high-intensity events leading to large flow rates into the pond. The rain event data are tabulated in Table 2.

Table 2. Rainfall data of the events in 2007.

Date	Rainfall Depth (mm)	Start of the Event	Duration
28 May	4.8 mm	11:05 p.m. (May 27)	2 h
6 June	31.8 mm	5:55 a.m.	7 h 10 min
26 August	12.3 mm	9:25 a.m.	5 h 55 min
12 September	30 mm	1:05 p.m.	7 h 45 min
20 September	4 mm	8:00 p.m.	1 h 40 min

An autosampler with 24 pre-sterilized bottles was placed inside the skimming manhole and collected samples for bacteriological analysis once triggered by a rain gauge. The autosampler was programmed to collect 12 samples (two bottles per sample) during long events, and less than 12 samples were analyzed for short duration events. The samples were collected at unequal intervals (3 to 50 min intervals), and much attention was paid to collect samples more frequently in the early stages of the rain event in order to catch the first flush effect. The water samples were recovered right after rain events or very early the next morning in case rain events occurred at night. The samples were packed with ice and analyzed at the Alberta Provincial Laboratory for Public Health by assaying microorganisms using the membrane method [20].

3. Methods

The model is comprised of a hydrological component that simulates stormwater runoff feed into the CFD, which simulates the hydrodynamics and predicts the bacteria levels within a water body. The modeling approach was formulated in a way in which *E. coli* are primarily discharged into the pond through contaminated stormwater runoff (water fowl droppings from the surface were not considered) and *E. coli* concentrations were affected by the biological fate of *E. coli* and pond hydrodynamics, which were in turn affected by meteorology and pond bathymetry. Studies in the literature have revealed that the concentration of bacteria is highly dependent on land use-in particular, the percentage of imperviousness [21]. Developed areas, which have more impervious areas, had elevated increases in *E. coli* concentrations in runoff [22] as opposed to less developed areas. In order to provide insight into the relationship between land use and bacteria concentration in runoff, bacteria concentrations were measured during storms in different watersheds with different land uses [23]. They demonstrated the variability in fecal coliform concentrations for different land uses. In the present study, the fraction of bacteria in stormflow corresponding to each land use was calculated using the findings of Schoonover and Lockaby [23]. However, here, all of the land uses except for newly graded land and farmland were considered as urban. Thus, by knowing the fraction of bacteria that could be generated from the area draining into inlet I5, the bacteria load of different inlets was estimated based on the measured data of inlet I5.

3.1. Hydrological Model, Calibration and Validation

The Hydrologic Modeling System (HEC-HMS 4.2.1) (Hydrologic Engineering Center, Davis, CA, USA) [24] was used to simulate stormwater runoff generated by each subbasin and drained into the pond in storm events. In the HEC-HMS, the Soil Conservation Service (SCS) curve number (CN) loss method [25], which is a very common and simple method for runoff estimation [26] was adopted. The SCS-CN model considers precipitation excess as a function of cumulative precipitation, soil cover, land use, and antecedent moisture. Precipitation excess was estimated by Equation (1):

$$P_e = \frac{(P - I_a)^2}{P - I_a + S} \tag{1}$$

where P_e is the accumulated rainfall excess (mm) at time t; P is accumulated rainfall depth (mm) at time t; S is the potential maximum retention (mm); I_a is the initial abstraction (mm), where $I_a = 0.2S$. Here, S is a function of the CN, which is a function of watershed characteristics, and was calculated by Equation (2) [27]:

$$S = \frac{25{,}400 - 254CN}{CN} \tag{2}$$

The CN number of each subbasin was initially estimated based on the land uses within each subbasin. The time step of the HEC-HMS model was set at 5 min. The runoff at I5, measured in the event on 6 June 2007, was employed to adjust the initial CN of this subbasin aiming to reproduce the runoff well. Accordingly, the CN of other subbasins was modified using the difference between the adjusted CN number for I5 subbasin and its initial CN.

3.2. CFD Simulations

This paper adopted Reynolds averaged Navier Stokes (RANS) equations as follows:

$$\rho\left[\frac{\partial \overline{u}_i}{\partial t} + \frac{\partial}{\partial x_j}\left(\overline{u}_i\overline{u}_j\right)\right] = -\frac{\partial p}{\partial x_i} + \rho\frac{\partial}{\partial x_j}\left(v\frac{\partial \overline{u}_i}{\partial x_j}\right) - \frac{\partial}{\partial x_j}\left(\rho\overline{u'_i u'_j}\right) \tag{3}$$

$$\overline{u} = \frac{1}{T}\int_0^T u\,dt \tag{4}$$

where the velocity decomposed into mean velocity and velocity fluctuation ($u_i = \overline{u}_i + u'_i$); T is period of time (s); ρ is density (kg/m^3); t is time; p is pressure (pa); v is kinematic viscosity (m^2/s); x_j is an axis of the Cartesian coordinate system, and i,j = 1,2,3 are indices indicating the three axes in the coordinate system. The term $-\rho\overline{u'_i u'_j}$ is called Reynold's stress term [28]. Since the popular k-ε turbulence model has been successfully applied to simulate water body flow fields [29,30] in this study the k-ε turbulence model [31] was used to calculate the Reynold's stress terms. The turbulence model is not detailed here. The fundamentals of CFD and turbulence modeling can be found in Versteeg and Malalasekera [32].

Kunkel et al. [33] studied the attachment of *E. coli* to particles in stormwater. The study showed that *E. coli* appeared to attach predominantly to fine particles (<4 μm), while a further study on attachment to smaller particles was recommended [33]. Another study on the attachment of *E. coli* showed that most of them attached to particles smaller than 2 μm [34]. The City of Calgary characterizes the particles in stormwater and suggests the settling velocity of 0.00592 (mm/s) for small particles (<10 μm) [1]. Using Stokes relation for settling velocity [35], the settling velocity of finer particles (<2 μm) can be estimated using:

$$w_s = \frac{g\left(\rho_p - \rho_w\right)}{18\rho_w v}d_p^2 \tag{5}$$

where w_s is settling velocity (m/s); ρ_p and ρ_f are particle and fluid density (kg/m^3), respectively; d_p is particle diameter (m); g is gravitational acceleration (m/s^2). This relation is valid for small Reynolds numbers (Re < 1) [36]. With regard to the settling velocity of small particles, this condition was certainly satisfied in the Inverness pond [1]. The calculated settling velocity for the very fine particles (<2 μm) was 0.001 mm/s using the Stokes relation. This suggests that the attached bacteria settle less than 9 cm in 24 h. Therefore, it was assumed that all the bacteria (i.e., free-floating and attached *E. coli*) remain in the water column during the simulation. Accordingly, bacteria (both attached and free-floating bacteria) were modeled as passive scalars, which are massless particles that do not affect the physical properties of the flow field.

The passive scalar transport equation was solved for the transport of both free-floating and attached *E. coli*. The transport equation for a passive scalar, ϕ_i is:

$$\frac{\partial}{\partial t}\int_V \rho\phi_i dV + \oint_A \rho\phi_i v.dA = \oint_A j_i.dA + \int_V S_{\phi_i}dV \tag{6}$$

where t is time (s); V is volume (m^3); A is area (m^2); i is the component index; j_i is diffusion flux (kg/m^2·s); S_{ϕ_i} is source term (kg/m^2·s).

The fate of bacteria has been modeled using a negative value for the source term in the passive scalar transport equation in previous studies on waste stabilization ponds [18,19]. This approach is, however, not appropriate for stormwater ponds due to two reasons. Firstly, the flow field and bacteria transport in stormwater ponds are intrinsically unsteady; thus, the location and strength of the sources cannot be determined. Secondly, bacteria concentrations in some spots of stormwater ponds are very low, so a general source, representing decay over the entire domain, might cause negative bacteria concentrations which are not physically feasible. Therefore, in this paper, the transport of bacteria was calculated with Equation (6), while the fate of bacteria was modeled separately.

A finite volume discretization scheme was used to discretize Equations (3) and (6), and the turbulence model equations over the domain to solve the flow field and the concentration of passive scalar. Then, a field function was implemented for the first-order kinetic decay (Equation (7)) for the free-floating portion of bacteria and added the concentrations of both attached and the remaining free-floating bacteria at each computational cell. Characklis et al. [13] found that 30–55% of bacteria attach to sediments in stormwater, whereas the other study [37] applied a higher value in the range (50%). Therefore, in this paper, it was assumed that 50% of bacteria are attached to sediments and do not decay. Decay is calculated using Equation (7).

$$N = N_0 \exp(-\mu t) \tag{7}$$

where N and N_0 are the numbers of indicator bacteria at time t and at $t = 0$, respectively; μ is the decay rate (s^{-1}) [38]. The decay rate of *E. coli* was calculated with the following relationship that has been used in several studies [14,39] in the literature:

$$\mu = k_{20} \frac{e^{\left(-\frac{(T-25)^2}{400}\right)}}{e^{\left(-\frac{25}{400}\right)}} \tag{8}$$

where k_{20} is the decay rate of *E. coli* at 20 °C and T is water temperature (°C). A value of 1.25×10^{-5} s^{-1} was proposed for k_{20} of *E. coli* in the water column; average measured data were used for temperature in this paper.

3.2.1. Boundary and Initial Conditions

The effect of wind on the flow field was not negligible considering the size of the study pond. Since the wind vector was not stationary with time, a dynamic boundary condition of wind on the pond surface was defined. The effect of the wind was applied as a velocity vector on the top layer of the pond. The velocity of the top layer of the pond was calculated from the wind velocity based on the experimental data from a study by Banner and Peirson [40], which measured velocity at the top layers of water bodies under windy conditions.

The boundary condition of the inlets was set as the "velocity inlet" with parameters changing with time including velocity and *E. coli* concentration. A time variable boundary condition was also defined which linearly interpolated the value of bacteria concentration and velocity between the interval that data were collected. "Pressure outlet" and "stationary wall" were set as the boundary conditions for the outlets and the bottom of the pond, respectively.

A steady-state simulation in no storm conditions and with an average wind was simulated and its result (flow field) was used as the initial condition for the main simulations. The main simulation is unsteady. Using the steady simulation's results also accelerated the convergence. The unsteady simulations started one hour before the storm in order to let the flow field form based on the real wind data before the storm.

3.2.2. Other CFD Settings

The domain of the CFD simulation was sketched based on the bathymetric data of the pond comprised of approximately 33000 GIS points. AutoCAD CIVIL 3D 2018 was used for sketching the domain that includes the water body of the pond and a few meters (10 m on average) of inlet and outlet pipes. A grid was made that is more highly clustered close to the inlets, outlets and water edges. In the CFD simulations, grid dependence was checked by comparing velocity from the simulation with different numbers of grids. An unstructured grid with hexahedral cells and a number of 1.5 million was found sufficiently fine and the corresponding results were presented. Using finite volume discretization, RANS equations with the k-ε turbulence model were solved in an unsteady-state with the time step of 0.5 s and 24 inner iterations between each time step. The time step was selected by an approach similar to the grid independence comparison but the number of inner iterations was chosen based on the convergence trend of the solution. The pond was simulated from 1 h before the start of an event until 24 h after the end of the event. The simulations were run on Cedar High-Performance Computer, located at Simon Fraser University, Vancouver, BC, Canada, with 576 GB of memory and 192 computational cores (each core has two CPUs of 2.1 GHz). The CFD simulations were performed by using commercial CFD code STAR-CCM+ 12.04.011 [41].

4. Results and Discussions

4.1. Hydrological Modeling Performance

The hydrological model was manually calibrated with the measured data in inlet I5 on 6 June 2007 and then validated on four other events: 28 May, 26 August, 12 September, and 20 September 2007. During the calibration, the error in the total volume of water between prediction and observation was adopted to evaluate model performance, as the total volume of water entering the pond was considered to be the key factor that determines bacteria loading.

The CN number for the subbasin draining to I5 was calculated to be 11% less than the initially calculated CN value in the calibration of the 6 June 2007 event. Noting this, all other CN numbers were decreased by 11%. Table 3 shows the CN numbers for the different subbasins before and after calibration of the model. Lag time was also calibrated at the same time, and a 30 min lag time was found to be optimal. In addition, an attempt was made to calibrate initial abstraction; however, the default setting ($I_a = 0.2S$) resulted in the lowest error and was, therefore, left unchanged. Figures 2–5 show the modeled and observed hydrographs for the validation events for subbasin I5. The Nash-Sutcliffe model efficiency (NSE), which is a reliable criterion for assessing the goodness of fit of hydrological models [42], and the error in total volume for all events, are tabulated in Table 4. The calibrated CNs (Table 3) were then used to model other subbasins.

Table 3. Curve numbers of the different subbasins draining to the Inverness pond inlets.

Inlet	I1	I2	I3	I4	I5	I6	I7
Estimated CN	91	86.49	87.33	87.99	78.34	81.21	83.39
Calibrated CN	80.99	76.97	77.72	78.31	70	72.27	74.21

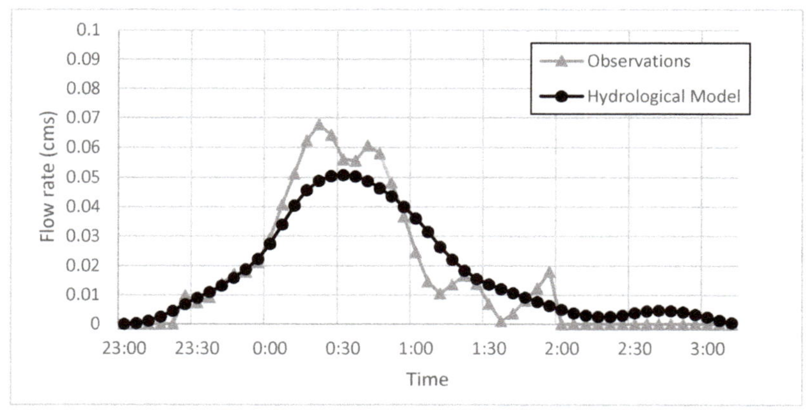

Figure 2. Modeled and observed hydrographs at I5 inlet of storm event on 28 May 2007.

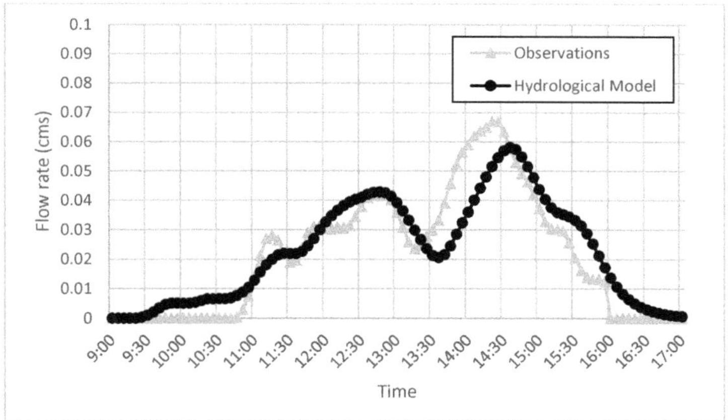

Figure 3. Modeled and observed hydrographs at I5 inlet of storm event on 26 August 2007.

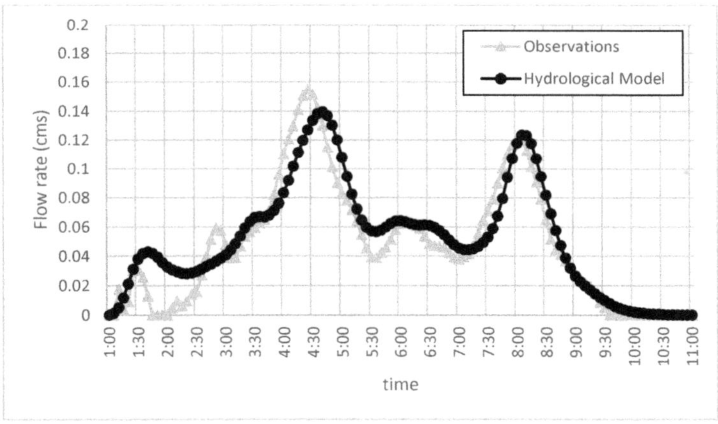

Figure 4. Modeled and observed hydrographs at I5 inlet of storm event on 12 September 2007.

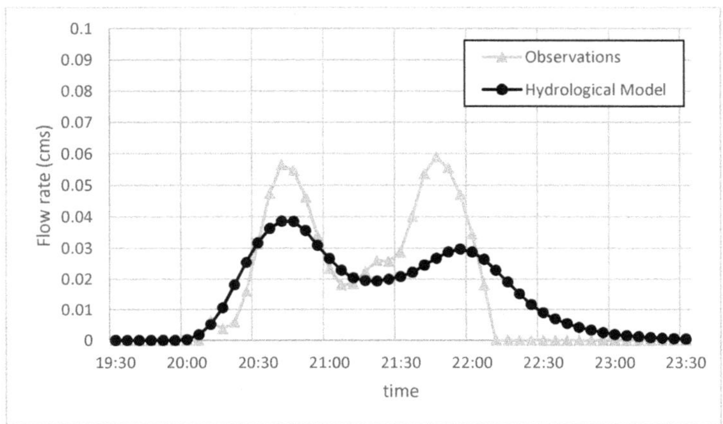

Figure 5. Modeled and observed hydrographs at I5 inlet of storm event on 20 September 2007.

Table 4. Model performance in model calibration and validation. The event on 6 June 2007 is the calibration event and all others are validation events.

Event	Total Volume Error	Nash-Sutcliffe Efficiency (NSE)
6 June 2007	4.8%	0.89
28 May 2007	2.8%	0.92
26 August 2007	4.1%	0.81
12 September 2007	7.4%	0.86
20 September 2007	−7.3%	0.66

Results show good agreement between the modeled and measured data. In all of the events except for 20 September 2007, the model slightly overestimated the total volume but the total volume error is less than 10%. The underestimation of the total volume for the event on 20 September 2007 may be due to the excessive moisture in the soil resulting from heavy rains that fell prior to the event; thus, potentially leading to greater runoff generation. However, the Nash-Sutcliffe efficiency in all of the modeled events is reasonable.

4.2. CFD Simulations

The simulation results for the event on 26 August 2007 are detailed and discussed in this section as an example. The concentrations of *E. coli* on the surface of the pond at different time steps after the rain event are shown in Figure 6. As time passes, bacteria entering from the inlets redistribute in the pond. The flow field was affected by the inlet velocities for the first few hours after the events, but afterwards, the wind was the only parameter affecting the flow field.

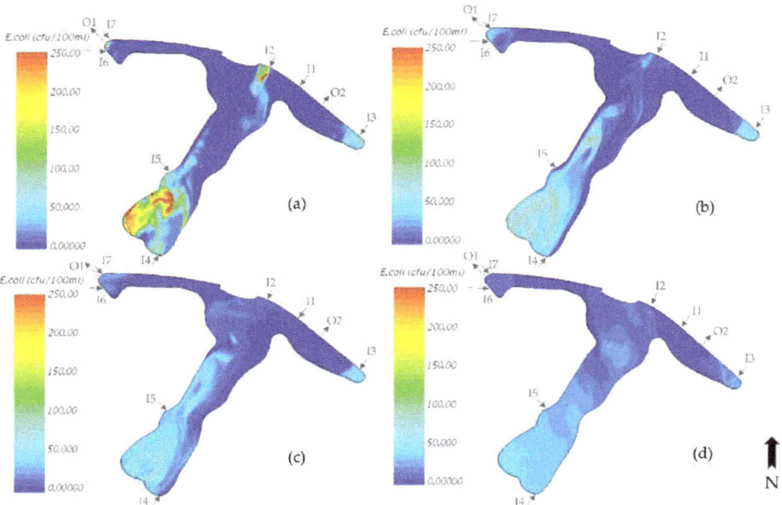

Figure 6. Contours of *E. coli* concentration (cfu/100 mL) on the surface of the Inverness pond on (**a**) 26 August 2007 at 5 p.m., (**b**) 26 August 2007 at 11 p.m., (**c**) 27 August 2007 at 5 a.m., and (**d**) 27 August 2007 at 11 a.m.

The bacteria concentrations in the south wing of the pond were greater as compared with the other two wings (the east and west wings). There are two inlets (I5 and I4) in the south wing. The *E. coli* concentration is higher in the stormwater runoff entering the pond from inlet I5 than all other inlets; while bacteria loading from I4 is highest as it drains the largest area (Table 1). For example, the *E. coli* concentration in the storm runoff from I5 on 26 August 2007 at 2 p.m. (during the storm) was

2100 cfu/100mL, while the storm runoff from I4 had a concentration of 1 038 cfu/100 mL. However, the flow rates of the inlets at that time were 0.045 m^3/s and 0.39 m^3/s for the I5 and I4 inlets, respectively. Therefore, more bacteria mass entered the pond from inlet I4 even though the concentration of bacteria was greater in I5. In addition, a relatively large amount of bacteria entered the pond from inlets I3, I7 and I6, but most of the bacteria that came from I7 and I6 immediately exited the pond through outlet O1, because these two inlets are in proximity to the outlet. Therefore, they do not have a significant effect on the bacteria level in the pond. In general, inlets I4, I3 and I5 had the most significant effect on the bacteria levels in the pond.

Figure 7 shows the vertical profile of *E. coli* distribution in the pond at 6 h after the end of the storm. It reveals that the bacteria concentration also changes with depth. As illustrated in Figure 7, the maximum concentration of *E. coli* on the surface barely reached 90 cfu/100 mL. However, the maximum *E. coli* concentration was more than 120 cfu/100 mL at 2 m depth. In addition, the bacteria in the middle of the pond (where the three wings join) was less distributed at the bottom compared to the surface. The reason is that the direction of the wind had been NE and NNE for the last few hours, so the bacteria escaping from the sediment forebay of inlet I2 could reach the other side where the south wing and the west wing join. In contrast, there was current flow toward the NE and NNE directions at the bottom simply because of mass conservation. This current potentially brought clean water close to the I2 sediment forebay. There was also a downwelling near the bank of the middle of the pond where the south and the west wings join. Generally, the wind causes differences in *E. coli* distribution in different layers of the water column. *E. coli* data collected at different depths on five random days between the year 2006 and 2007 [2] reveal that the bacteria concentration changed with depth. However, these data did not show any specific trend, and this is likely an indication of the influence of multiple environmental conditions on the bacteria distribution at various depths.

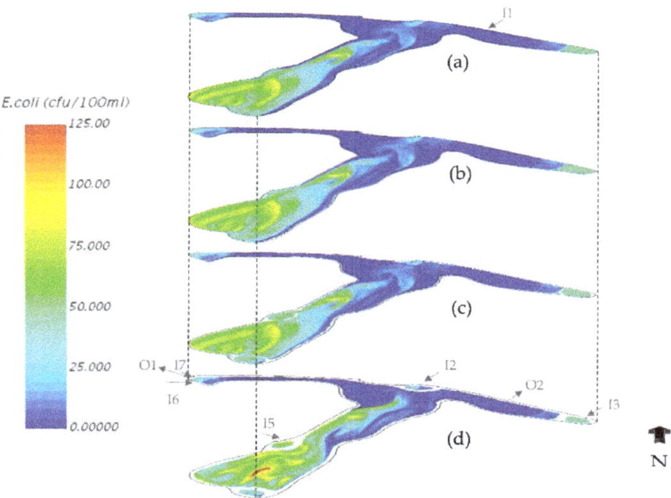

Figure 7. The vertical profile of *E. coli* concentration on 26 August 2007 at 11 p.m. (**a**) on the surface, (**b**) at 0.5 m below the surface, (**c**) at 1 m below the surface, and (**d**) at 2 m below the surface.

The sediment forebay corresponding to I3 was full of bacteria at all depths after the event (Figure 7). The sediment forebay was able to retain bacteria many hours after the event, which resulted in keeping the east wing relatively clean as compared with the south wing. Figure 8 magnifies the contour of *E. coli* concentration near the sediment forebays at the surface on 27 August 2007 at 2 a.m. It also confirms that the sediment forebay of I3 outperforms the forebay of other inlets. As illustrated in Figure 8, the concentration of *E. coli* in the sediment forebays of inlets I4 and I5 appears to be similar or even slightly less than that in the region outside the forebays. This reveals that the bacteria entering these forebays during the storm were promptly discharged out of the forebays and consequently increased the *E. coli* concentration in their nearby regions.

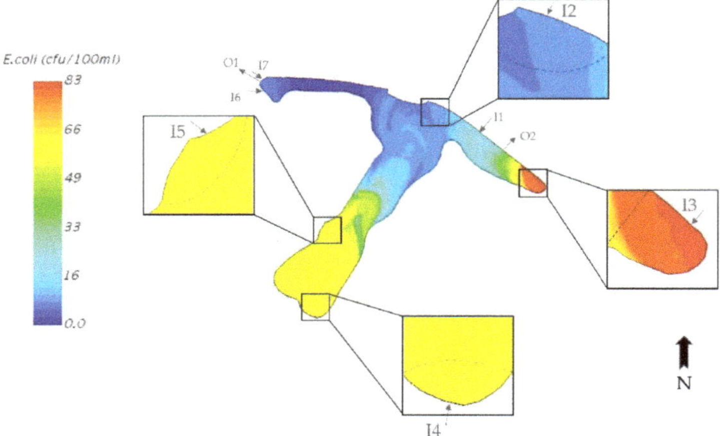

Figure 8. *E. coli* distribution on the surface 27 August 2007 at 2 a.m. (9 h after the end of the storm).

The design of sediment forebays, such as their configuration and size, determines their efficiency in trapping bacteria. Figure 9 shows streamlines coming out from the inlets and spreading throughout the pond during the storm. The streamlines from I4 continue straight out of the forebay, which means the bacteria were transported into the pond. One reason is that the size of the forebay was not large enough to fit a circulation proportional to the high flow rate of the I4 inlet. Another reason may be the configuration of the inlet and sediment forebay. The direction of the streamline from I4 and the corresponding sediment forebay were perpendicular to each other (i.e., they were in front of each other), so the water jet coming out of the inlet easily escaped the forebay without circulating. The situation was the same in the sediment forebays corresponding to the inlets I2 and I5. However, their size was proportional to their flow rate. The sediment forebay corresponding to the inlet I3 had the best configuration since the direction of streamlines coming out of the inlet was parallel to the forebay. Thus, the bacteria coming from the inlet had no way but to circulate and in the long term, they die, which kept the east wing relatively clean. In addition, the size of the forebay was large enough to fit two large eddies.

During the data collection campaign, surface grab samples were collected from six different sites at the pond over several days [2]. All of the samples were collected at an average depth of 15 cm. One of the sample collection days was after a day with heavy rain. On 10 September 2005, 68 mm of rain fell and on the next day, water samples were collected from the six sites. Although much greater rain fell on 10 September 2005 than in the simulated events from 2007, it provided an interesting case for validation due to the dominant effect of rain on the bacteria distribution. The initial bacteria level in the pond before an event can influence the final bacteria levels after an event, particularly for small, low rainfall events. However, for high rainfall events, the majority of bacteria is transported into the pond with the storm runoff, and the initial bacteria level is a far smaller proportion of the total bacteria level. In this paper, it was assumed that the initial bacteria level was negligible, and only the effect of the storm was studied. Thus, the modelling results can be compared with the collected data on 10 September 2005. Although it was noted that due to a lack of data it was difficult to validate the process-based FIB methods thoroughly [11], a comparison was done between the different events using a non-dimensional number. The non-dimensional number was computed as the ratio of *E. coli* concentration at a site to the maximum *E. coli* concentration among all of the six sites at a certain time. Thus, the non-dimensional number at the site where *E. coli* concentration was a maximum was one. The number was calculated for the individual modeled events, and then the average was taken for each site. Figure 10 shows the non-dimensional number at the six sites for the measured data and the average of the simulated events with its variation range. Simulation results show a good agreement with the measured data in recognizing hot spots (spots with a high concentration of bacteria) and spots with the lowest level of bacteria. The variation in the non-dimensional number in the east and west wings were higher than that in the south wing and in the middle of the pond. This may be due to a stronger influence from meteorological factors such as wind and rain at the tip of the east wing and the tip of the west wing. On the other hand, the low variation in the calculated data in the south wing and their high values show that the south wing always had the highest bacteria levels in the pond. Therefore, it is not recommended to extract water for reuse from the south wing.

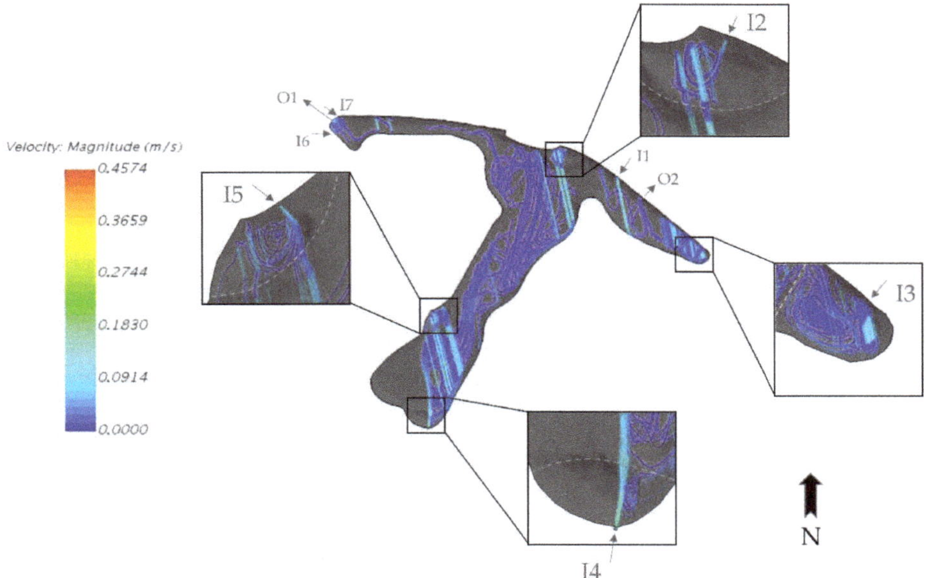

Figure 9. Velocity streamlines on 26 August 2007 at 3 p.m. (during the storm).

In addition, both measured and calculated data show that the middle of the pond, where the three wings join, had the lowest level of bacteria among the six sites; this site was a reliable spot for water extraction due to low concentrations of bacteria and low variability in the calculated data. There were no collected data for the middle of the west wing; however, simulation results show that this spot had the lowest velocity and *E. coli* concentrations in the pond. In a region with low velocity, fewer bacteria can enter. In addition, even if some bacteria enter a low-velocity region, most of them die due to high retention time in the region, which keep the bacteria concentration relatively low. In addition, the west wing has two inlets and an outlet all located at the tip of the wing. Thus, most of the bacteria that were discharged to the wing by the inlets exited the pond promptly. Therefore, the middle of the west wing can be considered as a potential location for water extraction. However, more data should be collected from the entire pond for a more comprehensive validation. In addition, further modeling studies after rain events are recommended when more data are available. More experimental and modeling studies need to be conducted in order to study the effect of sedimentation on bacteria transport in the pond, which was assumed to be negligible in this paper.

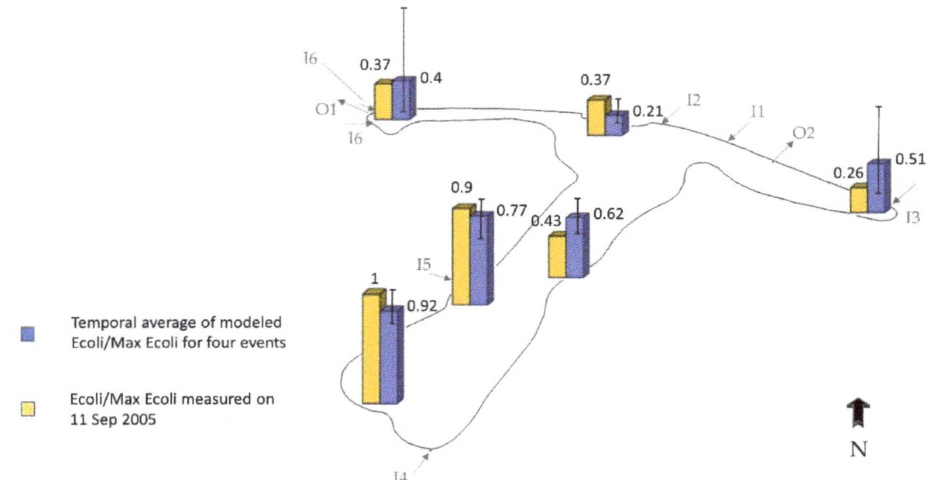

Figure 10. Non-dimensional *E. coli* concentration 15 cm below the surface.

5. Conclusions

An integrated hydrological-CFD model was developed to simulate bacteria concentrations within the Inverness stormwater pond in Calgary, Alberta, in order to locate the best (cleanest) locations within the pond to extract water for potential reuse. The integrated model was calibrated at one inlet with observed flow rate data in order to estimate the flow rates and bacteria levels at other inlets. The flow rate and bacteria concentration at the inlets were implemented as boundary conditions for an unsteady CFD simulation to determine the distribution of bacteria in the pond. The bacteria concentration distribution in Inverness pond was simulated both during and after four rain events using the integrated model. Results showed a good agreement with collected data from six different points in the pond, indicating that the model can be successfully used to model fate and transport of bacteria in stormwater ponds. It was found that the wind plays a crucial role in forming the flow field in the pond, which affects the bacteria distribution. In addition, one of the sediment forebays, located in the east wing of the pond, successfully trapped large amounts of bacteria until death occurred. However, other sediment forebays were found ineffective. The middle of the west wing was found to be the best location for extracting water because it showed the lowest levels of bacteria during the simulation of the pond both in and after the events.

Author Contributions: Conceptualization, modeling, simulation, writing, F.A.; supervision, investigation, reviewing, editing, and fund acquisition, C.V., J.H.; fund acquisition, project administration, and reviewing, N.F.N.

Funding: This research was funded by Alberta Innovates, grant number RES#0030866.

Acknowledgments: The authors would like to thank Mostafa Rahimpour for his comments on CFD modeling and Belaid Moa from WestGrid for his technical support on the parallel computation of the work.

Conflicts of Interest: The authors declare no conflict of interest.

References

1. The City of Calgary Water Resources. *Stormwater Management and Design Manual*; Urban Development Publications: Calgary, AB, Canada, 2011.
2. He, J.; Valeo, C.; Chu, A.; Neumann, N.F.; Tn, A. Water Quality Assessment in the Application of Stormwater Reuse for Irrigating Public Lands. *Water Qual. Res. J. Can.* **2008**, *43*, 93–107. [CrossRef]

3.   Ahilan, S.; Guan, M.; Wright, N.; Sleigh, A.; Allen, D.; Arthur, S.; Haynes, H.; Krivtsov, V. Modelling the long-term suspended sedimentological effects on stormwater pond performance in an urban catchment. *J. Hydrol.* **2019**, *571*, 805–818. [CrossRef]

4.   Clevenot, L.; Carré, C.; Pech, P. A Review of the factors that determine whether stormwater ponds are ecological traps and/or high-quality breeding sites for amphibians. *Front. Ecol. Evol.* **2018**, *6*, 40. [CrossRef]

5.   Gorgoglione, A.; Bombardelli, F.A.; Pitton, B.J.L.; Oki, L.R.; Haver, D.L.; Young, T.M. Role of sediments in insecticide runoff from urban surfaces: Analysis and modeling. *Int. J. Environ. Res. Public Health* **2018**, *15*, 1464. [CrossRef]

6.   Di Modugno, M.; Gioia, A.; Gorgoglione, A.; Iacobellis, V.; la Forgia, G.; Piccinni, A.; Ranieri, E. Build-up/wash-off monitoring and assessment for sustainable management of first flush in an urban area. *Sustainability* **2015**, *7*, 5050–5070. [CrossRef]

7.   St-hilaire, A.; Duchesne, S.; Rousseau, A.N. Canadian Water Resources Journal/Revue canadienne Floods and water quality in Canada: A review of the interactions with urbanization, agriculture and forestry. *Can. Water Resour. J./Rev. Can. Des Ressour. Hydr.* **2016**, *41*, 277–291.

8.   Borrego, J.J.; Figueras, M.J. Microbiological quality of natural waters. *Microbiologia* **1997**, *13*, 413–426.

9.   Leclerc, H.; Mossel, D.A.A.; Edberg, S.C.; Struijk, C.B. Advances in the bacteriology of the coliform group: Their Suitability as Markers of Microbial Water Safety. *Annu. Rev. Microbiol.* **2001**, *55*, 201–234. [CrossRef]

10. Tallon, P.; Magajna, B.; Lofranco, C.; Leung, K.T. Microbial indicators of faecal contamination in water: A current perspective. *Water Air. Soil Pollut.* **2005**, *166*, 139–166. [CrossRef]

11. De Brauwere, A.; Ouattara, N.K.; Servais, P. Modeling Fecal Indicator Bacteria Concentrations in Natural Surface Waters: A Review. *Crit. Rev. Environ. Sci. Technol.* **2014**, *44*, 2380–2453. [CrossRef]

12. Anna, H.; Jeng, C.; Englande, A.J.; Bakeer, R.M.; Bradford, H.B. Impact of urban stormwater runoff on estuarine environmental quality. *Estuar. Coast. Shelf Sci.* **2005**, *63*, 513–526.

13. Characklis, G.W.; Dilts, M.J.; Simmons, O.D.; Likirdopulos, C.A.; Krometis, L.H.; Sobsey, M.D. Microbial partitioning to settleable particles in stormwater. *Water Res.* **2005**, *39*, 1773–1782. [CrossRef]

14. Ouattara, N.K.; de Brauwere, A.; Billen, G.; Servais, P. Modelling faecal contamination in the Scheldt drainage network. *J. Mar. Syst.* **2013**, *128*, 77–88. [CrossRef]

15. Wu, J.; Rees, P.; Storrer, S.; Alderisio, K.; Dorner, S. Fate and transport modeling of potential pathogens: The contribution from sediments. *J. Am. Water Resour. Assoc.* **2009**, *45*, 35–44. [CrossRef]

16. Shilton, A. Potential application of computational fluid dynamics to pond design. *Water Sci. Technol.* **2000**, *42*, 327–334. [CrossRef]

17. Wu, B.; Chen, Z. An integrated physical and biological model for anaerobic lagoons. *Bioresour. Technol.* **2011**, *102*, 5032–5038. [CrossRef]

18. Shilton, A.; Harrison, J. Integration of coliform decay within a CFD (computational fluid dynamic) model of a waste stabilisation pond. *Water Sci. Technol.* **2003**, *45*, 205–210. [CrossRef]

19. Shilton, A.N.; Mara, D.D. CFD (computational fluid dynamics) modelling of baffles for optimizing tropical waste stabilization pond systems. *Water Sci. Technol.* **2005**, *51*, 103–106. [CrossRef]

20. He, J. Reducing the Vulnerability of Water Supply under a Changing Climate: An Assessment of Stormwater Reuse. Ph.D. Thesis, University of Calgary, Calgary, AB, Canada, 2009.

21. Mallin, M.A.; Williams, K.E.; Esham, E.C.; Lowe, R.P. Effect of human development on bacteriological water quality in coastal watersheds. *Ecol. Appl.* **2000**, *10*, 1047–1056. [CrossRef]

22. Chen, H.J.; Chang, H. Response of discharge, TSS, and *E. coli* to raifall events in urban, suburban, and rural watersheds. *Environ. Sci. Process. Impacts* **2014**, *16*, 2313–2324. [CrossRef]

23. Schoonover, J.E.; Lockaby, B.G. Land cover impacts on stream nutrients and fecal coliform in the lower Piedmont of West Georgia. *J. Hydrol.* **2006**, *331*, 371–382. [CrossRef]

24. U.S. Army Corps of Engineers. *Hydrological Modeling System HEC-HMS User's Manual: Version 4.2*; Hydrologic Engineering Center: Davis, CA, USA, 2016.

25. Mockus, V. Hydrology. In *National Engineering Handbook*, 2nd ed.; Natural Resources Conservation Service: Washington, DC, USA, 1972; pp. 21.2–21.49.

26. Teegavarapu, R.S.V.; Chinatalapudi, S. Incorporating Influences of Shallow Groundwater Conditions in Curve Number-Based Runoff Estimation Methods. *Water Resour. Manag.* **2018**, *32*, 4313–4327. [CrossRef]

27. U.S. Army Corps of Engineers. *Hydrologic Modeling System HEC-HMS Technical Reference Manual*; Hydrologic Engineering Center: Davis, CA, USA, 2000.

28. Graebel, W.P. *Advanced Fluid Mechanics*, 1st ed.; Academic Press: Oxford, UK, 2007; pp. 233–250.
29. Abbasi, A.; Annor, F.O.; van de Giesen, N. Investigation of Temperature Dynamics in Small and Shallow Reservoirs. Case Study: Lake Binaba, Upper East Region of Ghana. *Water* **2016**, *8*, 84. [CrossRef]
30. Shilton, A.; Kreegher, S.; Grigg, N. Comparison of Computation Fluid Dynamics Simulation against Tracer Data from a Scale Model and Full-Sized Waste Stabilization Pond. *J. Environ. Eng.* **2008**, *134*, 845–850. [CrossRef]
31. Launder, B.E.; Sharma, B.I. Application of the energy-dissipation model of turbulence to the calculation of flow near a spinning disc. *Lett. Heat Mass Transf.* **1974**, *1*, 131–138. [CrossRef]
32. Versteeg, H.K.; Malalasekera, W. *An Introduction to Computational Fluid Dynamics: The Finite Volume Method*; Longman Scientific and Technical: New York, NY, USA, 1995.
33. Kunkel, E.A.; Privette, C.V.; Sawyer, C.B.; Hayes, J.C. Attachment of *Escherichia coli* to fine sediment particles within construction sediment basins. *Adv. Biosci. Biotechnol.* **2013**, *4*, 407–414. [CrossRef]
34. Muirhead, R.W.; Collins, R.P.; Bremer, P.J. Interaction of *Escherichia coli* and Soil Particles in Runoff. *Appl. Environ. Microbiol.* **2006**, *72*, 3406–3411. [CrossRef]
35. Stokes, G. On the Effect of the Internal Friction of Fluids on the Motion of Pendulums. *Trans. Camb. Phillosophical Soc.* **1851**, *9*, 8–106.
36. Gu, L.; Dai, B.; Zhu, D.Z.; Hua, Z.; Liu, X.; van Duin, B. Sediment modelling and design optimization for stormwater ponds. *Can. Water Resour. J./Rev. Can. Des Ressour. Hydr.* **2017**, *42*, 70–87. [CrossRef]
37. Bai, S.; Lung, W.S. Modeling sediment impact on the transport of fecal bacteria. *Water Res.* **2005**, *39*, 5232–5240. [CrossRef]
38. Chick, H. An Investigation of the Laws of Disinfections. *J. Hyg.* **1908**, *8*, 92–158. [CrossRef]
39. De Brauwere, A.; Gourgue, O.; de Brye, B.; Servais, P.; Ouattara, N.K.; Deleersnijder, E. Integrated modelling of faecal contamination in a densely populated river-sea continuum (Scheldt River and Estuary). *Sci. Total Environ.* **2014**, *468–469*, 31–45. [CrossRef]
40. Banner, M.L.; Peirson, W.L. Tangential stress beneath wind-driven air–water interfaces. *J. Fluid Mech.* **1998**, *364*, 115–145. [CrossRef]
41. CD-adapco. *STAR-CCM+ 12.04.011 User's Manual*; Siemens Product Lifecycle Management Software Inc.: Melville, NY, USA, 2017.
42. McCuen, R.H.; Knight, Z.; Cutter, G. Evaluation of the Nash-Sutcliffe Efficiency Index. *J. Hydrol. Eng.* **2006**, *11*, 597–602. [CrossRef]

Article

Near-Wake Flow Structure of a Suspended Cylindrical Canopy Patch

Ayşe Yüksel Ozan [1],* and Didem Yılmazer [2]

[1] Civil Engineering Department, Aydın Adnan Menderes University, Aydın Menderes Derslikleri, Efeler, 09010 Aydın, Turkey

[2] Civil Engineering Department, Namık Kemal University, Çorlu Faculty of Engineering, Çorlu, 59869 Tekirdag, Turkey; didem_yilmazer@yahoo.com

* Correspondence: ayse.yuksel@adu.edu.tr; Tel.: +90-256-213-7503 (ext. 3550)

Received: 26 November 2019; Accepted: 21 December 2019; Published: 25 December 2019

Abstract: Urban stormwater is an important environmental problem, especially for metropolitans worldwide. The most important issue behind this problem is the need to find green infrastructure solutions, which provide water treatment and retention. Floating treatment wetlands, which are porous patches that continue down from the free-surface with a gap between the patch and bed, are innovative instruments for nutrient management in lakes, ponds, and slow-flowing waters. Suspended cylindrical vegetation patches in open channels affect the flow dramatically, which causes a deviation from the logarithmic law. This study considered the velocity measurements along the flow depth, at the axis of the patch, and at the near-wake region of the canopy, for different submerged ratios with different patch porosities. The results of this experimental study provide a comprehensive picture of the effects of different submergence ratios and different porosities on the flow field at the near-wake region of the suspended vegetation patch. The flow field was described with velocity and turbulence distributions along the axis of the patch, both upstream and downstream of the vegetation patch. Mainly, it was found that suspended porous canopy patches with a certain range of densities (SVF20 and SVF36 corresponded to a high density of patches in this study) have considerable impacts on the flow structure, and to a lesser extent, individual patch elements also have a crucial role.

Keywords: suspended vegetation; FTW; ADV; velocity profile; submerge ratio; SVF

---

1. Introduction

Increases in urban and agricultural development cause stormwater runoffs that contains nutrients that can have severe effects on aquatic ecosystems and human health [1]. The use of floating treatment wetlands (FTWs) is a new and innovative tool to remove nutrients in stormwater wet detention ponds [2]. FTWs represent a human-made ecosystem similar to natural wetlands [3]. A typical FTW is shown in Figure 1 [4].

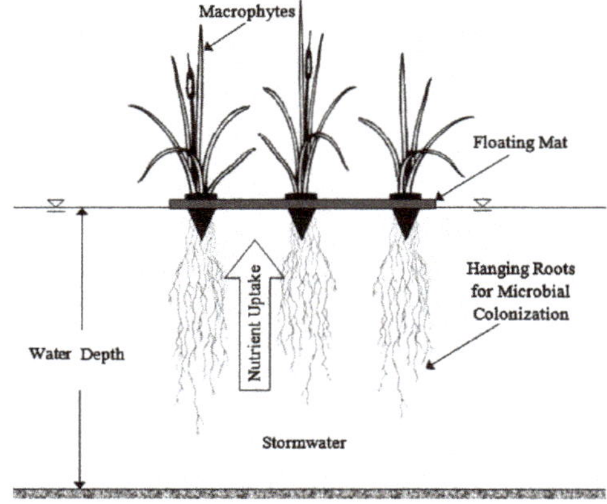

Figure 1. An example of floating treatment wetland [4].

Vegetation patches or layers are essential in aquatic environments in submerged, emergent, or suspended forms. Suspended canopies, which are the focus of this study, form on the water surface, with stems toward the bottom that do not touch the bottom, in contrast to submerged and emerged canopies that are attached to the bottom. *Lemnaceae, Eich- hornia crassipes, Pistia stratiotes, Salvinia molesta,* and *Laminariales* are typical examples of suspended canopies [5–7]. *Lemnaceae* are used extensively for the purpose of purifying wastewater and eutrophic water bodies [8]. Suspended canopies in flow have been investigated in terms of the ecological effects [9–12] and the mean velocity and turbulence structures of open-channel flows included in these types of canopies [6,7]. Fishing activities and marine navigation can be at risk because of these types of floating vegetation patches [8]. Bed friction has a substantial effect on the velocity and turbulence below and within floating vegetation patches [6,13], however, it has a small role on the flow structure inside submerged or emergent canopies [14–21]. Despite the increasing importance of suspended canopies, there is still a limited number of studies on them. Detailed flow structures in the wake regions of the suspended canopies should be examined to understand the transport of fluid and pollutants.

Liu et al. [22] studied longitudinal dispersion in flow along the floating vegetation patch. They also indicated that a large number of studies on the longitudinal dispersion in open-channel flows with submerged and emergent vegetation have been presented [16,23–25]. However, there is an insufficient number of studies on the effects of floating vegetation patches in the literature [22]. They proposed a four-zone model with which they defined the velocity profile and turbulent diffusion coefficient profile for the four zones in their study.

Huai et al. [7] studied the effect of suspended vegetation on velocity distribution at open-channel flows. They considered four different submergence ratios ($h_v/H = 0.125$–0.5, where h_v is submergence ratio of the suspended vegetation and H is the flow depth) with a constant density (defined as the canopy projected area per unit volume, $a = 1.272$ m^{-1}) of the canopy in their experiments. They defined four parts along the flow depth with the suspended vegetation (Figure 2) and determined the velocity profile for every part with the solution of the momentum equation. They called Zone 1 and Zone 2 the outer and internal vegetation layers, respectively. They explained that the drag force of vegetation is dominant in Zone 1 where the velocity is slow and almost constant. The velocity in this zone corresponds to that of the wake region in the submerged vegetated flow [7]. They defined the non-vegetated layer with two zones: Zone 3 and Zone 4. Kelvin–Helmholtz (K–H) vortices characterize the middle mixing layer (t_{ml}), which is placed near the bottom of the suspended vegetation [22]. The K–H vortices control the vertical mass transport in the lower region of the suspended vegetation and in the upper part of the gap, which is the non-vegetated region of the flow [26,27]. Zone 2 corresponds to the penetration depth, h_i, where the length of the K–H vortices penetrate the vegetation region [22].

The non-vegetated area in the mixing layer is defined as Zone 3, in which the thickness is $t_{ml}-h_i$ [22]. The resistance caused by suspended vegetation affects Zone 1, Zone 2, and Zone 3, while the friction resistance of the bed affects Zone 4, as it is located close to the bed [22]. The velocity has a maximum value at the junction of Zones 3 and 4. Liu et.al. [22] stated that because of the maximum velocity, the Reynolds shear stress has zero value at this point where the impacts of river bed and suspended vegetation are consistent.

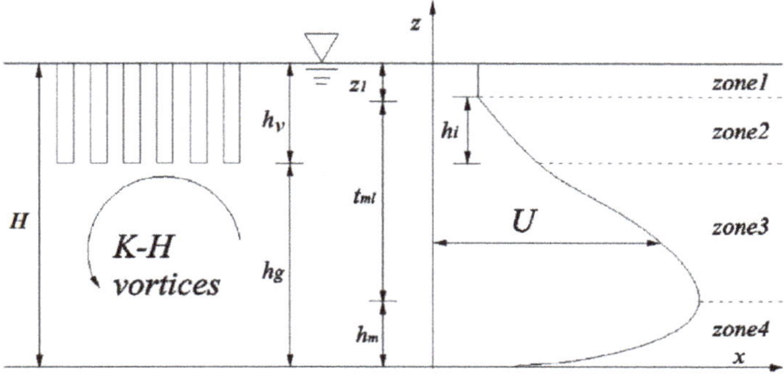

Figure 2. Schematic velocity distribution in suspended vegetated flow [22].

The wake structure downstream of a circular suspended porous obstruction, which represents a model of a suspended vegetation patch, was investigated in this paper. The main question of this study is how the submergence ratio affects the structure of the wake and flow close to the bed, downstream of the suspended patch. For this purpose, the velocity measurements were performed along with the flow depth downstream of the canopy for different submergence ratios with different patch porosities.

2. Experimental Methods

Experiments were performed in an 11 m long, 1.2 m wide, and 0.75 m deep re-circulating glass flume with a concrete horizontal bed at the Civil Engineering Laboratory of Aydın Adnan Menderes University (Figure 3). The suspended vegetation patch was of diameter D = 0.3 m and was formed by rigid plastic cylinders of diameter d = 0.01 m (Figure 4). The cylinders were arrayed in a staggered pattern (Figure 5). The suspended patch was placed at the center of the flume in the transverse direction. The number of cylinders per unit bed area, n (IP/cm^2), defines the density of the patch. Here, IP means individual plant number per square centimeter and area is the A = $\pi D^2/4$. Vegetation density is defined as the frontal area per unit volume, given with a = nd (cm^{-1}), and characterizes the blockage caused by the vegetation layer [28]. The average solid volume fraction is defined with ϕ = nπd^2/4 $\approx$ ad [29]. In the experiments, three different densities of the canopy patch and a solid cylinder were used to simulate the suspended vegetation (Table 1). The Plexiglas solid cylinder was composed as a closed cylinder with D = 30 cm to represent the suspended Solid Case (SC) in the flow area (Table 1). Additionally, a constant mean depth-averaged velocity (U_{dm} = 0.116 m/s) and constant flow depth (H = 0.3 m) were considered in the experiments. Here, velocity values were integrated along the flow depth to calculate mean depth-averaged velocity (U_{dm}). Mean depth-averaged velocity (U_{dm}) was used to calculate Reynolds number (Re = $U_{dm}H/\nu$) and Froude Number $\left(Fr = U_{dm}/\sqrt{gH}\right)$ for the non-vegetated case. Here, g is the gravitational acceleration, and ν is the kinematic viscosity. Patches were immersed in the flow at two different heights: h_{v1} = 0.1 m and h_{v2} = 0.2 m. The cases at the experiments are outlined in Table 1.

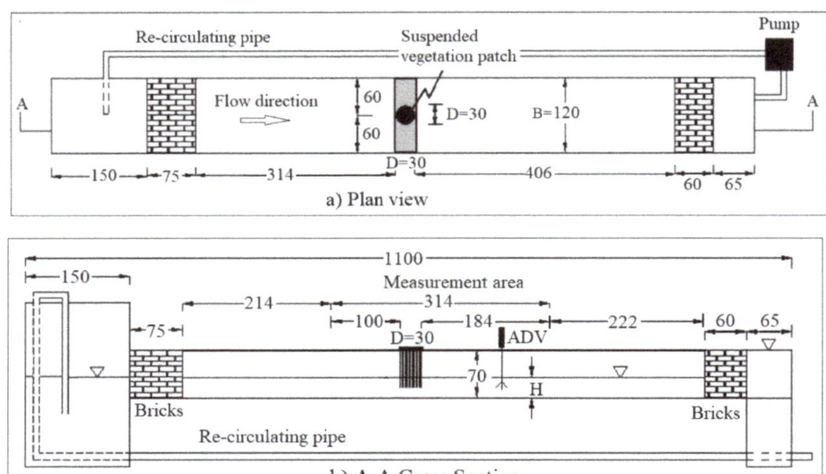

Figure 3. Experimental set-up: (**a**) plan view, (**b**) cross-section (units are in cm).

Figure 4. *Cont.*

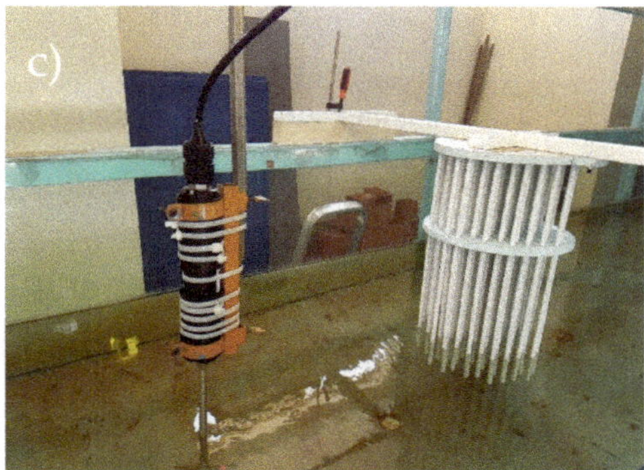

Figure 4. Photos from vegetation patches for different SVFs (Solid Volume Fractions): (**a**) side view, (**b**) top view, (**c**) view from an experiment with an Acoustic Doppler Velocimeter (ADV).

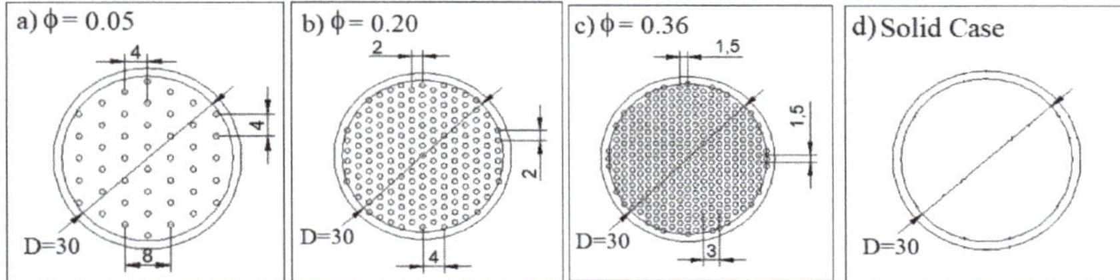

Figure 5. Schematic sketch of the plants of different densities; (**a**) $\phi = 0.05$, (**b**) $\phi = 0.20$, (**c**) $\phi = 0.36$, (**d**) Solid Case (units are in cm).

Table 1. Experimental conditions.

Cases	a (cm^{-1})	ϕ	h_v/H	U_{dm} (cm/s)	Re	Fr
SVF5	0.06	0.05	0.33 0.67			
SVF20	0.25	0.20	0.33 0.67	11.6	35,000	0.07
SVF36	0.46	0.36	0.33 0.67			
Solid Case (SC)	Solid	1.00	0.33 0.67			

In this study, x, y, and z are the longitudinal, transverse, and vertical coordinates, respectively. The origin of the coordinate system is at the upstream edge of the suspended vegetation for the x–y plane and the flume bed for the z direction (Figure 6). A 3-D Acoustic Doppler Velocimeter (ADV) SonTek 10 MHz with a 3-D down-looking probe was used to measure three instantaneous components of velocity (u, v, w), corresponding to x, y, and z directions shown in Figure 6, respectively, during 120 s at 25 Hz. The velocity range was ±0.30 m/s, with a measured velocity accuracy of ±0.5%, and the sampling volume diameter was 6 mm established 0.05 m down from the probe in the experiments. The velocity measurements were conducted at 11 points in the x-direction and at eight different levels along the flow depth (Figure 6). Measurement points along the x-direction were chosen by moving at intervals of 15 or 30 cm upstream and downstream of the vegetation patch (Figure 6c). Alongside this, measurement points along the Z direction were placed at different intervals which correspond to

$Z_1 = 0.5$ cm, $Z_2 = 1.0$ cm, $Z_3 = 1.5$ cm, $Z_4 = 2.0$ cm, $Z_5 = 3.0$ cm, $Z_6 = 4.5$ cm, $Z_7 = 6.0$ cm, $Z_8 = 8.0$ cm, $Z_9 = 11.0$ cm, and $Z_{10} = 15.0$ cm (Figure 6a,b). The ADV company suggests that a 15 db signal-to-noise ratio (SNR) and a correlation coefficient larger than 70% for high-resolution measurements [30]. For this reason, conditions where SNR <15 db and the correlation coefficient <70% are used to remove the low-quality samples in the measured data.

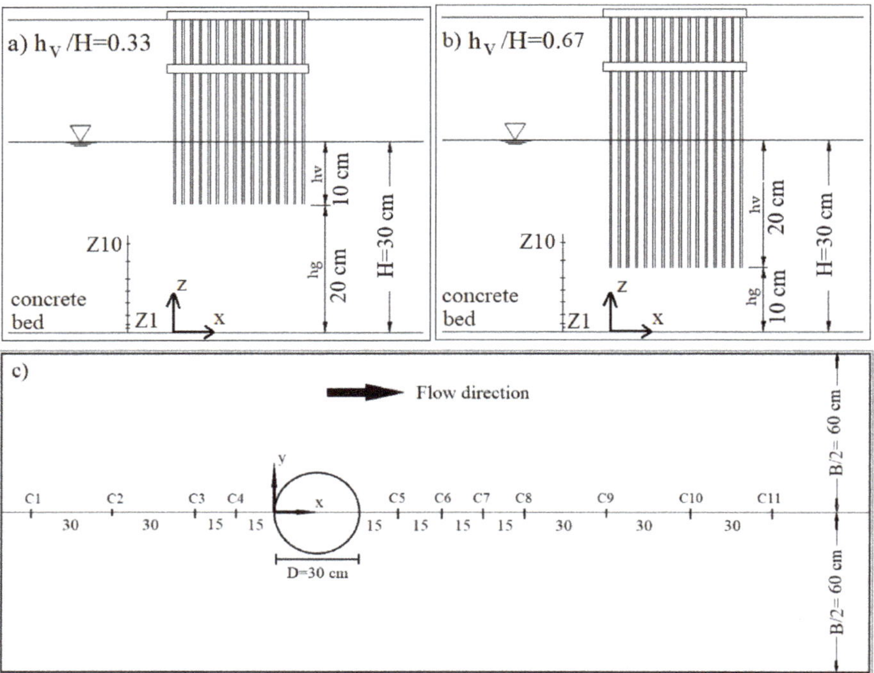

Figure 6. Velocity measurement points (**a**) along the flow depth for $h_V/H = 0.33$, (**b**) along the flow depth for $h_V/H = 0.67$, and (**c**) along the flow direction (units are in cm).

3. Results and Discussion

The longitudinal profiles of streamwise averaged velocity, U, for various different densities of patches at mid-depth ($z/H = 0.5$), for $h_V/H = 0.67$, and $h_V/H = 0.33$, are presented in Figure 7. Here, the vegetated part (VP) corresponds to the suspended vegetation canopy and the mid-depth corresponds to beneath the patch for $h_V/H = 0.33$. The diversion of flow begins almost one diameter upstream of the patch, which is consistent with Rominger and Nepf's [31] results (Figure 7a). There is a recirculation zone (U < 0) downstream of the flow, which extends to 1.5 D and 2 D distance from the edge of the patch for SVF20 and SVF36, respectively (Figure 7a). As the density increases, the recirculation zone moves downstream and the magnitude of negative streamwise velocity increases. Moreover, there is no recirculation zone for the lowest density (SVF5). Zong and Nepf [29] found that there is a recirculation region in the wakes behind denser patches ($\phi = 0.10$), but absent in the wakes of sparser patches ($\phi = 0.03$). This result is consistent with our results. A steady wake region was observed downstream of the patch, close to the patch for SVF5, which is not present at higher SVFs in this study. Here, steady wake corresponds to 2 D distance beyond the end of the patch, where U velocity is almost constant.

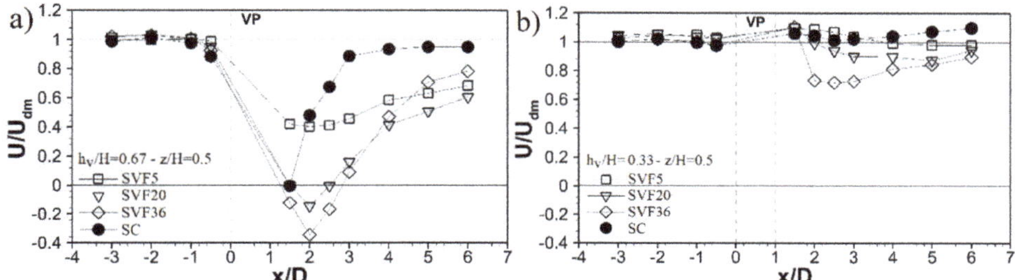

Figure 7. Longitudinal profiles of U velocity for different densities of patches at mid-depth (z/H = 0.5) for (**a**) h/H = 0.67 and (**b**) h/H = 0.33.

Zong and Nepf [29] mentioned that the wake can never perfectly recover since the drag-induced momentum deficit should be constant downstream of the obstruction. They further pointed out that the total length of the wake, L, is defined as the distance from the downstream end of the patch to a point where the velocity recovery rate is induced to $(\partial(U/U_{dm})/\partial(x/D) < 0.1)$. The total length of the wake (L) is almost 3 D at mid-depth for the solid body and for $Re_D = 3.5 \times 10^4$ (Figure 7a). This value is consistent with Zong and Nepf's [29] results, in which L = 3 D and L = 3.3 D for $Re_D = 2.2 \times 10^4$ and $Re_D = 4.1 \times 10^4$, respectively, in their study. Cantwell and Coles [32] also defined L = 2.5 D for a cylinder wake at $Re_D = 1.4 \times 10^5$ in their study. The recovery region extends to 5 D beyond the end of the patch in our study. Another noteworthy situation is that U increases rapidly to U_{dm} at the highest density (SVF36) compared to the lower densities (SVF5 and SVF20) (Figure 7a).

Considering the lower depth ratio (h/H = 0.33), there is almost no effect of the patch beneath the patch at mid-depth for the solid body and lowest SVF (Figure 7b). However, suspended vegetation still affects the flow beneath the patch for a higher density of patches (SVF20 and SVF36). The obstruction effect on the flow is higher at the highest patch density (SVF36).

Longitudinal profiles of time-averaged transversal velocity, V, for different densities of patches and depth ratios at mid-depth were also examined (Figure 8). A suspended vegetation patch does not have a significant effect on transversal velocity beneath the patch for a low solid volume fraction (SVF5) at a lower depth ratio (h_v/H = 0.33) (Figure 8a). As the suspended canopy become denser, the magnitude of the transversal velocity increases downstream until 2 D distance from the end of the patch for a lower depth ratio (h_v/H = 0.33). Similar to a lower depth ratio, there is almost no effect on transversal velocity downstream of the patch for low solid volume fractions at higher depth ratios (h_v/H = 0.67) (Figure 8a). There is diverging flow ($V > 0$) downstream of the patch which extends until 5 D distance for SVF20, 3 D distance for SVF36, and 1 D distance for the Solid Case. The position of diverging flow occurs closer to the patch with the increasing SVF. Transversal velocity varies around zero beyond the point where the diverging of the flow ends.

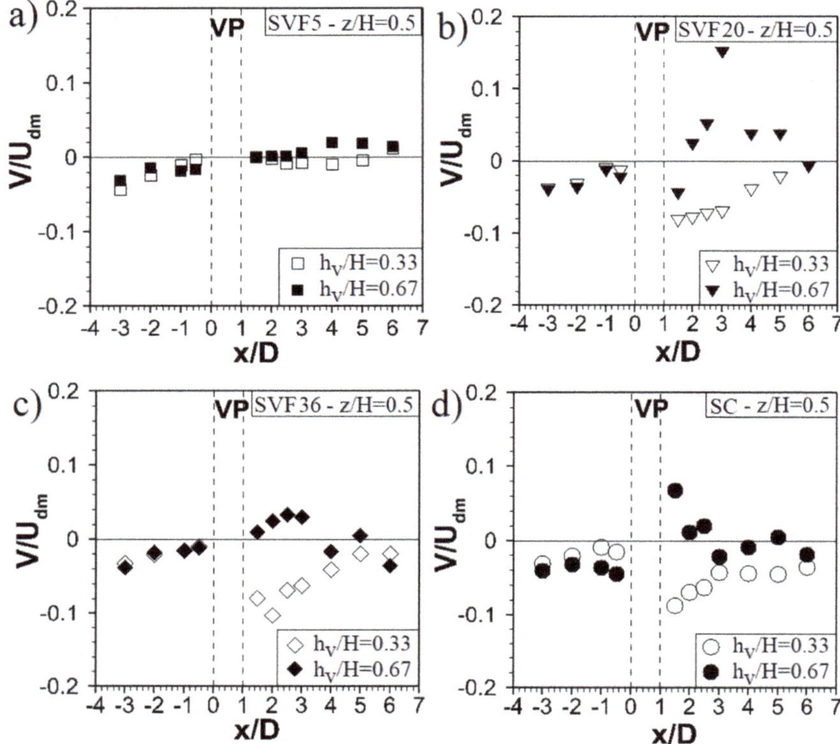

Figure 8. Longitudinal profiles of V velocity for different densities of patches and depth ratios at mid-depth (z/H = 0.5): (**a**) ϕ = 0.05, (**b**) ϕ = 0.20, (**c**) ϕ = 0.36, (**d**) Solid Case.

Figure 9 shows the vertical profiles of the normalized velocity U/U_{dm} at different upstream and downstream positions (x′/D) for h_v/H = 0.67. Here, x′ refers to the distance from the patch edge and NV and NVlog correspond to the normalized velocity (U/U_{dm}) distribution and normalized logarithmic velocity (U_{log}/U_{dm}) distribution, respectively, at non-vegetated flow conditions. You can find detailed information about the calculation of U_{log} from Yılmazer et al. [17]. A combination of a channel bed and suspended patch causes maximum velocity at the gap, with a flow similar to a wall jet beneath the vegetation patch. Here, the layer between the bed and the maximum velocity point is called the inner layer, where wall shear is effective; the area above the point of the maximum velocity is called the outer layer, where patch shear is effective. Flow has a maximum velocity of almost 1.45–1.5 U_{dm} beneath the patch and downstream for higher densities of patches (SVF20 and SVF36). Comparatively, the maximum velocity for the lowest density of patches (SVF5) is around 1.3 U_{dm}, which is similar to the Solid Case. It can be concluded that there is a suspended patch effect that increases the velocity beneath the suspended patch for a certain range of SVF, which corresponds to high densities of patches. Another important point is the placing of the maximum velocity, which is z/H $\cong$ 0.15 for SVF20 and SVF36, and z/H $\cong$ 0.3 for SVF5 and the Solid Case. Inner layer thickness decreases with the increasing SVFs. The placement and the magnitude of the maximum velocity at the gap region shows similar behavior at SVF5 and the Solid Cases, which is similar for higher densities (SVF20 and SVF36). It can be concluded that the individual patch elements have an important role for lower densities, which is SVF5 in this study. U velocity has negative values downstream of the patch for the densities SVF20 and SVF36, and its value increases with increasing patch density at the near-wake region (~x′/D ≤

1.5 D). There is no reversed flow region for the lowest density patch (SVF5) and solid patch. Moreover, the flow behavior at the Solid Case is similar to the lower densities of the patch, e.g., SVF5. Maximum velocity value bigger than U_{dm} is present along x′ < 3 D, downstream of the gap region, which also corresponds to the distance where flow behavior is similar to wall jet flow.

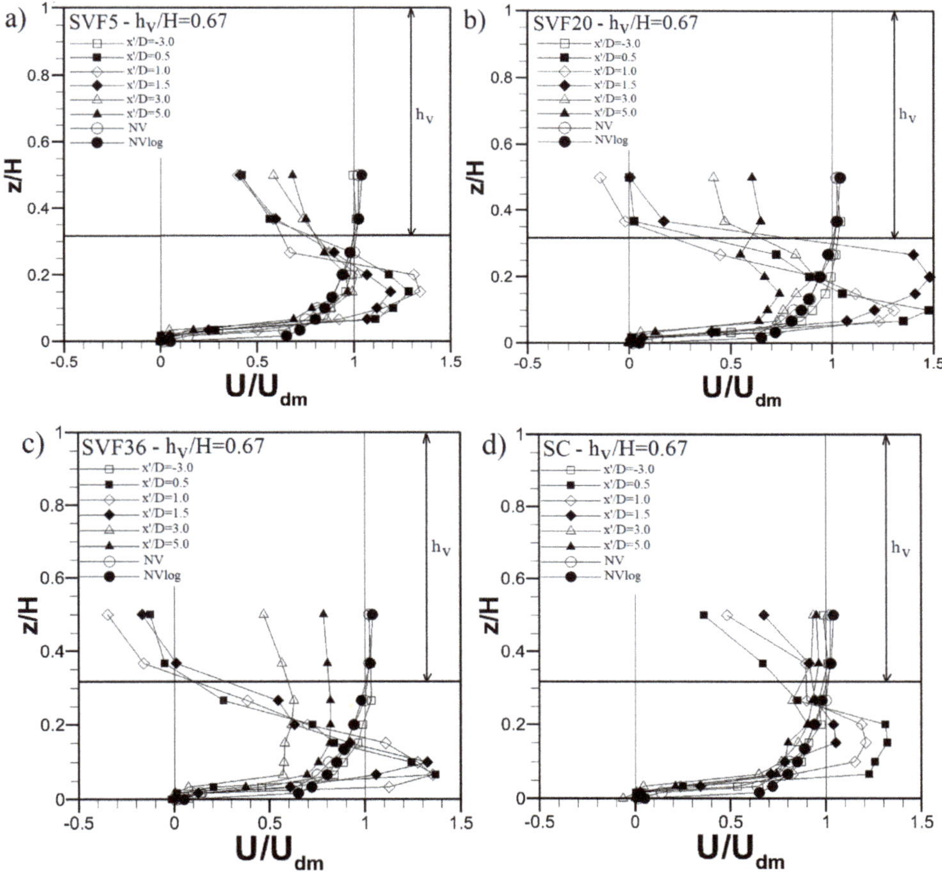

Figure 9. Vertical profiles of U velocity for different densities of patches along streamwise direction for h/H = 0.67: (**a**) ϕ = 0.05, (**b**) ϕ = 0.20, (**c**) ϕ = 0.36, (**d**) Solid Case.

The vertical profiles of U velocity for different densities of patches are shown in Figure 10 for h_V/H = 0.33. Velocity distribution and flow behavior deviate at x′ = 0.5 D distance downstream of the patch, and is totally different in the case of a higher submergence ratio. The patch does not have a big effect on the flow structure along with the entire flow depth for this submergence ratio. The behavior of the velocity distribution for SVF20 and SVF36 is similar, as is the case for SVF5 and the solid patch. Flow has a maximum velocity of almost ~1.25 U_{dm} for higher densities (SVF20 and SVF36) and around ~1.1 U_{dm} for the lowest density (SVF5) and Solid Case. This result shows us that it is still possible to see the effect of density for high densities (SVF20 and SVF36 in this study) downstream of the gap region for lower submergence ratios. Downstream of the gap region, a maximum velocity value bigger than U_{dm} is present along x′ < 3 D for SVF36 and SC, and x′ < 5 D for SVF5 and SVF20. It can be concluded that a high density of the submerged vegetation patches still has an influence on the flow structure beneath the patch, even at low submergence ratios.

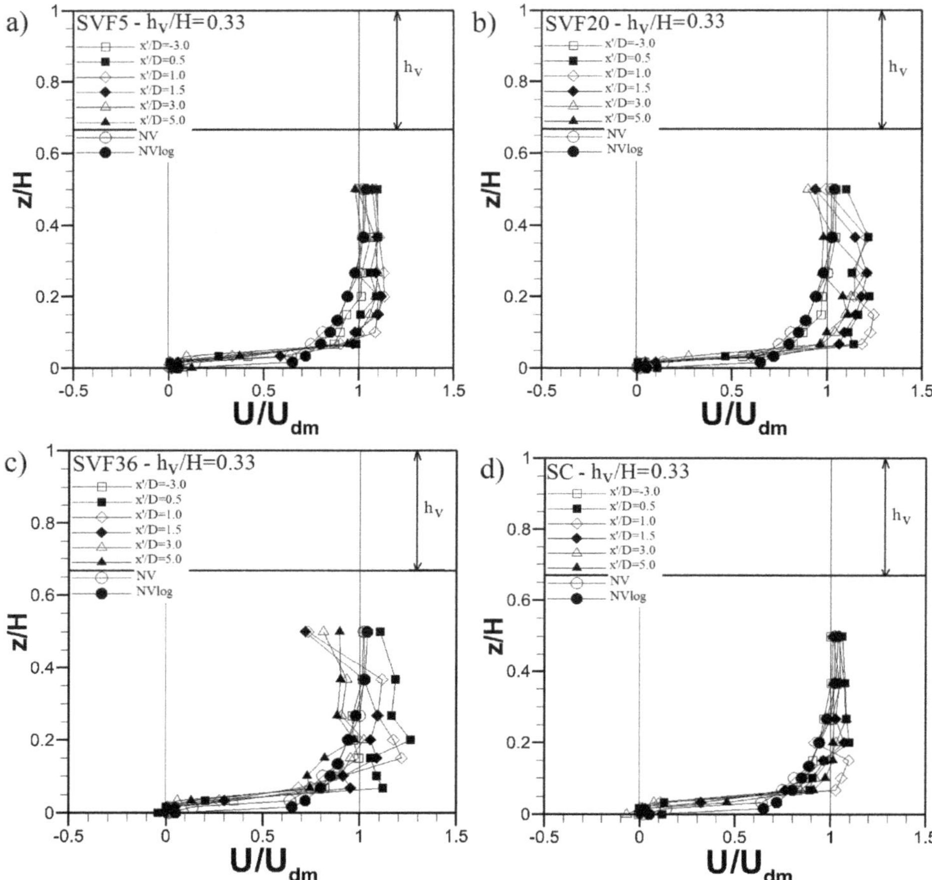

Figure 10. Vertical profiles of U velocity for different densities of patches along streamwise direction for h /H = 0.33: (**a**) φ = 0.05, (**b**) φ = 0.20, (**c**) φ = 0.36, (**d**) Solid Case.

Longitudinal profiles of u_{rms}/U_{dm} and v_{rms}/U_{dm} for different densities of patches at different depths (z/H = 0.2 and z/H = 0.5) along the streamwise direction for h_v/H = 0.67 are shown in Figure 11. Here, z/H = 0.2 corresponds to the gap region and z/H = 0.5 corresponds to the upstream and downstream parts of the patch. First of all, velocity fluctuations are higher upstream of the gap region than upstream of the patch for all cases. Moreover, there is a peak at x′= –0.5 D upstream of the gap region that is not present upstream of the patch. This shows us that the effect of a submerged obstruction starts in the upstream section, below the obstruction. Turbulence starts with a decaying trend downstream, and then it starts to increase for SVF5 and SVF20 cases at both levels. Specifically, the shear layer caused by the patch increases the turbulence level beneath the patch. Turbulence shows a decaying trend for SVF36, which is the highest density of patches, and the values of turbulence become closer at x′ = 3 D distance for different depths (z/H = 0.2 and z/H = 0.5). This shows us that the individual stem turbulence is not significant for the highest patch density (SVF36), but is significant at lower patch densities (SVF5 and SVF20). Individual cylinders cause small-scale turbulence, which results in a higher turbulence region behind the patch. The turbulence level has the highest value, especially for SVF20, in the near-wake region of the patch in this study. This is because of the significance of the individual cylinders for the patch intensity.

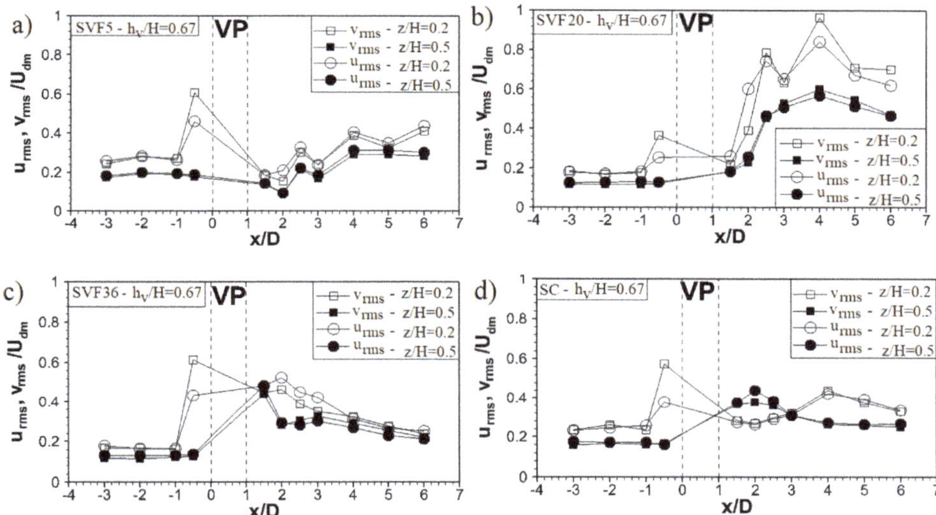

Figure 11. Longitudinal profiles of u_{rms}/U_{dm} and v_{rms}/U_{dm} for different densities of patches along the streamwise direction for $h_v/H = 0.67$: (**a**) $\phi = 0.05$, (**b**) $\phi = 0.20$, (**c**) $\phi = 0.36$, (**d**) Solid Case.

Comparing the cases of SVF20 and SVF36, the turbulence level decreases with increasing patch density (SVF36) to a level close to that of the solid body at mid-depth. This shows that the suspended patch with the highest porosity patch (SVF36) behaves like a solid body. The turbulence level is higher downstream of the solid body than it is downstream of the gap region along $x' = 2$ D distance, beyond which point it is reversed (Figure 11d). After this point, the magnitude of the turbulence is similar for the highest density (SVF36) of patches and the solid body at mid-depth. However, the magnitude of the turbulence is higher for the solid body than the highest density (SVF36) of patches in the recovery region. This may occur because of the wake region behind the solid body. There is only one point, which is the closest point to the patch, ($x = 0.5$ D) downstream of the highest density patch (SVF36), where turbulence is higher than the solid body at mid-depth. This point corresponds to the wake region of the individual stem of the highest density patch. This shows us that the highest density patches behave like a solid body at the wake region, with the exception of very close to the patch ($x \leq 0.5$D).

Figure 12a shows the spectral density variation of transverse velocity with frequency for Case SVF5 and the Solid Case. The eddy dominant frequency corresponds to the peak velocity spectra *Svvpeak*. Case SVF5 and the Solid Case are discussed as representative examples and for the comparison of the canopy and Solid Cases (Figure 12). The dimensionless Strouhal number ($St = fL/U_{dm}$) defines the relationship between the characteristic length of a vortex and the frequency. Here, f is the vortex shedding frequency (dominant frequency), L is the characteristic vortex length, and U_{dm} is the characteristic flow velocity. St is almost constant around $St = 0.2$ within a wide range of Reynolds numbers, between Re = 250 to 2×10^5 for flows passing a cylinder [33]. Zhao and Huai [34] assumed that this can also be used for the bundle of cylinders in their study. Therefore, we considered $St = 0.2$ for both Case SVF5 and the Solid Case. For Case SVF5 ($x'/D = 0.5$ and $z/H = 0.5$), the vortex shedding frequency is $f = 0.08$ Hz (Figure 12a) and the characteristic velocity U_{dm} is approximately 0.116 m/s. The characteristic vortex length L is almost 0.293 m, which is almost equal to the diameter of the patch

(D = 0.3 m). For the Solid Case (x/D = 0.5 and z/H = 0.5), the vortex shedding frequency is f = 0.11 Hz (Figure 12a) and the characteristic velocity U_{dm} is almost 0.116 m/s. The characteristic vortex length L is almost 0.211 m, which is smaller than the canopy diameter. When the magnitude of the peaks in S_{vv} is compared, it is clear that the turbulence intensity of the wake-scale vortices is greater for the solid body than for the patch, which is consistent with Zong and Nepf's [29] results. This magnitude difference becomes smaller moving further downstream of the patch.

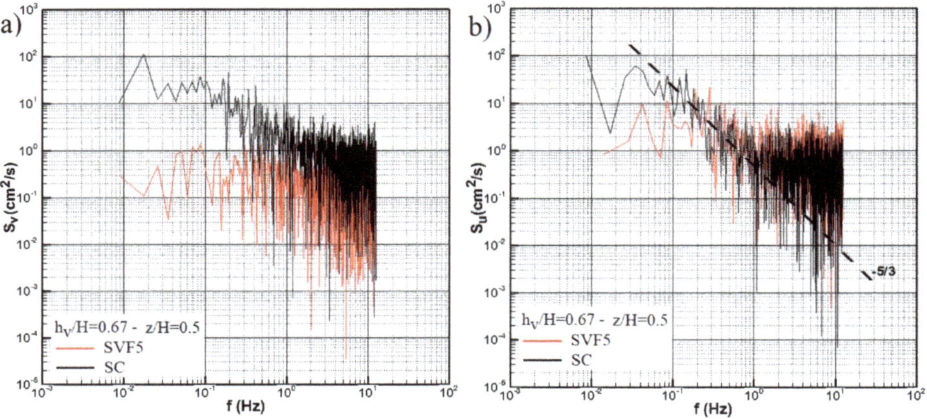

Figure 12. Spectral densities of the transverse velocity Sv located at (**a**) C5 and (**b**) C11 for Case SVF5 and Solid Case.

Figure 12b shows the change in longitudinal velocity spectral density with frequency for Case SVF5 and the Solid Case. When the frequency exceeds 0.16 Hz for the Solid Case and 0.04 Hz for SVF5, the linear decrease of spectral densities is in agreement with the Kolmogorov line, with a slope of −5/3, as shown in Figure 12b.

Channel blockage D/B = 0.25 (Figure 3) in this study may have some effect on the vortex shedding frequency in our study. Chen et al. [35] and Coutanceau and Bouard [36] stated that both Re_c and the Strouhal number ($St_D = fD/U_{dm}$) increase with increasing channel blockage. Here, Re_c is the critical Reynolds number at which unbounded steady flow past a circular cylinder becomes unstable. Sahin and Owens [37] found that Re_c and St_D were 285 and 0.44, respectively, for a blockage of 0.64. Zong and Nepf [29] considered D/B = 0.10, 0.18, and 0.35 for the emergent patch in their study and showed that Re_D ($=U_{dm}D/\nu$) is O (10^4), which is within the turbulent wake regime. In our study, Re_D is O (10^5), which is the within the turbulent wake regime. As a result, the channel blockage might have some effects on the vortex shedding frequency in our study.

Rominger and Nepf [31] defined deceleration through a long porous patch where length is much larger than width. They mentioned that the velocity at the centerline starts to decline from some distance upstream of the patch, which was also observed in this study (Figure 7a), and that the deceleration continues into the patch over a length x_D, after which the flow reaches a steady velocity, U_0—the final interior velocity. Rominger and Nepf [31] obtained that the interior adjustment length (x_D) and the final interior velocity (U_0) depend on the dimensionless parameter, $C_{Da}D$, which is called the patch flow-blockage. Here, C_D is the drag coefficient for the cylinders within the patch.

A low blockage effect is present for $C_{Da}D < 4$, where interior velocity is driven by turbulent stress; while in a high blockage effect, when $C_{Da}D > 4$, interior velocity is driven by a pressure gradient [29,31]. In our study, $C_{Da}D$ = 1.8 for ϕ = 0.05 (SVF5), $C_{Da}D$ = 7.5 for ϕ = 0.20 (SVF20), and $C_{Da}D$ = 13.8 for ϕ = 0.36 (SVF36), where C_D = 1 for simplicity, as Zong and Nepf [29] considered in their study. There is a low blockage regime for the case ϕ = 0.05 < 0.1 and at this regime $x_D \sim 2$ $(C_{Da})^{-1}$ [31] and

$x_D = 33.33$ cm in this study. Here, $C_D a$ is the canopy drag length scale, which is the length scale of flow deceleration associated with canopy drag. There is high blockage regime for the cases $\phi = 0.20 > 0.1$ and $\phi = 0.36 > 0.1$, which is consistent with Zong and Nepf's [29] experimental conditions. In these cases, the interior adjustment length $x_D \sim D = 30$ cm and diameter of the canopy patch provide us this length. As such, velocity just downstream of the patch is expected to be equal to the interior velocity for these cases.

4. Conclusions

The purpose of this study was to answer how the submergence ratio and density of vegetation patches affects the flow structure along the axis of the patch, downstream of the suspended cylindrical vegetation patch. For this purpose, the velocity measurements with ADV (Acoustic Doppler Velocimeter) were performed along the flow depth at the axis of the patch along different longitudinal distances downstream of the suspended patch for different submergence ratios ($h_{v1}/H = -0.33$ and $h_{v2}/H = 0.67$) and different patch porosities ($\phi_1 = 0.05$, SVF5; $\phi_2 = 0.20$, SVF20; $\phi_3 = 0.36$, SVF36; $\phi_4 = 1.00$, Solid Case).

It was found that there is a recirculation zone behind the obstruction, which is strongest for the highest density (SVF36) patch, while there is no recirculation zone for the lowest density (SVF5). It was also shown that U velocity increases more rapidly to the U_{dm} value at the highest density (SVF36) compared to the lower densities (SVF5 and SVF20). Moreover, a steady wake region was observed downstream of the patch for SVF5, which is not present at higher SVFs.

The flow becomes stronger downstream of the gap region with increasing submergence ratio and flow behavior becomes similar to wall jet flow. It was found that SVF has an important impact on this flow structure in terms of the magnitude and the placement of the maximum velocity.

The patch with the lower depth ratio ($h_v/H = 0.33$) affected the flow slightly downstream for the solid body and lowest SVF. However, suspended vegetation still affects the flow downstream of the gap region for a higher density of patches (SVF20 and SVF36) and it is concluded that high densities of submerged vegetation patches still have an influence on the flow structure downstream of the gap, even at low submergence ratios.

It was found that suspended vegetation patches do not have a significant effect on transversal velocity downstream of the patch for a low solid volume fraction (SVF5) at both depth ratios. As the suspended canopy becomes denser, the magnitude of transversal velocity increases, and the placement of diverging flow occurs closer to the patch, in the downstream section, with increasing SVF.

The effect of suspended vegetation on turbulence starts in the upstream section and causes a peak upstream of the gap region. It was found that the shear layer caused by the patch increases the turbulence level beneath the patch. It was also found that the individual stem turbulence is not significant for the highest patch density (SVF36), whereas it is significant at lower patch densities (SVF5 and SVF20). Individual cylinders cause the small-scale turbulence, which results in a higher turbulence region behind the patch. Moreover, the turbulence level decreases with increasing patch density (SVF36) and this turbulence level is close to the turbulence level for the solid body at mid-depth. This means that the suspended patch with the highest porosity patch (SVF36) behaves like a solid body in terms of turbulence and structure of the flow.

Author Contributions: A.Y.O. and D.Y. conceived and designed the experiments; A.Y.O. and D.Y. performed the experiments; A.Y.O. and D.Y. analyzed the data; A.Y.O. wrote the paper. All authors have read and agreed to the published version of the manuscript.

Funding: This research received no external funding.

Acknowledgments: The authors would like to thank Aydın Adnan Menderes University, Civil Engineering Department students Omer Botan YILDIZHAN, Ali CAMBOLAT and Ebru ERSIN who helped during the experiments.

References

1. Anderson, D.M.; Glibert, P.M.; Burkholder, J.M. Harmful algal blooms and eutrophication: Nutrient sources, composition, and consequences. *Estuaries* **2002**, *25*, 704–726. [CrossRef]
2. Hartshorn, N.; Marimon, Z.; Xuan, Z.M.; Chang, N.B.; Wanielista, M.P. Effect of floating treatment wetlands on control of nutrients in three stormwater wet detention ponds. *J. Hydraul. Eng.* **2016**, *21*, 8. [CrossRef]
3. Sample, D.J.; Laurie, J.F. Innovative Best Management Fact Sheet No. 1: Floating Treatment Wetlands. Available online: https://vtechworks.lib.vt.edu/bitstream/handle/10919/70627/BSE-76.pdf?sequence=1&isAllowed=y (accessed on 19 November 2019).
4. Wanielista, M.P.; Chang, N.B.; Chopra, M.; Xuan, Z.; Islam, K.; Marimon, Z. Floating Wetland Systems for Nutrient Removal in Stormwater Ponds; Final Report FDOT Project BDK78 985-01; 2012. Available online: http://frogenvironmental.co.uk/wp-content/uploads/2014/09/Wanielista-et-al-2012-FTW-for-nutrient-removal-in-stormwater-ponds-final-report.pdf (accessed on 19 November 2019).
5. Rosman, J.H.; Koseff, J.R.; Monismith, S.G.; Grover, J. A field investigation into the effects of a kelp forest (*Macrocystis pyrifera*) on coastal hydrodynamics and transport. *J. Geophys. Res.* **2007**, *112*, C02016. [CrossRef]
6. Plew, D.R. Depth-averaged drag coefficient for modeling flow through suspended canopies. *J. Hydraul. Eng.* **2010**, *137*, 234–247. [CrossRef]
7. Huai, W.; Hu, Y.; Zeng, Y.; Han, J. Velocity distribution for open channel flows with suspended vegetation. *Adv. Water Resour.* **2012**, *49*, 56–61. [CrossRef]
8. Zhao, F.; Huai, W.; Li, D. Numerical modeling of open channel flow with suspended canopy. *Adv. Water Resour.* **2017**, *105*, 132–143. [CrossRef]
9. Adams, C.S.; Boar, R.R.; Hubble, D.S. The Dynamics and Ecology of Exotic Tropical Species in Floating Plant mats: Lake Naivasha, Kenya. *Hydrobiologia* **2002**, *488*, 115–122. [CrossRef]
10. Scheffer, M.; Szabo, S.; Gragnani, A. Floating plant dominance as a stable state. *Proc. Natl. Acad. Sci. USA* **2003**, *100*, 4040–4045. [CrossRef]
11. Fang, Y.Y.; Babourina, O.; Rengel, Z. Ammonium and nitrate up-take by the floating plant *Landoltia punctate*. *Ann. Bot.* **2007**, *99*, 365–370. [CrossRef]
12. Nichols, P.; Lucke, T.; Drapper, D.; Walker, C. Performance Evaluation of a Floating Treatment Wetland in an Urban Catchment. *Water* **2016**, *8*, 244. [CrossRef]
13. Plew, D.R.; Spigel, R.H.; Stevens, C.L.; Nokes, R.I.; Davidson, M.J. Stratified flow interactions with a suspended canopy. *Environ. Fluid Mech.* **2006**, *6*, 519–539. [CrossRef]
14. Nepf, H.M.; Mugnier, C.G.; Zavistoski, R.A. The effects of vegetation on longitudinal dispersion. *Estuar. Coast. Shelf Sci.* **1997**, *44*, 675–684. [CrossRef]
15. Ghisalberti, M.; Nepf, H.M. The limited growth of vegetated shear layers. *Water Resour. Res.* **2004**, *40*, W07502. [CrossRef]
16. Lightbody, A.F.; Nepf, H.M. Prediction of velocity profiles and longitudinal dispersion in salt marsh vegetation. *Limnol. Oceanogr.* **2006**, *51*, 218–228. [CrossRef]
17. Yılmazer, D.; Ozan, A.Y.; Cihan, K. Flow Characteristics in the Wake Region of a Finite-Length Vegetation Patch in a Partly Vegetated Channel. *Water* **2018**, *10*, 459. [CrossRef]
18. Ben Meftah, M.; Mossa, M. Prediction of channel flow characteristics through square arrays of emergent cylinders. *Phys. Fluids* **2013**, *25*, 1–21. [CrossRef]
19. Ben Meftah, M.; De Serio, F.; Mossa, M. Hydrodynamic behavior in the outer shear layer of partly obstructed open channels. *Phys. Fluids* **2014**, *26*, 1–19. [CrossRef]
20. Ben Meftah, M.; Mossa, M. Partially obstructed channel: Contraction ratio effect on the flow hydrodynamic structure and prediction of the transversal mean velocity profile. *J. Hydrol.* **2016**, *542*, 87–100. [CrossRef]
21. Ben Meftah, M.; De Serio, F.; Malcangio, D.; Mossa, M.; Petrillo, A.F. A modified log-law of flow velocity distribution in partly obstructed open channels. *Environ. Fluid Mech.* **2016**, *16*, 453–479. [CrossRef]
22. Liu, X.; Huai, W.; Wang, Y.; Yang, Z.; Zhang, J. Evaluation of a random displacement model for predicting longitudinal dispersion in flow through suspended canopies. *Ecol. Eng.* **2018**, *116*, 133–142. [CrossRef]
23. Cheng, N.S. Representative roughness height of submerged vegetation. *Water Resour. Res.* **2011**, *47*, 1–18. [CrossRef]
24. Shucksmith, J.D.; Boxall, J.B.; Guymer, I. Determining longitudinal dispersion coefficients for submerged vegetated flow. *Water Resour. Res.* **2011**, *47*, 124–132. [CrossRef]

25. Liu, Z.; Chen, Y.; Zhu, D.; Hui, E.; Jiang, C. Analytical model for vertical velocity profiles in flows with submerged shrub-like vegetation. *Environ. Fluid Mech.* **2012**, *12*, 341–346. [CrossRef]
26. Stoesser, T.; Salvador, G.P.; Rodi, W.; Diplas, P. Large Eddy Simulation of Turbulent Flow Through Submerged Vegetation. *Transp. Porous Media* **2009**, *78*, 347–365. [CrossRef]
27. Nepf, H.; Ghisalberti, M. Flow and transport in channels with submerged vegetation. *Acta Geophys.* **2008**, *56*, 753–777. [CrossRef]
28. Kaimal, J.; Finnigan, J. *Atmospheric Boundary Layer Flows*; Oxford University Press: New York, NY, USA, 1994.
29. Zong, L.; Nepf, H. Vortex development behind a finite porous obstruction in a channel. *J. Fluid Mech.* **2012**, *691*, 368–391. [CrossRef]
30. Wahl, T.L. Analyzing ADV Data Using WinADV. In Proceedings of the Joint Conference on Water Resources Engineering and Water Resources Planning & Management, Minneapolis, MN, USA, 30 July–2 August 2000.
31. Rominger, J.; Nepf, H. Flow adjustment and interior flow associated with a rectangular porous obstruction. *J. Fluid Mech.* **2011**, *680*, 636–659. [CrossRef]
32. Cantwell, B.; Coles, D. An experimental Study of Entrainment and Transport in the Turbulent Near Wake of a Circular Cylinder. *J. Fluid Mech.* **1983**, *136*, 321–374. [CrossRef]
33. Schlichting, H. *Boundary-Layer Theory*; McGraw-Hill: New York, NY, USA, 1979.
34. Zhao, F.; Huai, W. Hydrodynamics of discontinuous rigid submerged vegetation patches in open-channel flow. *J. Hydro-Environ. Res.* **2016**, *12*, 148–160. [CrossRef]
35. Chen, J.H.; Pritchard, W.H.; Tavener, S.J. Bifurcation for flow past a cylinder between parallel planes. *J. Fluid Mech.* **1995**, *284*, 23–41. [CrossRef]
36. Coutanceau, M.; Bouard, R. Experimental determination of the main features of the viscous flow in the wake of a circular cylinder in uniform translation. Part 1. Steady flow. *J. Fluid Mech.* **1977**, *79*, 231–256. [CrossRef]
37. Sahin, M.; Owens, R.G. A numerical investigation of wall effects up to high blockage ratios on two-dimensional flow past a confined circular cylinder. *Phys. Fluids* **2004**, *16*, 1305–1320. [CrossRef]

Article

Experimental Hydraulic Investigation of Angled Fish Protection Systems—Comparison of Circular Bars and Cables

Heidi Böttcher [1,*] and Roman Gabl [1,2,*] and Markus Aufleger [1]

[1]   Unit of Hydraulic Engineering, University of Innsbruck, Technikerstraße 13, 6020 Innsbruck, Austria; markus.aufleger@uibk.ac.at

[2]   School of Engineering, Institute for Energy Systems, FloWave Ocean Energy Research Facility, The University of Edinburgh, Max Born Crescent, Edinburgh EH9 3BF, UK

*   Correspondence: heidi.boettcher@gmail.com (H.B.); roman.gabl@uibk.ac.at or roman.gabl@ed.ac.uk (R.G.)

Received: 23 April 2019; Accepted: 14 May 2019; Published: 21 May 2019

Abstract: The requirements for fish protection at hydro power plants have led to a significant decrease of the bar spacing at trash racks as well as the need of an inclined or angled design to improve the guidance effect (fish-friendly trash racks). The flexible fish fence (FFF) is a new developed fish protection and guidance system, created by horizontally arranged steel cables instead of bars. The presented study investigated experimentally the head loss coefficient of an angled horizontal trash rack with circular bars (CBTR) and the FFF with identical cross sections in a flume (scale 1:2). Nine configurations of different bar and cable spacing (blockage ratio) and rack angles were studied for CBTR and FFF considering six different stationary flow conditions. The results demonstrate that head loss coefficient is independent from the studied Bar–Reynolds number range and increases with increasing blockage ratio and angle. At an angle of 30 degrees, a direct comparison between the two different rack options was conducted to investigate the effect of cable vibrations. At the lowest blockage ratio, head loss for both options are in similar very low ranges, while the head loss coefficient of the FFF increases significantly compared to the CBTR with an increase of blockage. Further, the results indicate a moderate overestimation with the predicted head loss by common head loss equations developed for inclined vertical trash racks. Thus, an adaption of the design equation is proposed to improve the estimation of head loss on both rack options.

Keywords: fish protection; head loss; intake; hydraulics of renewable energy systems; hydraulic structure design and management; scale model test

---

1. Introduction

1.1. General

Fish protection and downstream migration measures are considered essential in the design, construction or retrofitting and operation of hydro power plants. Since downstream migrating fish, particularly juveniles, tend to swim within the main current, they will consequently pass the turbines without appropriate measures at the water intake [1]. Depending on site-specific conditions such as turbine type, total head, operation mode of the power plant or fish size, fish are injured, followed by a high risk of mortality [2]. Fish that refuse to enter the turbines remain in the reservoir area, sometimes for several days [3]. Physical barriers, particular trash racks with horizontal or vertical bars that are angled to the flow direction (angled trash racks) or inclined to the bottom (inclined trash racks), are one technical solution to prevent fish from turbine-passage and guide them to a bypass [4].

For these devices, bar spacing b has to be designed considering the target fish species and minimum fish sizes, while values of 10–30 mm are recommended [4–6]. Furthermore, a slight angle

($\leq 45°$) of the barrier to the flow direction is recommended, if an efficient bypass is located at the downstream end of the barrier [4,5,7,8].

Nevertheless, these solutions often lead to a great challenge for energy operators in consideration of increased operational and financial efforts, particularly due to debris related blockage and high head losses [8,9]. Particularly for run-of-river plants with low design heads, the head loss through a fish-friendly trash rack can cause significant relative energy production losses and hence should be part of the optimisation process of the intake structure [10]. Furthermore, for larger plants (design discharge ≥ 90 m^3/s), these solutions are not yet feasible [11].

Therefore, an interdisciplinary research programme on a new developed fish protection system, called the flexible fish fence (FFF), was initiated. The FFF is a physical barrier, created by horizontal oriented steel cables instead of bars [9,12–14]. To divert fish towards the bypass, the flexible structure can be positioned in various (slight) angles to the flow direction. Thus, the fish protection effect of the FFF is similar to an angled trash rack with horizontal bars, but allows a favourable mode of operation: Local clogging at the FFF during normal operation (e.g., small branches leaves, grass) is mobilised and cleaned by releasing individual steel cables or cable clusters [9]. Since the FFF works only as a fish protection device, an additional thrash rack downstream of the FFF for turbine protection is necessary (if not already existing in the case of a retrofitting).

The presented study compared a trash rack with horizontal circular bars (geometrically similar to the FFF) to the new concept with the cables in an experiment in the flume. Therefore, the same cross section of the structure (diameter s is 5 mm) was chosen and the spacing b between them as well as the angle α of the installation in relation to the main flow direction was investigated. This allowed identifying the effect of the comparable flexible structure made of cables in relation to circular bars.

1.2. Basic Equations

The aim of the experimental study presented here was to investigate local head loss h_v through a physical barrier similar or equal the flexible fish fence (FFF). In particular, it was evaluated how the specific geometry including the bar spacing, rack angle, bar shape and hydraulic conditions affect head loss through the FFF. The evaluation of the local head loss h_v is based on Bernoulli's equation and the comparison of two cross sections, which are numbered in flow direction. By assuming that the kinetic energy correction factor is equal to 1 (-), the head loss $h_{v,total}$ between two sections can be calculated as following [15]:

$$h_{v,total} = z_1 - z_2 + \frac{p_1 - p_2}{\rho \cdot g} + \frac{v_1^2 - v_2^2}{2 \cdot g} = h_v + h_{v,cont.} \tag{1}$$

For this analysis, the elevation $z_{1,2}$, pressure $p_{1,2}$ and velocity $v_{1,2}$ were needed for each cross section as well as two constants, namely density of water ρ and gravity acceleration g. Furthermore, the loss $h_{v,cont}$ based on surface roughness between the two measurement positions had to be eliminated to identify the local head loss h_v due to the structure. For the current case, the flume was used in its horizontal position ($z_1 = z_2$) and the water depth $h_{1,2}$ could replace the formulation of the pressure to

$$h_v = h_1 + \frac{v_1^2}{2 \cdot g} - h_2 - \frac{v_2^2}{2 \cdot g} - h_{v,cont.} \tag{2}$$

For most of the applications in turbulent flow conditions, the local head loss coefficient ζ can be introduced. This assumes a linear connection between h_v and the velocity height $v_{ref}^2/(2 \cdot g)$:

$$h_v = \zeta \cdot \frac{v_{ref}^2}{2 \cdot g}. \tag{3}$$

A clear definition of the reference velocity v_{ref} was needed [16,17]. For the quantification of head loss at trash racks, the undisturbed cross upstream the structure was used (mean approach velocity) [18]. It was assumed that the velocity distribution over the complete cross section is homogeneous. For some applications and under specific circumstances, the velocity of the complete cross section of the trash rack can change significantly, which could be part of future investigations [19]. For the presented results, the velocity v_1 was used as v_{ref}, hence this is a section that is in most cases easily accessible and does not include the influence of the trash rack (Section 3.2).

A differential pressure transmitter was applied in this study, which allowed directly measuring the differential pressure $\Delta p = p_1 - p_2$ and increasing the accuracy of this value [15]. For free surface applications, at least one water depth is need in addition to Δp to describe the flown through area. In the following case, the upstream water depth h_1 was used as well as the continuity equation. Consequently, the local head loss coefficient ζ of the trash rack can be calculated as followed:

$$\zeta^* = \frac{\Delta p}{\rho} \cdot \frac{2 \cdot B^2 \cdot h_1^2}{Q^2} + \left(h_1^{-2} - \left(h_1 - \frac{\Delta p}{\rho \cdot g} \right)^{-2} \right) \cdot h_1^2 - \zeta_{v,cont}. \tag{4}$$

This evaluation was only based on the three measurements of the differential pressure Δp, upstream water depth h_1 and the discharge Q. The coefficient $\zeta_{v,cont}$, which includes roughness and the influence of the support structure for the trash rack in the experimental set-up, had to be evaluated separately (Section 3.3).

1.3. Literature Values

A wide range of investigations are conducted to quantify the head loss (coefficient) through water intakes at hydro power plants in the past. However, most of them address conventional trash racks with vertical bars and a comparable big bar spacing. In the last years, the requirements changed since the awareness of the need for fish protection and downstream migration measures is increasing. Smaller spacing between the bars is needed as well as new developments.

One of the first fundamental investigations was carried out by Kirschmer [20] nearly a century ago. He studied the head loss on trash racks with vertical bars of several bar shapes, bar widths, bar spacings and rack inclinations. According to Kirschmer [20], the head loss coefficient ζ_K is given by

$$\zeta_K = k_F \cdot \left(\frac{s}{b} \right)^{\frac{4}{3}} \cdot k_\beta, \tag{5}$$

where k_F is the bar shape coefficient and $k_\beta = sin(\beta)$ considers the rack inclination β in relation to the initial vertical orientation of the bars. The geometry of the trash rack includes the bar width s and the bar spacing b. Those constants are connected with the velocity height using v as the approach velocity (comparable to the v_{ref} in Equation (3)). Extensive experiments on trash racks with vertical bars were also carried out by Meusburger [18], who expanded the basic equation introduced by Kirschmer [20]. The blockage ratio through the rack structure (including spacers and supporting elements) and through debris or trash clogging is added. Furthermore, the flow angle θ is taken in account based on adjusted calculation approaches [21–23]. Based on his results, the head loss coefficient ζ_M is given by:

$$\zeta_M = \zeta_p \cdot k_\beta \cdot k_\delta \cdot k_v, \tag{6}$$

with ζ_p defined by the head loss coefficient resulting from blockage, which depends on the blockage ratio p and the bar shape coefficient k_F:

$$\zeta_p = k_F \cdot \left(\frac{p}{1-p} \right)^{1.5}, p = \frac{A_b + A_s}{A_t}, \tag{7}$$

where A_b is the area of bars; A_s is the area through supports, spacers, etc.; and A_t is the total area of the trash rack in the flow. The constant k_δ is a factor considering the approach flow angle δ and k_v the blockage by clogging. Both factors were further described by Meusburger [18], but not considered in the current investigation.

Clark et al. [24] studied head loss on various trash configurations under pressurised flow conditions. However, research on hydraulic loss more and more focuses on fish friendly trash racks in recent years. Raynal et al. [25] tested inclined vertical trash racks with rectangular and hydrodynamic bar shapes, small bar spacings ($b \leq 30$ mm) and a wide range of inclinations ($\beta = 15°$ to $90°$). Based on their results, they proposed a new head loss equation, where particular low angles and the separation of the blockage ratio due to bars and transversal elements are taken into account. However, their head loss equation is restricted to investigated bar shapes. Other fish-friendly solutions such as angled trash racks with vertical angled bars (bar racks) are investigated within various experiments [8,26–28]. With respect to the flow conditions at these trash rack types, where flow is deflected twice through the angled slats, the results and developed head loss formulas are not transferable to geometries similar to the FFF. Head loss on trash racks with horizontal bars, which are more similar to studied rack configurations, has still not been sufficiently investigated for a comparable wide parameter range. Szabo-Meszaros et al. [27] studied head losses and hydraulic conditions at six angled trash rack configurations (with vertical angled, vertical streamwise bars and horizontal bars of rectangular and hydrodynamic bar shape). The results show that head loss at horizontal trash racks is comparably low (and lowest with a hydrodynamic bar shape) and the hydraulic conditions at the bypass entrance are favourable from an ecological point of view [27]. Berger [29] studied head loss at horizontal trash racks with a rectangular bar shape for a wider range of blockage ratios ($p = 0.34$–0.57) and rack angles ($\alpha = 30°$–$70°$), but similar to the above results are not directly transferable to the current study due to differences in the experimental design (e.g., they included a bypass and different hydraulic conditions).

According to the circular bar shape, length L and small bar spacings, the studied racks are highly vulnerable to flow-induced vibrations, which additionally affect head loss. In general, the vertical amplitudes of these vibrations are dependent on the frequency, the preload forces and cable length as well as the flow velocity [14]. This study evaluated basically the effect of blockage and rack angle on head loss on angled horizontal trash racks with circular bars (CBTR), where vibrations were firstly mitigated by an additional spacers. Further, the same parameters were studied at the FFF and compared with the former results to estimate the effect of cable vibrations on head loss. Finally, it was evaluated if the common formulas for head loss estimation (Equations (5) and (6)) at (inclined) trash racks with vertical bars are transferable to the rack configurations.

2. Materials and Methods

2.1. Experimental Setup

The experiments were carried out in a rectangular channel at the Hydraulic Engineering Laboratory at the University of Innsbruck and split into two different main configurations: (a) circular bars (CBTR); and (b) cables (FFF) (Figure 1). The flume is 20 m long and has a width B of 0.8 m. On both sides, 1 m high glass side walls allow observing the experiments. The water supply from an elevated tank fills up a tank and a flow straightener homogenises the flow further (Figure 2). At the outlet of the flume, the water flows over a flap gate into the underground tank, where it is pumped back to the elevated tank. It is noted that water temperature did not changed through this circulation process because of the high volume of both tanks. The discharge Q was varied from 50 to 230 L s^{-1}, which corresponds to approach velocities from 0.16 to 0.72 m s^{-1}, and bar Reynolds numbers Re_b from 750 to 3500. The Bar–Reynolds number is thereby defined by

$$Re_b = \frac{v_1 \cdot s}{\nu} = \frac{Q \cdot s}{B \cdot h_1 \cdot \nu}. \tag{8}$$

All experiments were conducted with turbulent flow conditions (see Reynolds numbers in Table 1). In this definition, the velocity v_1 in front of the installation was used and consequently the water depth h_1 multiplied by the width B of the flume allowed connecting the discharge Q. With the kinematic viscosity v, the properties of water were taken into account and, for the characteristic linear dimension, the diameter s of the bars/cables was used (Table 1).

Table 1. Overview of the parameter—varied ones are marked with a *.

Parameter	Rods	Cables
Bar diameter s (mm)	5	5
* Spacing b (mm)	5, 10, 15	5, 10, 15
* Angle α (°)	90, 45, 30	40, 30, 20
* Discharge Q (l s^{-1})	50–200	80–230
Bar shape coefficient k_F (-)	1.79	1.79
s/b (-)	0.33, 0.5, 1,0	0.33, 0.5, 1.0
Blockage ratio p (-)	0.25, 0.33, 0.5	0.25, 0.33, 0.5
Bar length l (m)	0.80, 1.24, 1.60	1.25, 1.60, 2.34
Approach velocity v (m s^{-1})	0.16–0.63	0.25–0.72
Bar–Reynolds-No. Re_b (-)	750–3000	1250–3500
Reynolds-No. Re (-)	31,000–125,000	50,000–144,000
Froude F (-)	0.08–0.3	0.13–0.36

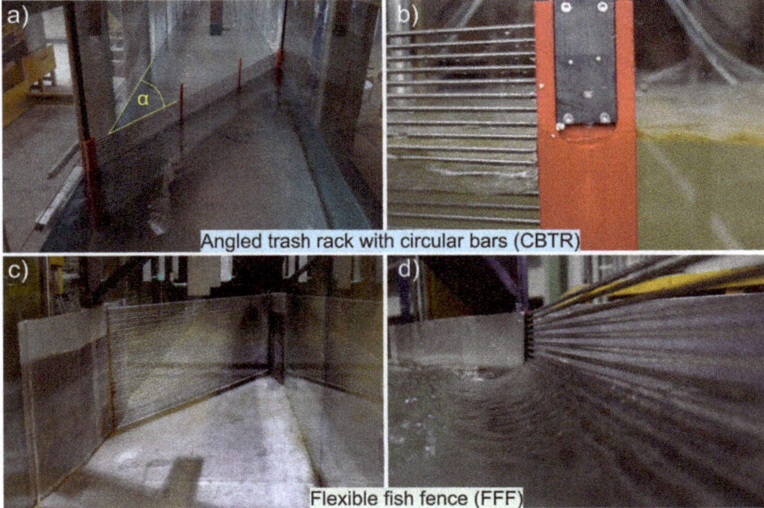

Figure 1. Angled trash rack with circular bars (CBTR) with: $\alpha = 30°$ and $b = 5$ m (upstream view) **(a)**; detailed side view of bar option from the side of the flume **(b)**; Flexible Fish Fence (FFF) with $\alpha = 30°$ and $b = 5$ mm **(c)**; and detailed side view of the flow pattern at the tail water of the FFF **(d)**.

A flap gate at the end of the flume allows controling the tailwater depth. In all experiments and independent of the discharge Q, the water depth of 0.4 m was maintained at the flap gate. This was controlled with a point gauge (PG) at $x = 17$ m, which is 0.3 m upstream of the ultrasonic sensor (US) 8.

The investigated trash racks were installed 9.6 m downstream from the channel inlet. The FFF requires high preload forces to tension the cables. A solid self-supporting structure had to be designed to install this in the glass-sided flume (Figure 1c). The influence of the supporting structure was minimised as far as possible. Separate experiments were conducted without the bars or cables to allow identifying head loss through the supporting structure at the flume walls together with the frictional loss. Based on the dominant influence of the free surface flow, an upscaling should be conducted according to Froude similarity [30]. All components were designed in a scale λ of 1:2, considering the ratio of water depth to bar width $h/s \geq 60$, which ensured that the resistance of the bars does not depend on the Froude number [21].

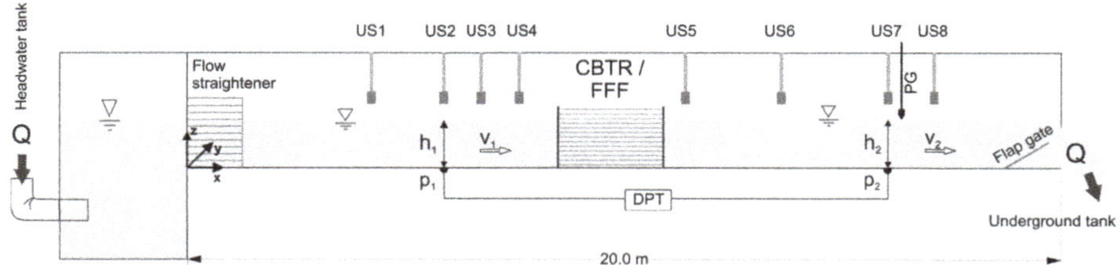

Figure 2. Schematic side view of the experimental flume including the local coordinate system—the locations of the points are given in Table 2.

Table 2. Location of the measurement points of the ultrasonic sensors (US) and the point gauge (PG) shown in Figure 2.

	US1	US2	US3	US4	US5	US6	US7	PG	US8
x (m)	5.3	6.5	7.5	8.3	11.3	13.7	16.1	17	17.3

2.2. Measurement

In each single experiment, the same procedure was applied. In a first step, the discharge Q through the flume was changed by the main inlet valve and then we waited until it was stable. Furthermore, the downstream water level at the end of the flume was adjusted to a water depth of 0.4 m. After steady flow conditions were reached, the measurement was conducted. Each condition was observed for approximately 10 min. At eight measuring points along the flume centre axis, the water levels were measured by ultrasonic sensors (US) (Figure 2) with an accuracy of ±1 mm and measurement frequency of 5 Hz. The point gauge was randomly used to control the measurement of the US8. Additionally, the pressure head loss through the CBTR and the FFF, respectively, was measured with a differential pressure transmitter (DPT) with a higher accuracy (±0.2 mm). Therefore, the two measuring points were located approximately 3 m upstream (measurement point US2) and around 6 m downstream of the racks (US7). Based on the measurements of the other US, it could be determined that US2 and US7 were in a homogeneous flow condition and not influenced by the racks or the flap gate (Figure 2).

The discharge Q was derived by two pipes from the headwater tank of the laboratory and measured with a magnetic inductive discharge-meters (MID) of ±1% accuracy. A plausibility check was conducted based on the needed weir height to reach the steady tail water height of 0.4 m. Therewith, all measurements, namely discharge and water depth, were redundant and could be checked independently.

2.3. Investigated Parameters

The circular cross section of the circular bars and the cables were identical with a diameter s of 5 mm, which allowed identifying the influence of cable vibrations. A supporting structure holds the rack at the side walls over a rack height of 0.5 m. Flow-induced vibrations were comparably strong based on the circular bar shape. These vibrations were intended to be excluded or even mitigated for the investigated option CBTR. Thus, additional spacers were applied (one for $\alpha = 90$ and 45° and two for $\alpha = 30°$, Figure 1a), which consist of rounded bars of a width of 5 mm and were tested as part of preliminary experiments.

In a first step, the bar spacing was varied from $b = 5$ mm to 15 mm, which reproduced full-scale bar spacing between 10 and 30 mm. Consequently, the ratio s/b [20] was 0.33–1.0 and blockage ratio p [18] ranged from 0.25 to 0.5, respectively. It is noted that the ratios s/b [20] and $p/(1-p)$ [18] were equal for both rack options since the few transversal elements installed at the CBTR had no significant effect on p. The second main parameter for the investigation is the rack angle α related to the flow direction (Figure 1). For the CBTR, three angles were studied from fish-friendly ($\alpha = 30°$ and $45°$) to the conventional ($\alpha = 90°$) resulting in variable lengths of the bars. For the FFF, fish-friendly solutions with smaller rack angles were the focus of the study. Thus, α was varied in a smaller range with $\alpha \leq 40°$ (Table 1).

In total, nine different geometries were investigated under six steady discharges, as summarised in Table 1. Each measurement was repeated in a different order so that independence of the previous experiments could be ensured. Additional reference experiments allowed determining the head loss $h_{v,cont}$, which includes the surface friction as well as the supporting structures (trash rack without bars or FFF without cables, respectively). Those were conducted for the whole discharge range and all observed angles α. Consequently, the main study was based on 216 basic experiments and 36 reference tests.

3. Results

3.1. Overview

The investigated head loss was comparably small and close to the measurement accuracy as well as the nearly unavoidable changes in the water level in the flume with higher discharge. Therefore, the verification process of the measurement process is of great importance and is presented in Section 3.2. The constant losses due to the supporting structure and friction are quantified in Section 3.3 and later excluded from the main results of the investigated different options. For each of the two options, namely the trash rack (CBTR) and the flexible fish fence (FFF), the effect on the head loss based on the spacing between bars or cables, respectively, as well as the angle α are considered in Section 3.4. A direct comparison was conducted for an angle α equal to $30°$. Further, the measured head loss coefficients were compared with literature equations and a regression model is fitted to the data to adapt Equation (6) to the studied rack options. This would allow implementing the findings in future applications with comparable settings.

3.2. Measurement Accuracy and Data Verification

As a first step of the verification process, each individual measurement was analysed independently of the investigated option and parameter combination. It was assumed that the investigated rack geometry had no effect on the individual analysis. Consequently, all runs were simultaneously checked to verify the chosen analysis method for the measurements. The second part of the verification process focused on the calculation of the total head loss coefficient ζ^*, which includes the constant losses due to the support structure and friction (Section 3.3). Generally, it was assumed that ζ^* is a constant value and only depending on the geometry. This allowed connecting the local head loss h_v with the velocity height in front of the installation, as presented in Equation (3). For the verification of the calculation, the total head loss $h_{v,total}$ based on Equation (1) was applied for the two main sections US2 and US7 (Figure 2).

Figure 3 presents all measurements standardised by the mean value of each individual parameter combination for all eight ultrasonic sensors (US) as well as the two different approaches to evaluate the pressure difference between the two sections. The found deviations of the mean values are in the range of a few millimetres and can be caused by the measurement accuracy but more likely by waves. The scattering of the measured water depths downstream of the trash racks (US5–US8) is generally higher than for measurements upstream of the investigated structures. Increased turbulence causes this effect, as expected. Based on this analysis, the velocity v_1 was chosen as a reference velocity v_{ref} in Equation (3) for the calculation of the total local head loss coefficient ζ^*, respectively, ζ for the main investigations.

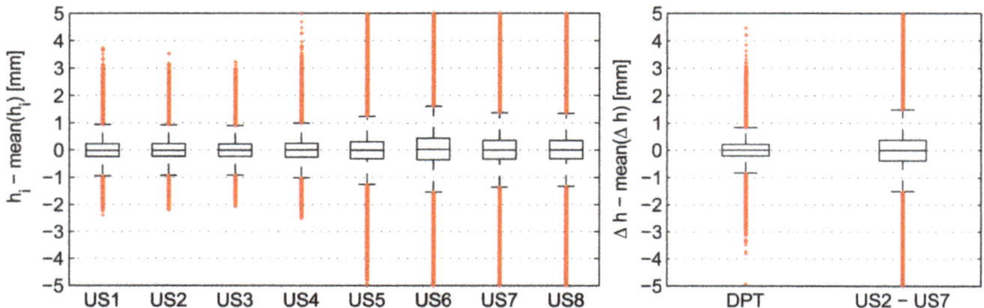

Figure 3. Analysis of the individual measured value standardised by the mean value of each geometry set-up: water depth at measurement points US1–US8 with ultrasonic sensors (**left**); and differential pressure transmitter (DPT) and difference between US7 and US2 (**right**).

A similar analysis was conducted for the pressure difference between the two cross sections US2 and US7 (Figure 3). The direct measurement with the differential pressure transducer (DPT) shows a comparable result to the single US measurements upstream of the structure. The uncertainty would increase, if instead the second US measurement were used to evaluate the pressure difference (US2–US7 in Figure 3). To further investigate this, the parallel measurement with the differential pressure transmitter Δh_{DPT} can be put in opposition with the ultrasonic sensors difference $\Delta h_{US} =$ US2–US7, as shown in Figure 4. Ideally, all points of each individual measurement would lay on the blue line of equality for this analysis. Obviously, the pressure head values differ considerably and partially strong outliers are produced by the ultrasonic measurement, which can also be seen in Figure 3. If the individual measured values are averaged for each measurement (Figure 4, right), it can be seen that the single outliers do not have such a significant influence on the mean value. Nevertheless, the previous comparison highlights the importance of the DPT to measure the differential pressure Δp_{DPT} with a very high accuracy, as known from previous studies [15,19]. The DPT was further chosen as input values for the ζ-calculation based on Equation (4).

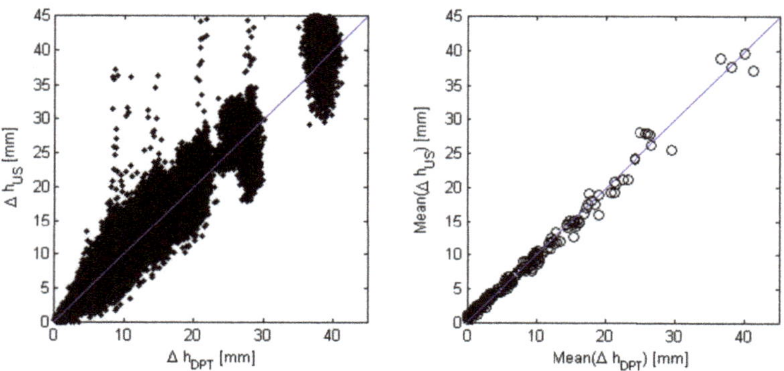

Figure 4. Comparison of pressure head loss Δh_{US} and Δh_{DPT} including all measuring points (**left**); and as a mean value for each geometrical set-up (**right**).

In the second part of the verification, the influences of the chosen measurement values h_1, Δp_{DPT} and Q on the local head loss coefficient was investigated based on Equation (4). For this, the variable ζ^* was used, which is based on the total loss including also those separately investigated (Section 3.3), $h_{v,cont}$ or $\zeta_{v,cont}$. The coefficient ζ_{ij}^* was calculated based on each of the 2700 single measured values (j) for each parameter combination and rack option (i). Those results were standardised by the corresponding mean value of the measurement and presented in three classes depending on the Bar–Reynolds Number Re_b in Figure 5. The fluctuations are in the range of 0.1 (-) and decrease with a higher discharge. In a previous step, the influence of the averaging was investigated. The difference between the mean value of ζ_i^* and a single calculation of the coefficient ζ_{ij}^* based on the previously averaged measurement values is negligible.

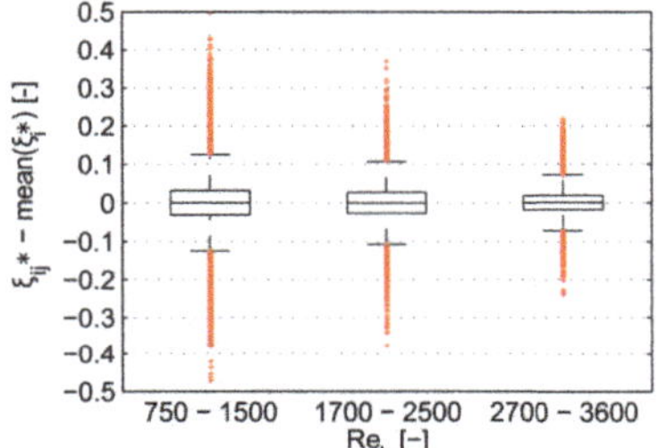

Figure 5. Head loss coefficients ζ_{ij}^* calculated for each data point ($j = 1$–2700) subtracted by the mean value of ζ_i^* of each measurement for three ranges of Bar–Reynolds number Re_b.

In the analysis, the assumption is made that the local head loss coefficient is independent from the discharge Q under turbulent conditions. Hence, the model scale of 1:2 is expected to have no influence on the studied Bar–Reynolds numbers. The resistance coefficient of circular bar shapes is nearly independent in the range of Re_b of 500–20,000 [31,32]. Nevertheless, recent studies on trash racks or bar racks with rectangular, rounded and hydrodynamic bar shapes recommend Bar–Reynolds numbers higher than 1500 [18,25,26,28]. Since the experiments of this study covered the range of Re_b of 750–3500, the scale effect of the physical model was checked for the studied cylindrical bar shapes in order to ensure the transferability of the results to nature conditions. Figure 6 demonstrates that ζ^* is roughly independent from Re_b for both trash rack types. Thus, the results can be transferred to applications under full-scale conditions.

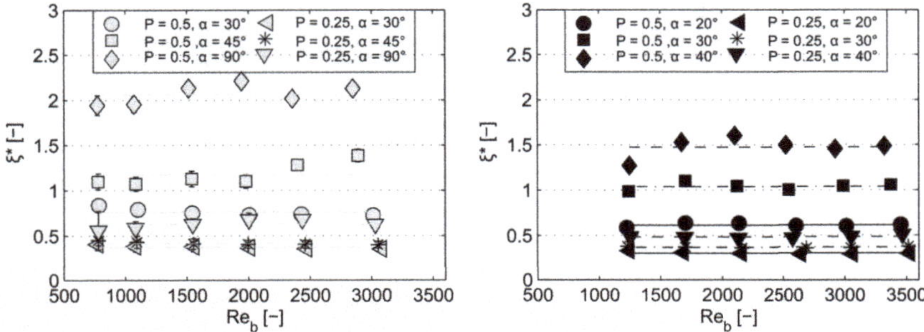

Figure 6. Head loss coefficient ζ^* versus Bar–Reynolds number Re_b for selected CBTR (**left**) and FFF configurations (**right**).

3.3. Head Loss Through Supporting Structures and Surface Friction

The installation of the trash rack (CBTR) and the flexible fish fence (FFF) in the experiments required supporting structures on both sides on the channel wall (Figure 1). For the FFF, this differs to real applications where these supporting structures are not exposed to the flow [9]. However, these supporting structures were optimised regarding their size and shape, but they still caused an additional form resistance to the flow. Therefore, the reference head loss $h_{v,cont}$ was measured without bars/cables and only remaining supporting structures. Thus, $h_{v,cont}$ includes the influence of the supporting structure as well as the friction of the flume boundary. The bar plots in Figure 7 demonstrate that head loss coefficients $\zeta_{v,cont}$ for both rack types and the studied rack angles are all in a similar range of $\zeta_{v,cont} \approx 0.2$, which is a substantial part of the measured loss in comparison to those due to the investigated structure. For the perpendicular CBTR ($\alpha = 90°$), $\zeta_{v,cont}$ tends to be slightly lower, which is probably due to the fact that flow separation is concentrated at one location. Obviously, the proportion of $\zeta_{v,cont}$ on ζ^* is relatively high depending on the rack configurations up to 2/3 of ζ^*.

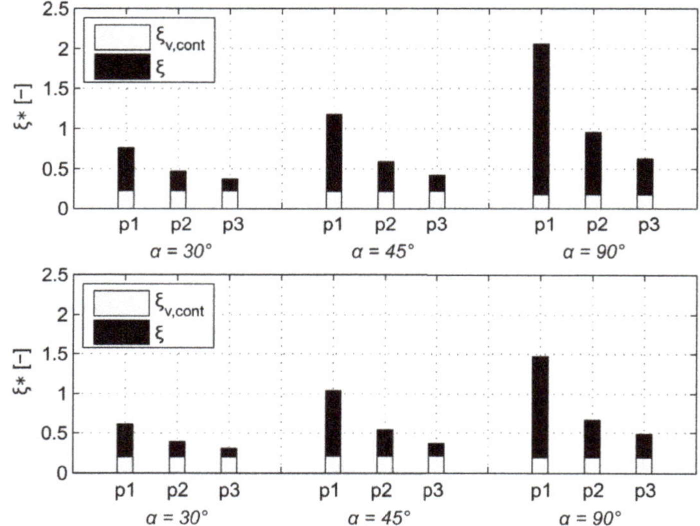

Figure 7. Proportion of head loss coefficient $\zeta_{v,cont}$ through the supporting structure and friction and ζ for rack option CBTR (**above**) and FFF (**below**) on the total head loss coefficient ζ^*.

3.4. Head Loss Coefficients of the Rack Configurations

The following presented results focus on the varied parameters spacing b and angle α (Table 1). All ζ-values measured at the trash rack with circular bars (CBTR) and flexible fish fence (FFF) are summarised in Table 3, in which similar configurations of both options are highlighted. Figure 8 demonstrates the effect of blockage ratio p—introduced in Equation (7)—on the head loss coefficient ζ, calculated by subtracting $\zeta_{v,cont}$ (Section 3.3) from the total head loss coefficient ζ^*. As expected, ζ increases disproportionally with increasing blockage ratios for both investigated rack options. Thereby, the increase of ζ with p positively depends on the angle (Figure 8). Furthermore, the comparison of the results with α equal 30° in Figure 8 reveal that the influence of p on ζ is more pronounced for the FFF.

Table 3. Local head loss ζ measured for the trash rack with circular bars (CBTR) and the flexible fish fence (FFF) depending on the blockage ratio p and rack angle α—differences between the two rack options for $\alpha = 30°$.

CBTR			FFF			Difference	
p (-)	α (°)	ζ_{CBTR} (-)	p (-)	α (°)	ζ_{FFF} (-)	$\Delta\zeta = \zeta_{FFF} - \zeta_{CBTR}$ (-)	$\Delta\zeta/\zeta_{CBTR}$ (%)
0.25	45	0.198	0.25	20	0.075	-	-
0.33	45	0.37	0.33	20	0.162	-	-
0.50	45	0.957	0.50	20	0.385	-	-
0.25	30	0.146	0.25	30	0.148	0.002	1.4%
0.33	30	0.242	0.33	30	0.322	0.08	33.1%
0.50	30	0.533	0.50	30	0.818	0.285	53.5%
0.25	90	0.449	0.25	40	0.305	-	-
0.33	90	0.775	0.33	40	0.488	-	-
0.50	90	1.884	0.50	40	1.295	-	-

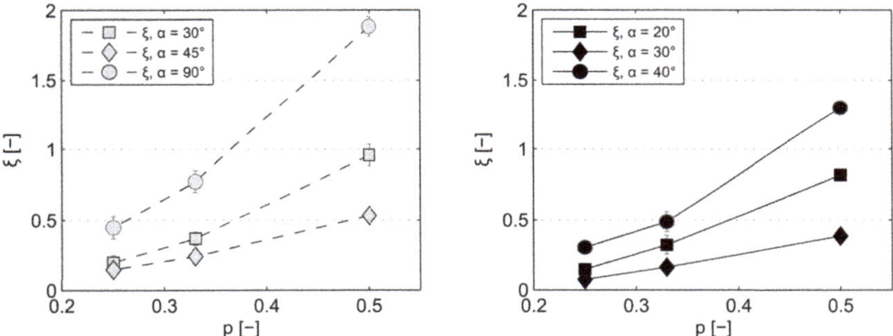

Figure 8. Head loss coefficient ζ versus blockage ratio p for the CBTR (**left**) and FFF option (**right**).

The effect of rack angle α on ζ is shown in Figure 9, where the measured data at the CBTR and FFF with all investigated p are all illustrated in one plot. Head loss coefficient ζ increases with increasing α and it is generally stronger for higher blockage ratios p. The comparison of the ζ-values revealed a stronger impact of α on ζ at the FFF than at the TR (Figure 9) in the range of $\alpha \leq 45°$.

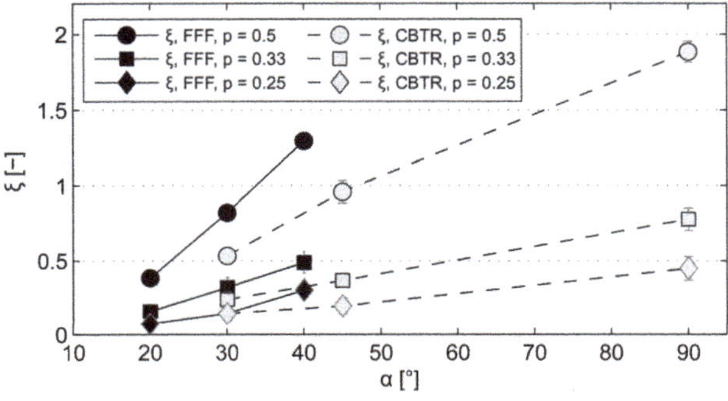

Figure 9. Head loss coefficient ζ versus rack angle α for both rack options (CBTR and FFF).

Besides the stronger trend of ζ with α observed at the FFF, the value of ζ at all rack configurations with $\alpha = 30°$ is up to 53% higher ($p = 0.5$, Table 3) for the FFF than for the CBTR. However, the magnitude of the head loss coefficient ζ is generally low for both rack options and for certain rack configurations of similar order than $\zeta_{v,cont}$ (Figure 7). In this respect, it is worth noting that $\zeta_{v,cont}$ of the FFF is probably lower for full-scale conditions, since the supporting structures are there usually not exposed to the flow.

3.5. Empirical Relations to Predict Head Loss of Angled Racks

An attempt was made to estimate the head loss of both rack options by applying the methods according Equation (5) provided by Kirschmer [20] and Equation (6) modified by Meusburger [18]. Both methods are developed for vertical (inclined) rack types and it is known that the approaching velocity at the inclined trash rack can change significantly over the cross section with a peak at the far end of the inclined structure [25,33]. Nevertheless, it was assumed that the vertical inclination β of Equation (5) and Equation (6) can be replaced by the horizontal angle α. Thus, the modified equations used in this study are defined by:

$$\zeta_K = k_F \cdot \left(\frac{s}{b}\right)^{\frac{4}{3}} \cdot sin(\alpha), \tag{9}$$

$$\zeta_M = k_F \cdot \left(\frac{p}{1-p}\right)^{\frac{3}{2}} \cdot sin(\alpha). \tag{10}$$

Herein, k_F was set to a constant value of 1.79, representing the bar shape coefficient for circular bars. It is noted that the $\frac{s}{b}$ and $\frac{p}{1-p}$ ratios were similar for both options, since the influence of transversal elements were negligible small for CBTR and absent for FFF. Figure 10 demonstrates the percentage deviation of the predicted head loss coefficients ζ_p with Equation (10) (left) and Equation (9) (right) from the measured ζ values with the $\pm 25\%$ and $\pm 75\%$ lines. Around 80% of the measured ζ_m values at the FFF and 70% at the CBTR deviate from the predicted ζ_p coefficients with Equation (10) (less than $\pm 25\%$), while the proportion is slightly lower for ζ_p with Equation (9). The comparison of predicted and measured head loss coefficients also revealed a positive overestimation of ζ_m for both formulae. Again, this is slightly more pronounced for the formula of Kirschmer (Equation (9)). Additionally, the deviation of predicted and measured ζ is correlated with the angle. There is an overestimation with the smallest angles and particularly for the highest blockage ratio $p = 0.5$ and a slight underestimation with the highest angles ($\alpha = 90°$ for CBTR and $\alpha = 40°$ for FFF). To compensate this overestimation, ζ_{fitted} was introduced based on a modified Equation (10), in which the measured head loss coefficients ζ_m are used to obtain the coefficients k_0 to k_2:

$$\zeta_{fitted} = k_0 \cdot \left(\frac{p}{1-p}\right)^{k_1} \cdot (sin(\alpha))^{k_2}. \tag{11}$$

Therefore, a multiple linear regression analysis with the log-transformed parameters $\frac{p}{1-p}$ and $sin(\alpha)$ was performed. Table 4 shows the corresponding coefficients k_0 to k_2, R^2 and $RMSE$ (root-mean-square error) for the measured head loss coefficients at the CBTR and the FFF. The exponent of $\frac{p}{1-p}$ is defined by k_1 and is 1.30 for the CBTR and 1.44 for the FFF, respectively, and in a similar range of both equations. In contrast, the exponent k_2 of $sin(\alpha)$ with 1.70 for the CBTR and 1.96 for the FFF option varies widely from the proposed value of Kirschmer [20]. At least, the coefficient k_0 with $exp(0.59) = 1.8$ for the CBTR matches well with the bar shape coefficient of 1.79 for circular bar shapes given by Equation (9). For the FFF configurations, k_0 is comparably higher with $exp(1.16) = 3.19$ (Table 4). Further tests are needed to confirm and refine this analysis.

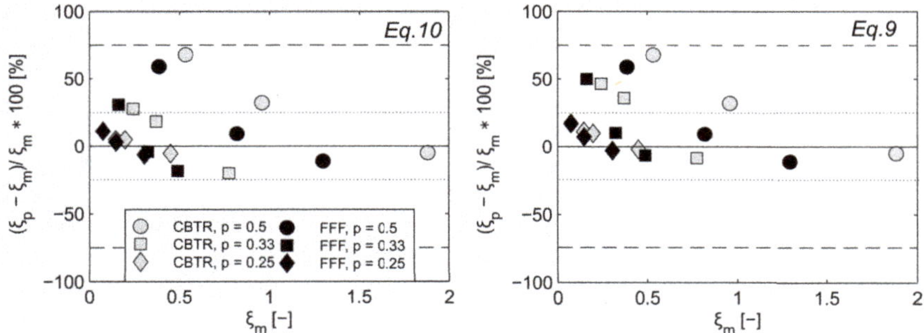

Figure 10. Percentage deviation of the predicted head loss coefficients ζ_p by Equation (10) (**left**) and Equation (9) (**right**) from the measured head loss coefficients ζ_m. The two rack options are highlighted in grey (CBTR) and black (FFF) and differentiated by the related blockage ratios of the rack configurations. The dashed and dotted lines represent the $\pm 75\%$ and $\pm 25\%$ deviation of ζ_p from ζ_m.

Table 4. Results of the multiple linear regression analysis for Equation (11).

	k_0	k_1	k_2	R^2	*RMSE*
CBTR	1.80	1.3	1.7	0.9904	0.0414
FFF	3.19	1.44	1.96	0.9861	0.0006

4. Discussion

4.1. Accuracy and Scale Effects

The measured ζ values are for all investigated rack options in a comparably small range from 0.075 to 1.884 (-) (Table 3, $v_{ref} = v_1$). Consequently, the accuracy of the measurements and the analysis are crucial. The verification analysis, presented in Section 3.2, clearly indicated the use of the water height in front of the trash rack as well as the differential pressure transducer instead of a second ultrasonic measurement downstream of the investigated structures. The deployment of a discharge measurement with a better accuracy would be desirable but would have been required to bypass the existing fix installation of the flume. All used measurements were independently checked by a redundant system continuously or randomly manually. The long observation period for each run of 10 min allowed finding a very stable mean value of the local head loss.

Scale effects of the 1:2 model were also analysed for the studied Bar–Reynolds Numbers of 750–3500, since a recent study by Albayrak et al. [28] on louvers and angled bar racks shows that ζ values for $Re_b \leq 1500$ are slightly lower for a 1:2 than for a 1:1 scale model. A similar phenomenon was observed by Raynal et al. [25], in which higher variations of ζ for lower Re_b occur, mostly pronounced for the inclined trash racks with a hydrodynamic bar profile. However, this is not confirmed for the head loss coefficients at the CBTR and FFF, where ζ is almost constant for the whole Re_b range (Figure 6). These differences are probably attributed to the different bar shapes of the studied trash racks. For rack configurations with a circular bar shape, ζ is at least nearly independent from $Re_b \geq 750$, which is more consistent with the results of Blevins [31] of drag coefficients on a single circular cylinder.

4.2. Effect of Blockage and Angle

As expected by past studies [18,20,24,25], head loss coefficient is a function of the blockage ratio p and bar spacing, respectively. Thereby, ζ disproportionally increase with increasing blockage ratios for both rack options. In comparison, this increase of ζ with p is stronger for the FFF than for the CBTR (Figure 8 and Table 1). This is likely due to the different characteristics of flow induced vibrations of cables and bars. In general, both rack options are very vulnerable to flow-induced vibrations due to their circular bar shape [34]. The vertical movement of the vibrating circular bars or cables leads to a significant reduction of the free flow area, which results in a higher form resistance and head loss. For the basic experiments with the CBTR, it was intended to exclude or reduce the effect of bar vibrations, thus additional spacers were installed to increase the stiffness. However, vibrations were mitigated but could not be fully avoided at the CBTR (in dependence of Re_b), but the vertical amplitudes of the individual bars were comparably small. Instead, the whole trash rack area was oscillating for some specific CBTR configurations (high blockage ratios and high angles) with higher Bar–Reynolds numbers. In comparison, cable vibrations at the FFF were very pronounced and occurred over the whole range of Re_b for the configurations with highest blockage ratio $p = 0.5$ (vibration amplitudes increase with increasing velocity and Re_b, respectively). This observation corresponds to Naudascher and Rockwell [34], who stated that smaller spacing of bars leads to a higher intensity of the buffeting due to wake turbulence and interference [34]. For lower blockage ratios, the cables oscillated at least above a certain threshold of Re_b. In summary, the phenomenon of flow-induced vibrations was significantly dominant at the FFF and, further, it was amplified by small bar spacing. For the experiments with $p = 0.5$ at the FFF (corresponding with the bar spacing being equal the bar diameter), the vertical movement of the vibrating cables reached the same magnitude of spacing b. According to Tanida et al. [35], the resistance coefficient can increase up to a double, if the amplitudes are high enough.

Besides the blockage, head loss coefficients are also affected by the rack angle, whereby ζ increases with increasing α. This probably results from a reduction of the total rack area with increasing α and a corresponding increase of the mean flow velocity at the rack. The same phenomenon was shown by Raynal et al. [25] at vertical inclined trash racks with $\beta = 15$–$90°$, in which lower angles significantly reduce velocities in front and downstream the racks and head loss coefficients [25]. Furthermore, Berger [29] measured the head loss Δh at horizontal trash racks with rectangular bars and observed also an increase of Δh with increasing angle α. In contrast, this effect of rack angle α was not observed at angled streamwise (vertical) bar racks [36]. Raynal et al. [36] showed that ζ as well as the velocity distribution in front of the rack is not affected by the rack angle at a trash rack with vertical bars. It is assumed that the vertical trash racks deflects the incoming flow along the full length of the rack and therewith has as a higher influence on the direction of the flow. A direct comparison between vertical and horizontal racks under identical boundary conditions including an intensive velocity measurement should be part of further research to clarify quantify the influence of the orientation of the racks.

4.3. Prediction of Head Loss Coefficients of Angled Racks with Empirical Equations

The head loss coefficients for the CBTR and the FFF estimated by Equation (10) roughly correspond with the measured head loss coefficients, but systematic deviations arise for configurations with the highest blockage ratio ($p = 0.5$) and low angles (particularly for CBTR and $\alpha = 30°$ or FFF and $\alpha = 20°$). For those configurations, the estimated ζ values are too high. It is noted, that the corresponding original approach of Meusbuger [18] in Equation (6) uses the vertical inclination of the rack β rather than the horizontal angle α. However, the influence of β on ζ in Equation (6) is based on measurements of preceding studies [20] with a range of $\beta = 60°$ to $90°$. In contrast, Raynal et al. [25] observed a similar overestimation of Equations (9) and (10) at lower inclinations of vertical trash racks and derived the effect of β on ζ on blockage through bars with $\sin^2(\beta)$ in their newly developed formula [25].

For the current results, the regression in Equation (11) similar to Equation (6) was fitted to the measured head loss coefficients to improve the estimation and to quantify the dependency of rack angle and blockage on ζ (Table 4). As expected, the effect of rack angle α is not sufficiently described by the standard sinus function applied in Equation (6). With $\sin^{1.70}(\alpha)$ for CBTR and $\sin^{1.96}(\alpha)$ for FFF configurations (k_2 in Table 4), it is obvious that rack angle α has a stronger effect on head loss for the observed rack options. Furthermore, the fitted functions correspond more to the approach of

Raynal et al. [25] for vertical inclined trash racks with low values of β [25]. Moreover, the exponents of the blockage term $\frac{p}{1-p}$ with 1.3 and 1.44 for CBTR and FFF, respectively (k_1 in Table 4), are slightly below the value of 1.5 in Equation (6). Particularly, for the CBTR configurations, it fits better to Kirschmer's description with $\frac{s}{b}^{(4/3)}$, where transversal elements of trash racks are unattended [20]. However, the higher exponent of the blockage term for FFF compared to CBTR seem to be a result of the cable vibrations, which are intensified particularly at higher blockage ratios. The last regression coefficient k_0 corresponds to the bar shape coefficient k_F proposed by Kirschmer [20]. For the CBTR configuration, it is 1.8 (k_0 in Table 4), which matches very well with the bar shape coefficient k_F of 1.79 for circular bar shapes given by Kirschmer [20]. For the FFF configurations, k_0 is comparably higher (3.19, Table 4), which may be again due to flow-induced vibrations and a stronger interaction between the cables.

4.4. Transferability of the Results to Technical Applications and Outlook

The multiple regression model of the results should not be considered as a new developed formula, since the studied parameter range of p (spacing b, respectively) and α is limited. Moreover, other influences at real hydro power applications, e.g., blockage through debris clogging, angled approach flow or additional structures such as a bypass, are not considered. Thus, a wider parameter set should be further analysed in laboratory but also in real application in order to validate or enhance the proposed model into a formula. Meanwhile, the effect of flow-induced cable vibrations on head loss in quantitative terms is not directly transferable to full-scale applications of the FFF. The amplitudes of cable vibrations are dependent on parameters such as the preload forces and frequency, cable length and flow velocities on-site [14]. Again, further research is necessary to include the effect of vibrations in the formula. Nevertheless, the results show that the additional effect of flow-induced cable vibrations on head loss is not negligible.

The transferability of the results to conventional horizontal trash racks is difficult to assess, since only a few studies about head loss on these trash racks are published. In this respect, two investigations about head loss at horizontal trash racks can be roughly compared with the rack option CBTR of this study. Szabo-Meszaros et al. [27] investigated the head loss of two angled horizontal trash racks with $\alpha = 30°$ and a volume based blockage ratio (Equation (1) in [27]) of 0.35 and 0.32 with rectangular and hydrodynamic bar profiles, respectively [27]). In addition, Berger [29] investigated head loss at horizontal trash racks with a wider parameter range of p and α and rectangular bar profiles [29]. In both studies [27,29], the total head loss coefficients (either reported [27] or calculated from reported head loss and velocity [29] according to Equation (3)) are noticeably higher than the results presented herein at similar geometric configurations. However, the experimental designs in [27,29] differ considerably from this study (e.g., different bar shapes). Additionally, the reported head loss in [27,29] includes head losses through deflection into a bypass and partly through additional transversal elements. The latter refers to spacers and supporting structures, which are necessary at conventional horizontal trash racks. In this respect, the additional effect on head loss depends also on their geometry (e.g., circular spacers are independent from the rack angle [25], while other bar shapes have an adverse effect on ζ with decreasing α. In summary, it reveals the need to investigate a wider parameter set and to address each component separately.

5. Conclusions

The presented experimental study investigated the flexible fish fence (FFF), a physical barrier created by horizontally arranged steel cables, and an angled horizontal trash rack with circular bars (CBTR). Thereby, particularly the effect of bar spacing and rack angle was examined and a comparison of both rack options at an angle of 30° allowed assessing the influence of flow-induced cable vibrations (at the FFF) on the head loss coefficient ζ. The observed ζ-values were compared with the head loss (Equations (9) and (10)) of Kirschmer [20] and Meusburger [18] and a modified approach for angled horizontal trash racks is introduced. Therefore, additional coefficients to the original Equation (10) were fitted based on the observed data. This allowed quantifying the relationship between the independent variables bar spacing and rack angle to the head loss coefficient based on Equation (11). The following conclusions summarise the findings of this experimental investigating:

- Head loss coefficient ζ is independent from the Bar–Reynolds number in the studied range of Re_b of 750–3500 and scale effects can be neglected.
- The coefficient ζ is significantly affected by the blockage ratio and the rack angle (Section 3.4, Table 3). The strong increase of head loss with decreasing bar spacings, which are necessary for fish protection, can be countered by designing lower rack angles ($\alpha \leq 45°$).
- With increasing blockage ratios, the head loss coefficient at the FFF is up to 53% higher compared to the CBTR. This phenomenon is likely resulting from the effect of flow-induced cable vibrations and hence a further increase of blockage. Since amplitudes and frequencies of the vibrations are depending on parameters such as preload forces, cable length or flow velocity, the transferability to full-scale applications is limited.
- Head loss at the CBTR and FFF can be roughly estimated with a modified version of Equation (10) originally published by Meusburger [18], where the horizontal angle is used instead of the vertical rack inclination. However, the comparison of measured and estimated head loss revealed a systematic bias, which is more pronounced for rack options with low angles and high blockage.
- An adaption of Equation (11) with the coefficients given in Table 4 allows estimating the head loss coefficient for the investigated options CBTR and FFF. It better takes the specific characteristics of both rack options (low rack angles, high blockage ratios, and bars vs. cables) into account.

Author Contributions: H.B. and M.A. were responsible for the conceptualisation of the hydraulic investigation. H.B. designed and measured the data in the scale model test. H.B. and R.G. analysed the data and wrote the paper.

Funding: This publication was funded by Austrian Science Fund (FWF) grant number J3918. The experimental work was funded by the Austrian Research Promotion Agency (FFG) within the framework of the Energy Research Programme 2014. The overall research project Flexible Fish Fence was carried out in collaboration with the University of Innsbruck (Unit of Hydraulic Engineering, Unit of Applied Mechanics), University of Natural Resources and Life Sciences, Vienna (Institute of Hydrobiology and Aquatic Ecosystem Management) and the company Albatros Engineering GmbH.

Acknowledgments: The authors want to thank the Master-students Sebastian Ritsch and Christian Goessel for their contribution to the experimental work.

Conflicts of Interest: The authors declare no conflict of interest.

Notation

A	= area (m^2)	α	= rack angle in relation to the vertical wall (°)
b	= spacing between the bars (m)	β	= rack angle in relation to the ground plane (°)
B	= width of the flume (m)	λ	= scale factor (-)
F	= Froude number (-)	ρ	= mass density of water $\approx$ 997 (kg m^{-3})
g	= gravity acceleration (m s^{-2})	ν	= kinematic viscosity (m^2 s^{-1})
h	= water depth (m)	ζ	= head loss coefficient (-)
h_v	= head loss (m)	ζ^*	= total head loss coefficient (-)
k_F	= bar shape coefficient (-)	ζ_p	= predicted head loss coefficient
k	= constant	ζ_m	= measured head loss coefficient
l	= bar length (in cross section) (m)	$\zeta_{v,cont}$	= ζ due to supports and surface friction (-)
$p_{1,2}$	= pressure (Pa)	Δp	= differential pressure $= p_1 - p_2$ (Pa)
p	= blockage ratio (-)	CBTR	circular bar trash rack
Q	= discharge (m^3 s^{-1})	DPT	differential pressure transducer
Re	= Reynolds Number (-)	FFF	Flexible Fish Fence
Re_b	= Bar–Reynolds Number (-)	PG	point gauge
s	= diameter of the bar/cable (m)	US	ultrasonic sensor
$v_{1,2}$	= velocity (m s^{-1})		
z	= elevation (m)		

References

1. Williams, J.G.; Armstrong, G.; Katopodis, C.; Larinier, M.; Travade, F. Thinking like a fish: A key ingredient for development of effective fish passage facilities at river obstructions. *River Res. Appl.* **2012**, *28*, 407–417. [CrossRef]
2. Cada, G.F. The development of advanced hydroelectric turbines to improve fish passage survival. *Fisheries* **2001**, *26*, 14–23. [CrossRef]
3. Schilt, C.R. Developing fish passage and protection at hydropower dams. *Appl. Anim. Behav. Sci.* **2007**, *104*, 295–325. [CrossRef]
4. Larinier, M.; Travade F. Downstream migration: problems and facilities. *Bulletin Français de la Pêche et de la Pisciculture* **2002**, *364*, 181–207. [CrossRef]
5. Ebel, G. *Fischschutz und Fischabstieg an Wasserkraftanlagen-Handbuch Rechen-und Bypasssysteme. Ingenieurbiologische Grundlagen, Modellierung und Pronose, Bemessung und Gestaltung [Fish Protection and Downstream Passage at Hydro Power Stations-Bioengineering Principles, Modelling and Prediction, Dimensioning and Design]*, 4th ed.; Mitteilungen aus dem Büro für Gewässerökologie und Fischereibiologie: Halle, Germany, 2018.
6. Dumont, U. Zum Stand der Technk Einer Ökologisch Angepassten Wasserkraftnutzung [State of the Art about An Ecologically Sustainable Use of Hydro Power]; In Proceedings of the Seminar Gewässer-verträglicher Wasserkraftausbau, Renexpo, Salzburg, Austria, 28 November 2013; RENEXPO INTERHYDRO: Salzburg, Austria, 2013.
7. Cuchet, M. Fish Protection and Downstream Migration at Hydropower Intakes—Investigation of Fish Behavior under Laboratory Conditions. Ph.D. Thesis, Technical University of Munich (TUM), Munich, Germany, 2014; ISBN 978-3-943683-08-0.
8. Kriewitz-Byun, C.R. Leitrechen an Fischabstiegsanlagen: Hydraulik und fischbiologische Effizienz [Guidance Screens at Fish Protection Facilities—Hydraulics and Fishbiological Efficiency]. Ph.D. Thesis, ETH Zürich, Zürich, Switzerland, 2015. [CrossRef]
9. Boettcher, H.; Brinkmeier, B.; Aufleger, M. Flexible Fish Fences. In Proceedings of the Norwegian University of Science and Technology 10th International Symposium on Ecohydraulics, Trondheim, Norway, 23–27 June 2014.
10. Gabl, R.; Innerhofer, D.; Achleitner, S.; Righetti, M.; Aufleger, M. Evaluation criteria for velocity distributions in front of bulb hydro turbines. *Renew. Energy* **2018**, *121*, 745–756. [CrossRef]
11. Umweltbundesamt, Bundesministerium für Umwelt, Naturschutz, Bau und Reaktorschutz, Ecologic Institute. *Forum Fischschutz und Fischabstieg—Empfehlungen und Ergebnisse des Forums "Fischschutz und Fischabstieg" [Forum Fish Protection and Downstream Migration—Recommendations and Results]*; Umweltforschungsplan des Bundesministeriums für Umwelt, Natuschutz, Bau und Reaktorsicherheit: Bau, Germany, 2015.
12. Böttcher, H.; Unfer, G.; Zeiringer, B.; Schmutz, S.; Aufleger, M. Fischschutz und Fischabstieg–Kenntnisstand und aktuelle Forschungsprojekte in Österreich [Fish protection and downstream migration: current state of knowledge and research projects in Austria]. *Österreichische Wasser-und Abfallwirtschaft* **2015**, *67*, 299–306. [CrossRef]
13. Böttcher, H.; Gabl, R.; Ritsch, S.; Aufleger, M. Experimental study of head loss through an angled fish protection system. In Proceedings of the 4th IAHR Europe Congress, Liege, Belgium, 27–29 July 2016; Dewals, B., Ed.; CRC Press: Liege, Belgium, 2016; pp. 637–642.
14. Unit of Applied Mechanics, University of Innsbruck. *Projektbericht—Berechnung der Seilschwingungen beim Seilrechen [Research Project Report—Calculation and Modelling of Cable Vibrations at the Flexible Fish Fence]*; University of Innsbruck: Innsbruck, Austrian, 2017.
15. Gabl, R.; Achleitner, S.; Neuner, J.; Aufleger, M. Accuracy analysis of a physical scale model using the example of an asymmetric orifice. *Flow Meas. Instrum.* **2014**, *36*, 36–46. [CrossRef]
16. Adam, N.J.; De Cesare, G.; Nicolet, C.; Billeter, P.; Angermayr, A.; Valluy, B.; Schleiss, A.J. Design of a Throttled Surge Tank for Refurbishment by Increase of Installed Capacity at a High-Head Power Plant. *J. Hydraul. Eng. ASCE* **2018**, *144*, 05017004. [CrossRef]
17. Gabl, R.; Righetti, M. Design criteria for a type of asymmetric orifice in a surge tank using CFD. *Eng. Appl. Comput. Fluid Mech.* **2018**, *12*, 397–410. [CrossRef]

18. Meusburger, H. Energieverluste an Einlaufrechen Von Flusskraftwerken [Head losses at intakes of run-of-river hydropower plants]. Ph.D. Thesis, Mitteilungen der Versuchsanstalt fur Wasserbau, Hydrologie und Glaziologie an der Eidgenossischen Technischen Hochschule Zürich, Zürich, Switzerland, 2002.

19. Gabl, R.; Gems, B.; Birkner, F.; Hofer, B.; Aufleger, M. Adaptation of an Existing Intake Structure Caused by Increased Sediment Level. *Water* **2018**, *10*, 1066. [CrossRef]

20. Kirschmer, O. *Untersuchungen über den Verlust an Rechen [Study on Head Loss At Trash Racks]*; Mitteilungen Hydraulisches Institut München: München, Germany, 1926; Nr. 1.

21. Zimmermann, J. *Widerstand Schräg Angeströmter Rechengitter [Resistance of Trash Racks Caused by Oblique Inflow]*; Mitteilungen der Universität Fridericana Karlsruhe, Theodor-Rhebock-Flußbaulaboratorium: Karlsruhe, Germany, 1969; Volume 157.

22. Spangler, J. *Untersuchung über den Verlust an Rechen bei schräger Zuströmung [Study about head Loss at Obliquely Approached Trash Racks]*; Mitteilungen des Hydraulischen Instituts der TH München: München, Germany, 1929.

23. Idelchik, I.E. *Handbook of Hydraulic Resistance Coefficients of Local Resistance and of Friction*; U.S. Department of Commerce National Technical Information Service (NTIS): Springfield, VA, USA, 1960.

24. Clark, S.P.; Tsikata, J.M.; Haresign, M. Experimental study of energy loss through submerged trashracks. *J. Hydraul. Res.* **2010**, *46*, 113–118. [CrossRef]

25. Raynal, S.; Courret, D.; Chatellier, L.; Larinier, M.; David, L. An experimental study on fish-friendly trashracks—Part 1. Inclined trashracks. *J. Hydraul. Res.* **2013**, *51*, 56–66. [CrossRef]

26. Raynal, S.; Chatellier, L.; Courret, D.; Larinier, M.; David, L. An experimental study on fish-friendly trashracks—Part 2. Angled trashracks. *J. Hydraul. Res.* **2013**, *51*, 67–75. [CrossRef]

27. Szabo-Meszaros, M.; Navaratnam, C.U.; Aberle, J.; Silva, A.T.; Forseth, T.; Calles, O.; Fjeldstad, H.-P.; Alfredsen, K. Experimental hydraulics on fish-friendly trash-racks: an ecological approach. *Ecol. Eng.* **2018**, *113*, 10–20. [CrossRef]

28. Albayrak, I.; Kriewitz, C.R.; Hager, W.H.; Boes, R.M. An experimental investigation on louvres and angled bar racks. *J. Hydraul. Res.* **2018**, *46*, 59–75. [CrossRef]

29. Berger, C. Rechenverluste und Auslegung von (elektrifizierten) Schrägrechen Anhand Ethohydraulischer Studien [Screen Losses and Design of Inclined (and Electrified) Screens with Horizontal Bars on the Basis of Ethohydraulic Studies]. Ph.D. Thesis, Technische Universität Darmstadt, Darmstadt, Germany, 2018.

30. Heller, V. Scale effects in physical hydraulic engineering models. *J. Hydraul. Res.* **2011**, *49*, 293–306. [CrossRef]

31. Blevins, R.D. *Applied Fluid Dynamics Handbook*; Van Nostrand Reinhold Company, Inc.: New York, NY, USA, 1984; ISBN 978-1575241821.

32. Naudascher, E. *Hydraulik der Gerinnebauwerke [Hydraulics of Open Channel Flow Structures]*; Springer: Berlin/Heidelberg, Germany, 1992; ISBN 978-3211823668.

33. Krzyzagorski, S.; Gabl, R.; Seibl, J.; Böttcher, H.; Aufleger, M. Implementierung eines schräg angeströmten Rechens in die 3D-numerische Berechnung mit FLOW-3D [Implementation of an angled trash rack in the 3D-numerical simulation with FLOW-3D]. *Österreichische Wasser-und Abfallwirtschaft* **2016**, *68*, 146–153. [CrossRef]

34. Naudascher, E.; Rockwell, D. *Flow-Induced Vibrations: An Engineering Guide*; CRC Press: Boca Raton, FL, USA, 1993; ISBN 978-0486442822.

35. Tanida, Y.; Okajima, A.; Watanabe, Y. Stability of a circular cylinder oscillating in uniform flow or in a wake. *J. Fluid Mech.* **1973**, *61*, 769–784. [CrossRef]

36. Raynal, S.; Chatellier, L.; Courret, D.; Larinier, M.; David, L. Streamwise bars in fish-friendly angled trashracks. *J. Hydraul. Res.* **2013**, *52*, 426–431. [CrossRef]

Article

Exploring Explicit Delay Time for Volume Compensation in Feedforward Control of Canal Systems

Wenjun Liao [1], Guanghua Guan [1,*] and Xin Tian [2]

[1] State Key Laboratory of Water Resources and Hydropower Engineering Science, Wuhan University, Wuhan 430072, China; LWJ@whu.edu.cn

[2] Department of Water Management, Delft University of Technology, 2600 GA Delft, The Netherlands; X.Tian@tudelft.nl

* Correspondence: GGH@whu.edu.cn; Tel.: +86-027-68776389

Received: 5 May 2019; Accepted: 21 May 2019; Published: 23 May 2019

Abstract: In the open channel control algorithm, good feed-forward controllers will reduce the transition time of the canal and improve performance. Feedforward control algorithms based on active storage compensation are greatly affected by delay time. However, there is no literature comparing the three most commonly used algorithms, namely volume step compensation, dynamic wave principle and water balance models, under the operation mode of constant water level downstream. In order to compare the existing three algorithms, and to avoid storage calculation by calculating the constant non-uniform water surface line or identification of relevant parameters, combined with the open channel constant gradient flow theory with the storage compensation algorithm, a delay time explicit algorithm is proposed in this study. Tested on the first canal pool of the American Society of Civil Engineers (ASCE) Test Canal 2, the performance of the delay time explicit algorithm is assessed and compared to that of the three conventional algorithms. In the current water intake plan, i.e., in the second hour, the intake begins to take 1.2 m^3/s, and the upstream flow of the canal pool changes from 6 m^3/s to 7.2 m^3/s, among the three existing algorithms, the volume step compensation algorithm has better performance in terms of time to achieve stability, i.e., 1.25 h. The actual adjusted storage accounts for 99.6% of the target adjusted storage, which can basically meet the requirement of compensated storage of the canal pool. The delay time explicit algorithm only needs 1.47 h to stabilize the regulation system. The fluctuation of water level and discharge in the regulation process is small. The actual adjusted storage accounts for 99.6% of the target adjusted storage, which can basically meet the requirement of compensated storage for the canal pool. The delay time calculated by explicit algorithm can provide references for the determination of delay time in feedforward control.

Keywords: canal pool; delay time; volume compensation; feedforward control; downstream constant water level

---

1. Introduction

In China, the total annual water consumption for agriculture is 3.7×10^{11} m^3, accounting for 61% of the total water consumption for industry, agriculture, life and ecology [1]. However, the effective utilization coefficient of irrigation water is only 0.5 [2], and the potential of increasing agricultural production and saving water is still huge. One of the main ways to improve the water utilization efficiency of water conveyance is the automatic operation of the channel system. In the channel control algorithm, feedforward control algorithms adjust the input of the system in advance by predicting the possible state deviation in the future operation of the control system. This open-loop algorithm does not need to compare with the actual monitoring data of the channel operation. In order to solve the feedforwardcontrol problem of canals, gate stroking was proposed in 1969 [3]. According to the offtake discharges' schedules, this method could determine the discharge variations of check structures by inverse solutions of unsteady open-channel flow equations. However, it was difficult to get a reasonable solution sometimes because it needs some extreme or unrealistic inflow variation, or no solution at all when calculating [4]. Nowadays, the open-loop algorithm of canals mainly considers the difference of stable storage in different flow states, and actively compensate the difference of storage through the opening and closing lag between gates [5]. Compared with gate stroking, this algorithm has an advantage in calculation, i.e., there is basically no extreme or unrealistic solution in the solution process. However, if the opening and closing lag time between gates is too small and the surge wave has not arrived when the intake gate is opened, the water supply will be insufficient, otherwise the water will be wasted and the excess water will be discharged through the downstream canal pool. Combined with the actual project, Wei simulated the influence of different feedforward control time on the water level fluctuation at the intake, and proposed a feedforward control time calculation method to effectively reduce the water level fluctuation at the intake [6]. In order to solve the nonlinear optimization problem with constraints on the gate movements in feedforward control, the sequential quadratic problem (SQP) method is used [7]. In order to shorten the time necessary to stabilize the new flow rate at the buffer reservoir in a traditional automated upstream controlled canal, the method is proposed which requires calculated, remote manual adjustments to all the canal check structure gate positions in addition to two flow rate changes made at the head of the canal, followed by a return to automated upstream control [8].

When designing feedforward control algorithm, the appropriate gate opening and closing lag time will improve the response speed and control quality of the canal system. However, when water waves propagate in canal pools, there are many complex phenomena, such as reflection, superposition and energy attenuation. It is difficult to accurately estimate the time when water waves propagate downstream [5]. The delay time can be calibrated by experiments, but it is different in different sizes of canal pools. If the delay time of each canal pool in multi-canal pools is calibrated, the calibration workload will be large, so it is necessary to explore the feedforward control algorithm with better control effect. At present, there are three typical feedforward control algorithms commonly used to determine the opening and closing lag time between gates: (1) the volume step compensation algorithm: the delay time of water wave from upstream to downstream can be deduced according to the change of flow and storage required by water demand [5]; (2) dynamic wave principle algorithm: the delay time of water wave propagation is estimated by using the characteristics of dynamic wave, i.e., the velocity of the flow and the wave velocity in the initial state of the canal pool [9]; (3) water balance model algorithm: simplifying the water intake and supply process of the canal pool into a linear superposition process; according to the principle of water balance, the calculation formula of optimal intake time is obtained, and the optimal intake time is obtained by identifying the parameters of the calculation formula so as to ensure that the storage capacity of the canal pool is compensated by the intake opening [10].

Theoretically, the optimal water intake time calculated by water balance model algorithm is the same as the delay time calculated by volume step compensation algorithm. However, the water balance model algorithm obtains the optimal water intake time by identifying the parameters of the canal pool, and the volume step compensation algorithm calculates the delay time by estimating the required volume's change by assuming a constant non-uniform flow line relationship between the beginning and the end, in practice, it is necessary to discretize the canal system model to solve the water surface line. The accuracy of parameter identification and the error of surface line calculation will affect the results of the two algorithms. In general, the volume step compensation algorithm needs to calculate the storage of the canal pool according to the water demand, and then calculate the delay time. The calculation amount of storage will increase with the increment of the series canal pools. The dynamic wave principle algorithm needs to calculate the storage and determine the flow of the canal pool when the storage capacity is compensated according to the calculated lag time. When the volume compensation occurs, the problem of excessive flow regulation may be brought. The water balance model algorithm needs to separate the water intake and supply linearly according to the water demand, identify the corresponding parameter and then calculate the optimal water intake time. The workload of parameter identification will increase with the increase of the number of canal pools and the water demands. For most open channels, the change of water surface profile is small when the discharge changes. Moreover, most open channel sections are trapezoidal with regular shapes. Based on the theory of constant gradient flow in open channel and volume compensation algorithms, this paper intends to seek an algorithm, i.e., a delay time explicit algorithm, in order to calculate the delay time directly according to the geometric size and flow state of the canal pool by using linear formula, so as to avoid the workload caused by the calculation of storage and parameter identification of the above algorithms, and to provide a reference value for the calculation of delay time.

The downstream constant water level operation is widely used in channel control. It is the operation mode often adopted by many water transfer projects such as the middle line and the east line of China South-North Water Transfer Project. The control effects of the three algorithms (volume step compensation algorithm, dynamic wave principle algorithm, water balance model algorithm) in this mode of operation have not been compared by the literature. Further, the application of the delay time explicit algorithm needs to be discussed. Therefore, this paper will discuss the feedforward control algorithms for the downstream constant water level operation. Experiments can effectively measure the delay time of the canal pool. However, when using experiments to verify different delay time algorithms, the accuracy of experimental measurement will affect the comparison of different delay time algorithms. For example, when the bottom width of the canal pool is large, the measurement of flow and water level is difficult, and there are some errors in the measurement. However, the bottom width of the laboratory test canal is usually small, and the conclusion of the test should be further calibrated and discussed in the checking calculation of the large-scale canals. In the context of an engineering application, the numerical solution of one-dimensional unsteady flow in open channel is relatively mature and the calculation results are relatively reliable. The existing literatures exploring the delay time algorithm of canal pools mostly use numerical simulation to evaluate the effect of the algorithm [5–8,10]. Different delay time algorithms are discussed by numerical method in this paper. Based on the simulation and control software V1.0 [11], the article uses canal Pool 1 of the American Society of Civil Engineers (ASCE) Test Canal 2 [12] modeling to compare the differences of the four algorithms and recommends the algorithm with a better control effect.

2. Delay Time Algorithm

2.1. Volume Step Compensation Algorithm

The calculation of canal pool's storage is related to the geometric parameters, roughness, upstream and downstream boundary conditions of the canal pool, water intake plans and so on [5]. On this basis, the volume step compensation algorithm assumes that a series of intermediate stable states in the operation process exist. As shown in Figure 1, in the case of a single canal pool, according to the water intake plan, the flow should be changed by Δq_d at t_d. From the assumed initial state, it can be deduced that the storage of the canal pool needs to be adjusted by ΔV, i.e., the initial flow Q_0 is adjusted $Q_0 + \Delta Q_s$ at t_s. At t_d, the storage of the canal pool can be adjusted completely. The meaning of the volume step compensation is to adjust the discharge of the canal pool once in advance. When the intake begins to take water, the adjusted storage can satisfy the storage required by the current intake of the canal pool, and the flow need not be adjusted again. At this time, ΔQ_s is equal to Δq_d. The delay time $\Delta \tau$ is the difference between t_s and t_d. Its calculation is shown in Equation (1).

$$\Delta \tau = \Delta V / \Delta q_d \tag{1}$$

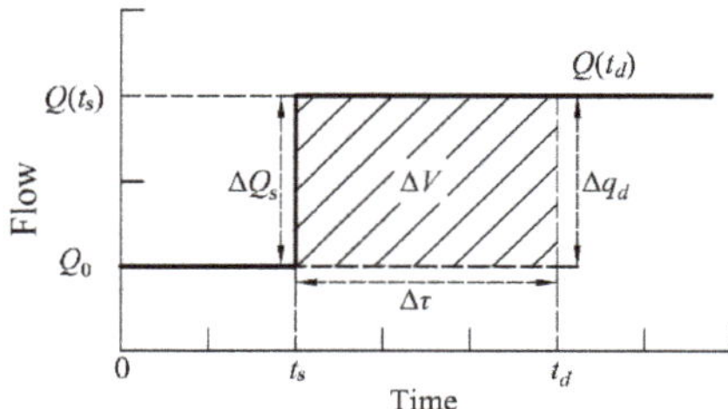

Figure 1. Schematic diagram of volume step compensation algorithm.

2.2. Dynamic Wave Principle Algorithm

The delay time of the volume step compensation algorithm is deduced according to the storage and flow required by currrent water demand of the canal pool. Even for a given canal pool, the delay time will vary with the different intake water. In the process of channel operation, when the downstream water demand plan is determined, the value of volume compensation is determined, but its realization is not the only way. For example, the storage compensation can be carried out with small flow change and a long delay time, and vice versa, with a large flow change and shorter delay time. The phenomenon of water wave lag is mainly related to the movement of water wave in the canal pool. Corriga [9] calculates the delay time by using the water wave characteristics of the initial state, as shown in Equation (2). Because the average velocity and wave velocity of the initial state are adopted without considering the attenuation of energy in the process of water wave propagation, the delay time calculated by the algorithm is the smallest [5] in the methods mentioned in this paper.

$$\Delta\tau_{DW} = L/(v_0 + c_0) \tag{2}$$

In Equation (2), $\Delta\tau_{DW}$ is the delay time of dynamic wave principle calculation; L is the length of canal pool, (m); v_0 is the initial average velocity of canal, (m/s); c_0 is the initial average velocity, (m/s).

In the dynamic wave principle algorithm, the flow regulation can be deduced according to the ideal delay time [9] on the basis of known storage variation. In fact, the flow regulation is not equal to the target value, so there is secondary regulation, i.e., to adjust the flow to the target value after reaching the target time. As shown in Figure 2, the initial discharge Q_0 of the canal pool is adjusted to $Q(t_s)$ at t_s, and from ΔQ_d to the target value $Q(t_d)$ at t_d. The purpose of secondary regulation is to avoid the waste of water when the intake opens and the inflow of the canal pool exceeds the flow required by the intake.

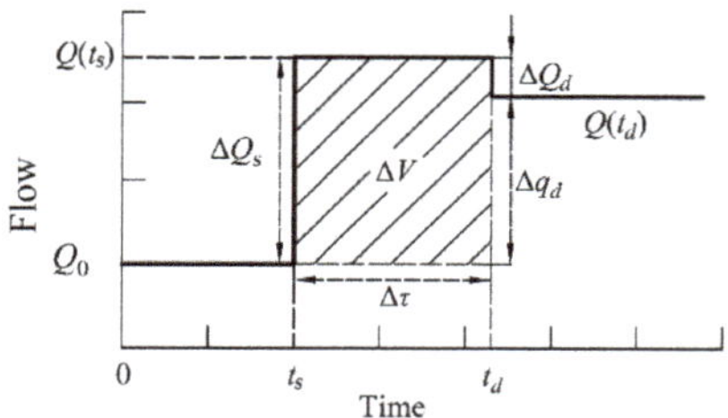

Figure 2. Schematic diagram of volume compensation algorithm with two adjustments.

2.3. Water Balance Model Algorithm

When the downstream users have water requirements, the upstream canal pools will transport the corresponding water to the downstream and be used by the downstream users to complete the whole water distribution process. In the channel operation, downstream water intake produces upstream precipitation wave and upstream water supply produces upstream water wave, which is superimposed by non-linear superposition. It is difficult to determine feedforward control time when unsteady flow is used for calculation. In order to calculate the delay time, the analytical formula of downstream canal pool response is introduced into the algorithm, as shown in Equation (3) [13].

$$q(x,t) = 1 - e^{-[t-\tau(x)]/K(x)} \tag{3}$$

In the formula, $q(x,t)$ is the flow at canal pool x at t, (m³/s); $\tau(x)$ is the delay time at canal pool x, (s); $K(x)$ is the time parameter at canal pool x, (s). In order to determine the time of water intake, the process of water intake and supply is simplified to a linear equation and superimposed [10], as shown in Equations (4)–(6).

When $t \leq \tau$,

$$q_d^{(t)} = 0 \tag{4}$$

When $\tau < t < T_w$,

$$q_d^{(t)} = \left(1 - e^{-(t-\tau)/K}\right)\delta Q_u \tag{5}$$

When $t \geq T_w$,

$$q_d^{(t)} = \left(1 - e^{-(t-\tau)/K}\right)\delta Q_u - q_{w,0}\left(1 - e^{-(t-\tau)/K_p}/(1 + k_d a)\right) \tag{6}$$

where K is the time constant introduced in the process of water wave propagation; K_p is the time parameter introduced by the change of downstream intake flow; $q_d^{(t)}$ is the change of downstream flow at t time, (m³/s); δQ_u is the upstream water supply, (m³/s); k_d is the flow coefficient, indicating the sensitivity of time parameter K to the influence of downstream flow and water level boundary, (m²/s); $q_{w,0}$ is the downstream intake flow, (m³/s); T_w is the optimal intake time; a is the sudden drop of water level caused by intake, (s/m²), τ is the delay time of parameter identification of water balance model. In order to minimize the downstream discarded water, the upstream water supply quantity δQ_u is equal to the downstream water intake quantity $q_{w,0}$. At this time, the canal pool does not produce discarded water, and the calculation formula of the optimal water intake time is demonstrated in Equations (7) and (8). In Equation (7), t_w is the difference between the optimal water intake time and the delay time of the canal pool. In the actual simulation process, K, τ, K_p, a and k_d parameters can be obtained by the parameter identification method.

$$t_w = K - K_p/(1 + k_d a) \tag{7}$$

$$T_w = t_w + \tau \tag{8}$$

3. Delay Time Explicit Algorithm

According to the volume step compensation algorithm, the storage capacity of the canal pool is calculated mainly by the relationship between the water surface of the constant and non-uniform flow, which requires numerical discrete calculation. In order to avoid the discrete calculation needed for deducing storage, this section seeks to find a formula that can directly calculate delay time according to canal pool geometry size and flow state, namely delay time explicit algorithm. In general, the compensatory storage of canal pool is affected by the flow variation of water demand. According to Equation (1), when the flow variation of water demand in canal pool is determined, the delay time $\Delta\tau$ of the canal pool is proportional to the compensation value ΔV of water demand in canal pool. In this paper, the open channel constant gradient flow theory [14] is used for analysis. When the canal pool is running at the downstream constant water level, the volume of storage change in the canal pool can be approximately considered as a wedge-shaped water body rotating around the control point

(Figure 3). For the artificial trapezoidal section canal pool, the hydraulic gradient J of the open channel steady gradient flow can be calculated by Equation (9).

$$J = \frac{Q^2}{K'^2} = \frac{v^2}{C^2 R} = \frac{Q^2 n^2}{A^2 R^{4/3}} \tag{9}$$

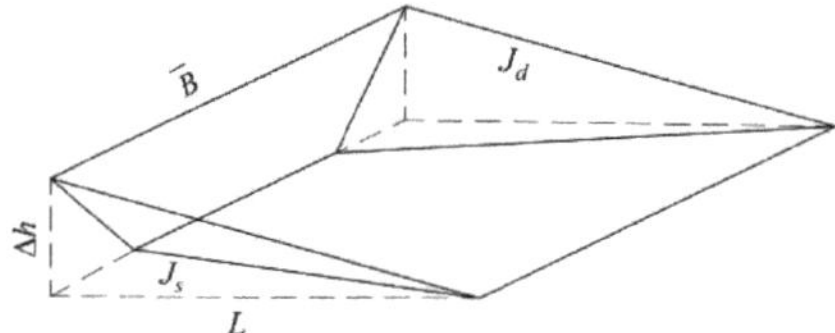

Figure 3. Schematic diagram of wedge-shaped water body.

In Equation (9), Q is the discharge of the canal pool, (m^3/s); K' is the flow modulus, (m^3/s); v is the average velocity of the canal, (m/s); C is the Chezy coefficient; R is the hydraulic radius, (m); n is the roughness; A is the area of the control point, (m^2). When the bottom slope of the canal pool is gentle and the water demand changes a little, the difference of A, R and B upstream before and after the volume compensation of the canal pool is small and the average values of the two points are calculated.

The actual process of solving J is often complex. To simplify the solution, J is expanded into Taylor polynomial near a certain flow to approximate it. When the bottom slope of the canal pool is gentle and the change of water demand is small, the difference of J under different discharge is small, so Taylor polynomial only expands to the second term in Equation (10). ΔJ is the second item of hydraulic gradient developed by the Taylor polynomial, i.e., the difference between the hydraulic gradient J_d and J_s before and after the volume compensation. The estimation of the required compensation value ΔV of the canal pool is shown in Equations (11)–(13).

$$J_d = \frac{n^2 (Q_0 + \Delta q_d)^2}{\left(\overline{A}\right)^2 \left(\overline{R}\right)^{4/3}} = J_s + \Delta J + \dots \tag{10}$$

$$\Delta h = L(J_d - J_s) \approx L\Delta J \tag{11}$$

$$\Delta J = \frac{n^2}{\left(\overline{A}\right)^2 \left(\overline{R}\right)^{4/3}} \left(\Delta q_d{}^2 + 2Q_0 \Delta q_d\right) \tag{12}$$

$$\Delta V = \frac{1}{2}\overline{B}\Delta h L = \frac{\overline{B}L^2 n^2}{2\left(\overline{A}\right)^2 \left(\overline{R}\right)^{4/3}} \left(\Delta q_d{}^2 + 2Q_0 \Delta q_d\right) \tag{13}$$

In Equation (13), $\overline{A}$ is the average flow area upstream of the canal pool before and after the volume compensation, m^2; $\overline{B}$ is the average water surface width upstream of the canal pool before and after the volume compensation, m; $\overline{R}$ is the average hydraulic radius upstream of the canal pool before and after volume compensation, m. According to Equation (1), the relationship between the delay time $\Delta \tau$ and the flow of the change of water demand Δq_d can be attained. However, in practical engineering, due to the influence of length L, initial discharge Q_0 and different water demand variation, the delay time calculated by $\overline{A}$, $\overline{B}$, $\overline{R}$ is smaller than actual delay time. That is to say, there exists amplification coefficient α_{am} to correct the calculation results. In Equation (13), when the initial and final discharge of the canal pool is determined, the main factors affecting the required volume are the geometric parameters of the canal pool, including the length, width of the canal bottom and so on. Combining with the geometric parameters of the canal pool, this paper introduce an empirical formula of α_{am} on the length and width of the canal bottom to correct the delay time (Equation (14)). The relationship between the delay time $\Delta\tau$ and the flow Δq_d of the change of water demand is shown in Equation (15). The coefficients a and c are calculated in Equation (16).

$$\alpha_{am} = 8 \times 10^{-5} L + 0.02b + 1.05 \tag{14}$$

$$\Delta\tau = \frac{\Delta V}{\Delta q_d} = \alpha_{am}\frac{\overline{B}L^2 n^2}{2\left(\overline{A}\right)^2\left(\overline{R}\right)^{4/3}}\left(\Delta q_d^2 + 2Q_0\Delta q_d\right) = a\Delta q_d + c \tag{15}$$

$$a = \alpha_{am}\frac{\overline{B}L^2 n^2}{2\left(\overline{A}\right)^2\left(\overline{R}\right)^{4/3}}; c = \alpha_{am}\frac{\overline{B}L^2 n^2 Q_0}{2\left(\overline{A}\right)^2\left(\overline{R}\right)^{4/3}} \tag{16}$$

4. Simulation Settings

Considering that the delay time of multi-channel pools in series is affected by the coupling effect of canal pools [15], i.e., the function of a single gate will cause the change of water level and discharge of several canal pools upstream and downstream, and the action of each gate will influence each other, this simulation model of single canal pool is built, i.e., canal pool 1 in ASCE Test Canal 2 [12]. The length of the canal pool is 7 km, the bottom slope is 0.0001, the roughness is 0.02, the slope is 1.5, the width of the canal bottom is 7 m, and the intake is located the most downstream of the canal pool. The total simulation time is 24 hours and the time step is 3 min. The downstream flow of the canal pool remains unchanged at 6 m³/s. As shown in Figure 4, in the second hour, the intake begins to take 1.2 m³/s, and the upstream flow of the canal pool changes from 6 m³/s to 7.2 m³/s (Figure 4).

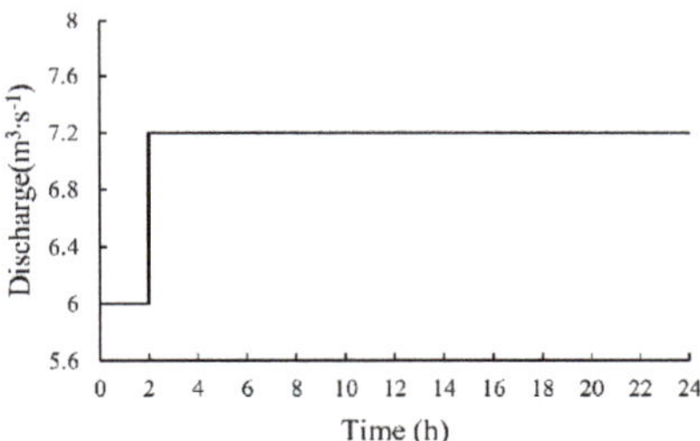

Figure 4. Change process of canal pool discharge according to intake plan.

The parameters identified by the water balance model algorithm [10] are shown in Table 1 [16]. The delay time calculated by four algorithms is shown in Table 2. Theoretically, the optimal intake time obtained by the water balance model algorithm should be equal to the delay time obtained by the volume step compensation algorithm, and the former algorithm needs to identify five parameters which are shown in Table 1. The delay, time calculated by the volume step compensation algorithm, needs to estimate the required volume by assuming a constant non-uniform flow line relationship between the initial and the final. It needs to discretize the canal pool in order to resolve the water surface line. The accuracy of parameter identification and the error of water surface line calculation will affect the results of the two algorithms. In the example, the delay time calculated by volume step compensation algorithm is 12.46 min less than the optimal intake time deduced by water balance model algorithm. According to the geometric parameters of the canal pool, the length and bottom width are substituted into Equation (14), and the amplification factor α_{am} of the canal pool is 1.75.

Table 1. Parameters identified by water balance model algorithm.

K/min	τ/min	K_p/min	a/(s·m⁻²)	k_d/(m²·s⁻¹)	T_w/min
67.6	29	63.8	0.026	15.9	51.46

Table 2. Delay time calculated by four algorithms.

Algorithm Name	Volume Step Compensation	Dynamic Wave Principle	Water Balance Model	Delay Time Explicit
Delay time/min	39	24	51.46	37

5. Results and Discussion

In order to compare the control effects of different algorithms, four performance indicators, i.e., transition time, maximum overshot flow, integral of absolute magnitude of error (IAE) and integrated absolute discharge change (IAQ) are selected. IAQ characterizes the fluctuation of the flow in the canal pool. The smaller the value, the more stable the system is, i.e., the smaller the flow fluctuation in the canal pool. IAE is the error accumulation of the water level deviating from the target water level. The smaller the value, the smaller the error range of the water level is. At this time, the smaller the gate actions of the channel control system are needed, and the faster the canal pool can reach the steady state [12]. The control performance indicators of the four algorithms are shown in Table 3. From the relationship of indicators, IAQ describes the flow fluctuation process in the regulation process of canal pool, and the range of flow fluctuation is affected by the maximum overshot flow. If the value of the maximum overshot flow is large, in order to make the flow of canal pool adjust to the steady flow state, the more regulations are needed, which causes the large the IAQ value. In Table 3, the maximum overshot flow of volume step compensation, water balance model and delay time explicit algorithm is close to the target flow of 7.2 m^3/s, and the IAQ value is small. However, the dynamic wave principle algorithm uses ideal delay time to compensate the storage by increasing the inflow of the canal pool. At this time, the maximum overshot flow exceeds the target flow by 0.644 m^3/s, which is reflected in the IAQ value. The IAQ value of the dynamic wave algorithm is about 43.67 times that of the volume step compensation algorithm, and the flow regulation fluctuates significantly.

Table 3. Performance indicators of canal pool.

Algorithm Name	Transition Time/h	Maximum Overshot Flow/($m^3 \cdot s^{-1}$)	IAE/%	IAQ/($m^3 \cdot s^{-1}$)
Volume step compensation	1.25	7.201	3.65×10^{-6}	0.030
Dynamic wave principle	2.55	7.844	1.49×10^{-5}	1.313
Water balance model	1.77	7.201	1.53×10^{-5}	0.033
Delay time explicit	1.47	7.201	3.62×10^{-6}	0.097

The simulation process of the canal pool discharge is shown in Figure 5. The volume step compensation and the delay time explicit algorithm can basically achieve the target flow in the canal pool by adjusting the flow only once. However, the dynamic wave principle uses the initial flow velocity and wave velocity of the canal pool to estimate the delay time, and the obtained delay time is small. In order to meet the demand of the current required storage volume, the flow of the canal pool is over-adjusted to 7.844m^3/s for compensation of the storage. After the water intake is opened, in order to avoid the waste of water, the flow of the canal pool is adjusted to the target flow for the second time. According to the compensated storage volume and the optimal intake time of the canal pool, the water balance model can calculate the flow value of the canal pool adjusted in advance. Affected by parameter identification, the optimal intake time of the calculation is slightly large. In order to meet the demand of the canal pool storage, the canal pool can adopt a small flow to compensate the storage capacity, i.e., the flow of the canal pool adjusted in advance is 6.843 m^3/s, which is less than the target flow (7.2 m^3/s). Then the flow of the canal pool increases to the target state, so as to avoid the phenomenon of insufficient water supply after opening the intake.

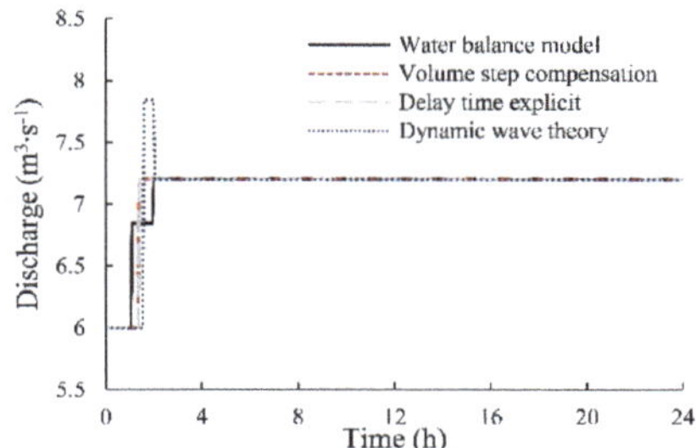

Figure 5. Simulation process of canal pool discharge.

The fluctuation of flow also affects the fluctuation of water level. The water level errors of downstream control points of the four algorithms are shown in Figure 6. Feedforward controllers are regulated according to plan and lacks real-time feedback mechanism. That is to say, the water level error of downstream control point of canal pool cannot converge to zero completely after adjustment, and the water level error will remain in a stable range. Taking the convergence of the water level error to zero as a reference, feedforward control algorithms compensate the storage of the canal pool by increasing the inflow ahead of time, and the water level at the downstream control point is higher than the target water level (the water level error is greater than 0). When the water intake begins, the water level at the downstream control point of the canal pool gradually falls and fluctuates. After stabilization, the water level error at the downstream control point remains within a stable range. Different feedforward control algorithms have different adjustment effects. After being adjusted by the volume step compensation or delay time explicit algorithm, the water level error of the downstream control point is stable at −0.0001 m, which is close to 0 basically. After being adjusted by the dynamic wave principle algorithm or water balance model, the water level error of downstream control point is stable at 0.002 m. IAE describes the accumulation of water level errors in the regulation process. In Table 3, the IAE of the water balance model algorithm is also the largest, and the IAE of the dynamic wave principle algorithm is slightly smaller than that of the water balance model algorithm, indicating that the errors accumulation of the water level deviating from the target water level is also greater. The IAE of the volume step compensation and the delay time explicit algorithm is smaller than the other two algorithms. The transition time of different control algorithms is affected by the fluctuation of flow and water level in the canal pool. When the IAQ and IAE of the canal pool are small, the time spent to reach the stable state of the canal pool is also less. In Table 3, the volume step compensation algorithm's IAE and IAQ are small, and the transition time is the shortest. The flow fluctuation of canal pool is large during the adjustment process of dynamic wave principle algorithm, which results in the longest transition time. When the fluctuation of flow in the canal pool is small (IAQ is small), the transition time is mainly affected by the water level fluctuation of canal pool. The IAE of the water balance model algorithm is slightly larger than that of the delay time explicit algorithm, so the transition time of the algorithm is slightly longer than that of the delay time explicit algorithm.

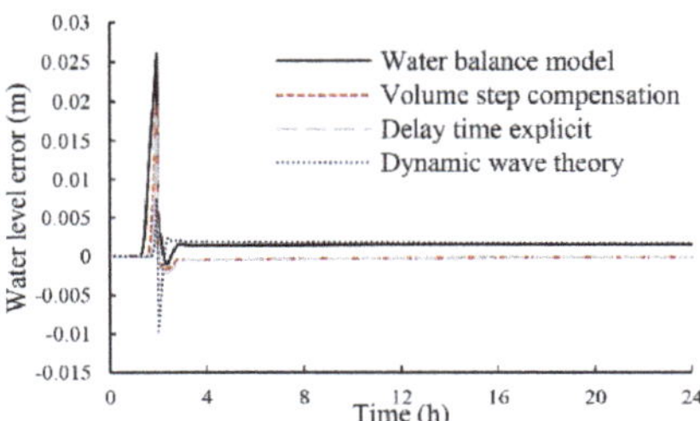

Figure 6. Water level error of downstream control point.

Feedforward control algorithms rely on water intake plan to adjust the input of control system in advance. If the amount of adjustment is too large in advance, it will cause unnecessary waste of water resources. The target compensation storage of the canal pool is 2628.4 m^3. The difference between the actual storage and the target storage of four algorithms are shown in Figure 7. Compared with the volume step compensation algorithm and delay time explicit algorithm, the actual compensation storage of the other two algorithms exceeds the target storage. The actual storage of the dynamic wave algorithm after water intake exceeds the target storage 121.79 m^3, the actual storage of the water balance model algorithm after water intake exceeds the target storage 114.64 m^3. In order to maintain the stability of the control system, the excess storage should be drained from the downstream. The dynamic wave principle algorithm is based on the velocity and velocity of the flow in the initial state of the canal pool, without considering the energy attenuation in the process of water flow propagation. The delay time calculated by this algorithm is small. In order to fully compensate the storage required by the canal pool, the flow of the storage compensation will be larger than the target flow. When the water intake is started, the flow of the canal pool will be adjusted to be equal to the target flow, so as to avoid the inflow of the canal pool being larger than the flow required by the canal pool, resulting in the waste of water resources. However, in the process of regulating the canal pool, the discharge has been adjusted twice, which results in the fluctuation of the water level and discharge of the canal pool, and the operation needed to stabilize the canal pool is more, which leads to a greater waste of water. Based on the principle of water balance in the canal pool, the water balance algorithm assumes that the canal pool does not produce discarded water, and derives the equation for calculating the optimal intake time. In practice, the optimal water intake time is obtained by identifying the K, τ, K_p, a and k_d parameters of the canal pool under water intake and supply processes. The identification accuracy of the five parameters will affect the optimal intake time. In the current example, the optimal intake time obtained is slightly large, and the actual adjusted storage exceeds the target.

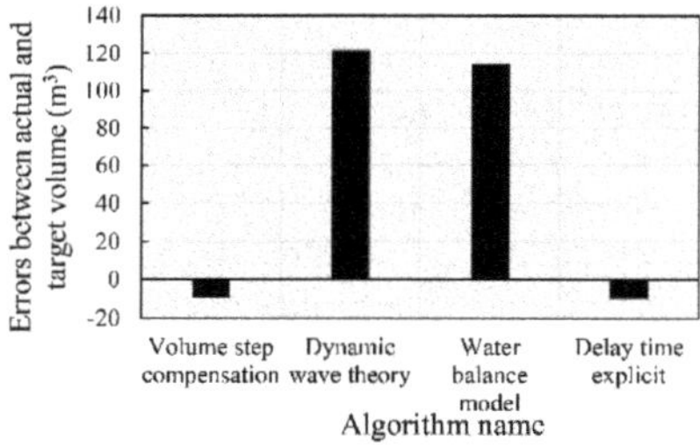

Figure 7. Differences between actual and target storage of canal pool after water intake.

While the actual storage of the volume step compensation algorithm is 9.68 m^3 less than the target storage, and the actual storage of the delay time explicit algorithm is 9.82 m^3 less than the target storage. The volume step compensation algorithm calculates the delay time by assuming the steady state at the beginning and the end. In order to calculate the storage of the canal pool, it is necessary to discretize the canal pool in space and obtain the water surface profile of the constant non-uniform flow. In order to ensure the continuity of the calculation of the water surface line, the water surface line will be smoothed at discrete points. At the same time, the algorithm ignores the intermediate state of the actual adjustment, so the actual storage adjustment value of the algorithm is slightly smaller than the target storage. The delay time explicit algorithm assumes that the change of the A, B and R of the canal pool before and after intake is small. Therefore, the delay time obtained by using the average value $\overline{A}$, $\overline{B}$, $\overline{R}$ is smaller than actual delay time and needs to be corrected by amplification coefficient. However, only considering the influence of the canal pool length and bottom width, this paper calculates the amplification factor by using a simple empirical formula (Equation (14)). The actual compensation storage is slightly smaller than the target storage. In practice, the measured data (flow or water level) and more geometric parameters of the canal pool should be taken into account to correct this empirical formula. The percentage of the actual compensation storage of the four algorithms to the target compensation storage is shown in Table 4.

Table 4. Percentage of actual compensation storage to target compensation storage.

Algorithm Name	Volume Step Compensation	Dynamic Wave Principle	Water Balance Model	Delay Time Explicit
Percentage/%	99.6	104.6	104.4	99.6

6. Conclusions

Under the operation mode of downstream constant water level of Canal Pool 1 in ASCE Test Canal 2, combining the theory of open channel constant gradient flow with volume compensation, this paper introduces a delay time explicit algorithm, and discusses the operation control effect of three existing algorithms of volume step compensation, dynamic wave principle, water balance model and delay time explicit algorithm. The conclusions are as follows:

(1) The delay time explicit algorithm establishes a linear formula between the delay time and the flow change required by water demand, which avoid the space discrete calculation of canal pool needed for storage estimation or identification of relevant parameters. Under the current water intake plan, only 1.47 h is needed to achieve stability of the canal pool. During the regulation process, the fluctuation of water level and flow in the canal pool is small. The actual adjusted storage accounts for 99.6% of the target adjusted storage, which basically meets the requirement of compensated storage required by the canal pool. The delay time calculated by the explicit algorithm can provide some references for the determination of delay time in feedforward control.

(2) Among the three existing algorithms, the volume step compensation algorithm has better control effect. Under the current water intake plan, the volume step compensation algorithm needs 1.25 hours to achieve stability, and the water balance model algorithm needs 1.77 hours to achieve stability. While the dynamic wave principle algorithm has secondary flow adjustment, the transition time is the longest among the three methods, and the flow fluctuates during the regulation process. The actual adjusted storage of volume step compensation algorithm accounts for 99.6% of the target adjusted storage, which basically meets the requirement of compensated storage for the canal pool. While the actual adjusted storage of the other two algorithms exceeds the target storage, which means the generation of discarded water.

However, in order to avoid the coupling effect of series canal pools, the case of single canal pools is only considered in this paper, and the case of multi-channel pools in series requires further discussion. At the same time, the paper only considers the influence of canal pool length and bottom width, and further calculates the amplification coefficient by using the simple empirical formula in the delay time explicit algorithm. In the actual process, the measured data (flow or water level) and the other geometric parameters of the canal pool should be taken into account to correct the empirical formula.

Author Contributions: All authors have contributed substantially to this work, as specified below: writing—original draft preparation, W.L.; writing—review and editing, G.G. and X.T.

Funding: This research was funded by National Key R&D Program of China (Grant No. 2016YFC0401810) and National Natural Science Foundation of China (No. 51439006).

Acknowledgments: The authors appreciate the comments from the anonymous reviewers.

References

1. Ni, W.J. Development and technology requirement of China rural water conservancy. *Trans. Chin. Soc. Agric. Eng.* **2010**, *26*, 1–8. [CrossRef]

2. Wang, H.; Wang, J.H. Sustainable utilization of China's water resources. *Bull. Chin. Acad. Sci.* **2012**, *27*, 352–358. [CrossRef]

3. Wylie, E.B. Control of transient free-surface flow. *J. Hydraul. Div.* **1969**, *95*, 347–362.

4. Cunge, J.A.; Holly, F.M.; Verwey, A. *Practical Aspects of Computational River Hydraulics*; Pitman Advanced Publishing Program: Boston, UK, 1980; ISBN 0273084429.

5. Bautista, E.; Clemmens, A. Volume compensation method for routing irrigation canal demand changes. *J. Irrig. Drain Eng.* **2005**, *131*, 494–503. [CrossRef]

6. Cui, W.; Chen, W.; Guo, X. Anticipation time estimation for feedforward control of canal. In Proceedings of the 33rd IAHR Congress, Vancouver, BC, Canada, 9–14 August 2009; pp. 1529–1536.

7. Soler, J.; Gomez, M.; Rodellar, J. GoRoSo: Feedforward Control Algorithm for Irrigation Canals Based on Sequential Quadratic Programming. *J. Irrig. Drain Eng.* **2013**, *139*, 41–54. [CrossRef]

8. Burt, C.M.; Feist, K.E.; Xianshu, P. Accelerated Irrigation Canal Flow Change Routing. *J. Irrig. Drain Eng.* **2018**, *144*, 1–9. [CrossRef]

9. Corriga, G.; Fanni, A.; Sanna, S.; Usai, G. A constant-volume control method for open channel operation. *Int. J. Simul. Model.* **1982**, *2*, 108–112. [CrossRef]

10. Belaud, G.; Litrico, X.; Clemmens, A.J. Response time of a canal pool for scheduled water delivery. *J. Irrig. Drain Eng.* **2013**, *139*, 300–308. [CrossRef]

11. Simulation and Control of Canal System. Available online: http://www.ccopyright.com.cn/ (accessed on 3 June 2011).

12. Clemmens, A.J.; Kacerek, T.F.; Grawitz, B.; Schuurmans, W. Test cases for canal control algorithms. *J. Irrig. Drain Eng.* **1998**, *124*, 23–30. [CrossRef]

13. Munier, S.; Litrico, X.; Belaud GMalaterre, P. Distributed approximation of open-channel flow routing accounting for backwater effects. *Adv. Water Resour.* **2008**, *31*, 1590–1602. [CrossRef]

14. Te Chow, V. *Open-Channel Hydraulics*; The Blackburn Press: West Caldwell, UK, 2009; ISBN 1932846182.

15. Cui, W.; Chen, W.; Mu, X.; Bai, Y. Canal controller for the largest water transfer project in China. *Irrig. Drain* **2015**, *63*, 501–511. [CrossRef]

16. Huang, K. Simulation for Time Model of Feedforward Control and Modified PID Feedbackalgorithm on Tandem Canal System. Master's Thesis, Wuhan University, Wuhan, China, 2016.

Article

Effect of the Area Contraction Ratio on the Hydraulic Characteristics of the Toothed Internal Energy Dissipaters

Ting Zhang [1], Rui-xia Hao [1,*], Xiu-qing Zheng [1] and Ze Zhang [2]

[1]   College of Water Resources and Engineering, Taiyuan University of Technology, Taiyuan 030024, China
[2]   Water conservancy engineering design Ltd. of Shanxi Qian Cheng, Taiyuan 030024, China
*   Correspondence: haoruixia@tyut.edu.cn; Tel.: +86-139-3516-1919

Received: 16 May 2019; Accepted: 3 July 2019; Published: 9 July 2019

Abstract: Toothed internal energy dissipaters (TIED) are a new type of internal energy dissipaters, which combines the internal energy dissipaters of sudden reduction and sudden enlargement forms with the open-flow energy dissipation together. In order to provide a design basis for an optimized body type of the TIED, the effect of the area contraction ratio (ε) on the hydraulic characteristics, including over-current capability, energy dissipation rate, time-averaged pressure, pulsating pressure, time-averaged velocity, and pulsating velocity, were studied using the methods of a physical model test and theoretical analysis. The main results are as follows. The over-current capability mainly depends on ε, and the larger ε is, the larger the flow coefficient is. The energy dissipation rate is proportional to the quadratic of Re and inversely proportional to ε. The changes of the time-averaged pressure coefficients under each flow are similar along the test pipe, and the differences of the time-averaged pressure coefficient between the inlet of the TIED and the outlet of the TIED decrease with the increase of ε. The peaks of the pulsating pressure coefficient appear at 1.3 D after the TIED and are inversely proportional to ε. When the flow is 18 l/s and ε increases from 0.375 to 0.625, the maximum of time-averaged velocity coefficient on the line of $Z/D = 0.42$ reduces from 2.53 to 1.17, and that on the line of $Z/D = 0$ decreases from 2.99 to 1.74. The maximum values of pulsating velocity on the line of $Z/D = 0.42$ appear at $1.57D$ and those of $Z/D = 0$ appear at $2.72D$, when the flow is 18 l/s. The maximum values of pulsating velocity decrease with the increase of ε. Finally, two empirical expressions, related to the flow coefficient and energy loss coefficient, are separately presented.

Keywords: toothed internal energy dissipaters (TIED); area contraction ratio; over-current capability; energy dissipation rate; time-averaged pressure; pulsating pressure; time-averaged velocity; pulsating velocity

---

1. Introduction

The internal energy dissipater effectively reduces the downstream flow speed, smoothly connects the downstream flow, and avoids the erosion of the river channel by a traditional energy dissipater in a water conservancy project with high water head and high flow. This is because it converts the large area effect of high-speed water flow into a local energy dissipation effect. The most common form of internal energy dissipation is the energy dissipaters with sudden reduction and sudden enlargement. The energy dissipaters with the sudden reduction and sudden enlargement forms are a pressure energy dissipation method that uses the sectional contraction water flow to adjust and dissipate excess energy [1], and its hydraulic characteristics are mainly affected by the geometric size of energy dissipaters.

The common energy dissipaters with sudden reduction and sudden enlargement forms were divided into orifice plates and plugs (thick plates) according to their thickness along the flow direction [2], and many researchers investigated their hydraulic characteristics and influencing factors. It was found that the contraction ratio was an important factor affecting the hydraulic characteristics of the plug [3], and the shape and thickness of the plug also had a great influence on the flow characteristics [4]. The affecting factors of head loss coefficient for orifice plate, such as relative thickness, area contraction ratio (ε), and Reynolds (*Re*), were studied by Wu [5], who proved that relative thickness and ε of orifice plate were important geometric factors affecting the head loss coefficient. Some researchers mainly focused on the energy dissipation rate and over-current capability of internal energy dissipaters with different body types and area contraction ratios. The energy dissipation ratios of single and two-stage plug energy dissipaters were obtained by the combination of a physical model and numerical simulation [6]. The energy dissipation ratio of a plug, with the combination of the vertical jet, straight curved hole plugs, and horizontal hole plugs was designed and analyzed [7]. The change of pressure was also one of the hydraulic characteristics concerned. The study concluded that the pressure pulsation decreased with the reduced cross-sectional area for the orifices plate [8]. The numerical simulation of pressure field for the thin and thick plates was conducted, and the change law of the sectional pressure drop was analyzed [9]. The distribution of wall pressure coefficient for the orifice dissipaters was discussed [10]. The distribution characteristics of mean flow velocity and pulsating velocity for the orifice plate or the plug were obtained by the simulated or measured flow fields [11,12]. Meantime, scholars have also studied the cavitation characteristics of orifice plates and plugs. The cavitation characteristics of orifice plates and plugs, the cavitation mechanism, and influencing factors of orifice dissipaters were studied by Ai [13]. Additionally, cavitation numbers decreased with the increase of ε for the internal energy dissipaters with the sudden enlargement and reduction forms in the work of Zhou [14] and Zhang [15]. To sum up, the area contraction ratio (ε) of internal energy dissipation with sudden enlargement and reduction was an important geometric factor to decide its hydraulic characteristics.

On the basis of previous studies, new internal energy dissipaters called toothed internal energy dissipaters (TIED) were proposed by our research team [16], which combines the internal energy dissipaters of the sudden reduction forms and sudden enlargement with the open-flow energy dissipation. The flow characters affecting the body factors of the TIED were discussed, and the reasonable body factors of the TIED were preliminary optimized [17–19]. In previous works, the body shape of internal energy dissipater was the key to analyze the variation of the hydraulic characteristics, and the area contraction ratio was the most critical body geometry parameter influencing the hydraulic characteristic of the TIED.

In order to study the effect of the area contraction ratio (ε) on the hydraulic characteristics of TIED, it is necessary to confirm the length, height, and angle of piers. Some researchers mainly focused on the length and height of the plug and TIED. The effect of the length of the plug [20] and piers [21] on the head loss coefficient was analyzed via numerical simulation. It was found that the head loss coefficient decreased sharply and then increased with the increase of their length, and its turning point was a length of 0.5*D*. When their length increased to 0.9*D*, the local head loss tended to be stable. The effect of the pier's height on the hydraulic characteristics of the toothed internal energy dissipate (TIED) was studied via a physical model test [22]. The results showed that the energy dissipating effect was better when the height of the piers (*h*) was no greater than 0.25*D*, and the flow coefficient was larger when the height of piers (*h*) was equal to 0.25*D*. According to the results, the length and height of piers was initially chosen to be 13.5 cm (0.9*D*) and 3.75 cm (0.25*D*), respectively.

In this paper, five different types of the TIED with four toothed piers (the length of each toothed piers is 0.9*D*), which have different area contraction ratios (ε), were designed and used to experimentally study the effect of the area contraction ratio (ε) on hydraulic characteristics.

2. Model Experiment

The test device consists of a flat water tank, a pipeline, an electromagnetic flow meter, a test section, and a valve. The test pipe is made of organic glass to observe the flow regime easily. The inner diameter (D) of the test pipe is 15 cm, and its total length is 370 cm. Figure 1 shows the layout of the test device.

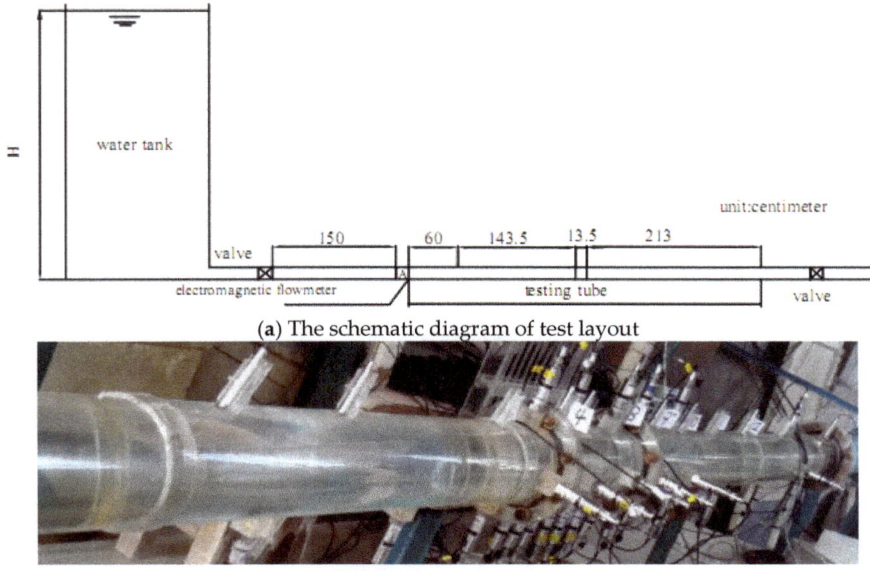

(**a**) The schematic diagram of test layout

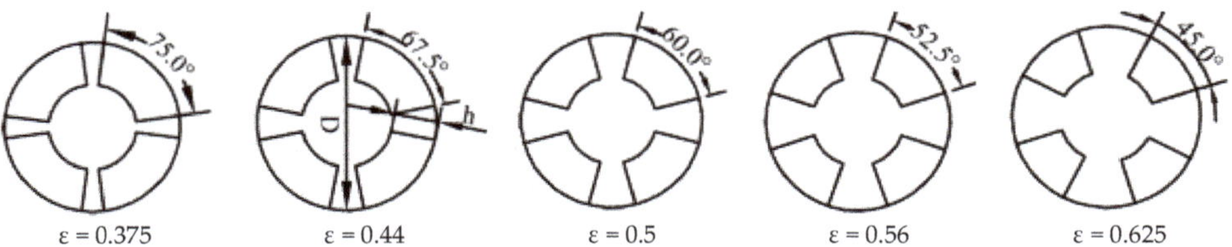

(**b**) The photo of test pipe

Figure 1. Layout of test pipe.

The TIED is located 143.5 cm away from the inlet of the test pipe, and its length (L) is 13.5 cm. For the TIED, the number of the piers (n) is 4, and the height of the piers (h) is 3.75 cm. The area contraction ratio (ε) can be defined as: $\varepsilon = A_0/A = 1 - \frac{\theta}{90°}\frac{4Dh-4h^2}{D^2}$. In equation, A_0 is the cross-sectional area of TIED, A is the cross-sectional area of test pipe, and θ is the angle of the piers. In order to study the effect of area contraction ratio (ε) on hydraulic characteristics of the TIED, five types of TIED were experimentally studied via the physical model. The cross-sectional areas of the TIED are displayed in Figure 2.

Figure 2. Cross-sectional areas of toothed internal energy dissipaters (TIED).

During the test, the upstream head (H) was kept constant, the flow was constant in the test pipe, the indoor temperature was 20 °C, and the range of flow was between 18 l/s and 42 l/s. In order to analyze the same flow condition, the flow of each group was about 6 l/s increased sequentially, and the flow measurement error of each experimental flow group was ±0.2 l/s.

The flow (Q) through the test pipe and the transient pressure (p_i) along the bottom of the test pipe were measured for each test group. The center point at the inlet of the TIED is the origin of the coordinate, the direction of water flow is positive for the X axis, and the direction for vertical upwards is the positive Z axis. Q was measured using an intelligent electromagnetic flow meter, and its measured accuracy was ±0.5%.

p_i was measured with a digital pressure sensor, and its measured accuracy was 0.1%, and its measuring frequency 100 HZ. Figure 3a shows the relative position of the pressure measuring points.

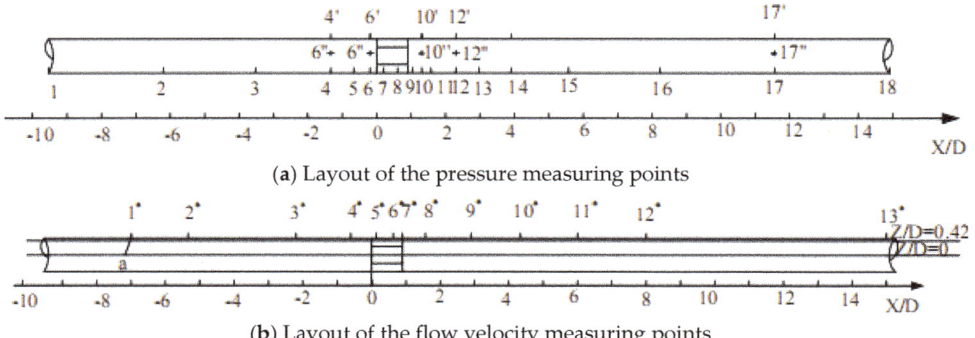

(a) Layout of the pressure measuring points

(b) Layout of the flow velocity measuring points

Figure 3. The distribution of measuring points along the test pipe.

When Q was 18 l/s, the transient flow velocity (u) was measured by the DOP3010 flow velocity meter for different ε. The sampling frequency of the DOP3010 flow meter was 1MHz and its resolution was 0.01 mm. The angle between the measuring probe of u and the pipe wall was 70°, and their gap was filled with coupling medium to make the measurement of u more precise. Taking the measuring point of 1* as an example, it is possible to obtain the value of u for measuring points at intervals of 1 cm on the line of a. $Z/D = 0$ is at the central axis of the test pipe and $Z/D = 0.42$ is 1.2 cm away from the upper side of test pipe. The measured value of u is stable in the position of $Z/D = 0.42$ and $Z/D = 0$ for every measuring point. The relative position of the measuring point is shown in Figure 3b.

3. Analysis of the Flow Characteristics Affected by the Area Contraction Ratio

3.1. Over-Current Capability

The flow coefficient (μ_c) reflects the over-current capability of the test pipe, written as:

$$\mu_c = \frac{Q}{A\sqrt{2g\Delta H}},\tag{1}$$

where ΔH is the head loss between the fifth and fourteenth measuring point.

ΔH is calculated by the following equation:

$$\Delta H = \left(z_{14} + \frac{\overline{p_{14}}}{\rho g} + \frac{v_{14}^2}{2g}\right) - \left(z_5 + \frac{\overline{p_5}}{\rho g} + \frac{v_5^2}{2g}\right) = h_w = \xi\frac{v^2}{2g},\tag{2}$$

where z_{14} and z_5 are the position head for the measuring points of 14 and 5, respectively, and the position head of z_{14} is equal to that of z_5; $\overline{p_{14}}$ and $\overline{p_5}$ are the time-averaged pressure for the measuring points of 14 and 5, respectively; v_{14} and v_5 are the average flow velocity for the measuring points of 14 and 5, respectively, and their value are equal to v; v is the averaged velocity of the test pipe; h_w is the head loss between the front and back of TIED; ξ is the head loss coefficient, and the sum of the resistance coefficient (λ) along the pipe and the local head loss coefficient (ζ).

Combining Equations (1) and (2) together, we can obtain:

$$\mu_c = \frac{1}{\sqrt{\xi}},$$

(3)

The influencing factors of λ are the flow regime and the roughness of the pipe wall, and the roughness of the pipe wall is affected by the geometric parameters of the TIED. When the water flow is laminar, λ is only affected by Re. When the water flow is turbulent and in the transition zone, λ is determined by the roughness of the pipe wall and Re. When the water flow is turbulent in the square zone of resistance, λ is determined by the roughness of the pipe wall and not affected by Re. When Re is between 1.5×10^5 and 3.5×10^5, the water flow is located in the square area of turbulent resistance, ξ is determined by the body type of the TIED, the wall roughness of the TIED, and the wall roughness of the testing pipe.

Figure 4 shows the change of the flow coefficients (μ_c) with ε for different Re. μ_c is little affected by Re and its relative errors are within 2% for the same ε when Re is between 1.5×10^5 and 3.5×10^5. For the same Re, μ_c increases from 0.4 to 0.9 with the increase of ε, so ε is the main influencing factor of over-current capability.

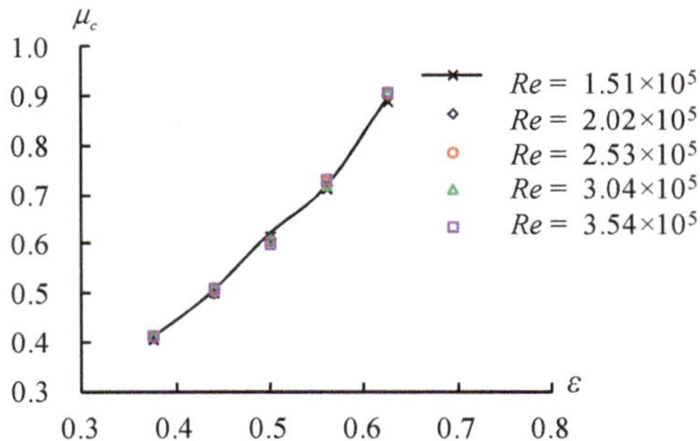

Figure 4. Relationship of the flow coefficient (μ_c) and area contraction ratio (ε).

In order to eliminate the error of μ_c caused by the Re in the experiment, the averaged value ($\overline{\mu_c}$) of μ_c for different ε is acquired in the testing flow range, and the relationship between $\overline{\mu_c}$ and ε is presented in Figure 5. It indicates that $\overline{\mu_c}$ increases with the addition of ε, and the empirical formula of μ_c is presented as:

$$\mu_c = 0.1213e^{3.1948\varepsilon} \quad (0.375 \leq \varepsilon \leq 0.625).$$

(4)

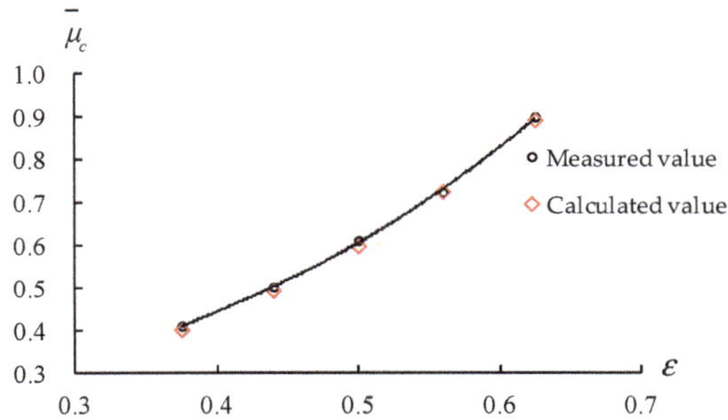

Figure 5. Relationship between the measured or calculated values of the averaged flow coefficient ($\overline{\mu_c}$) and the area contraction ratio (ε).

The calculated values of $\overline{\mu_c}$ obtained by Equation (4) and the measured values of $\overline{\mu_c}$ are shown in Figure 5. Comparing the calculated values with the measured values for the same ε, their errors are smaller than 5% and within the range of allowable error. Therefore, Equation (4) can calculate the flow coefficient of the TIED in the testing flow range.

3.2. Energy Dissipation Rate

The energy dissipation rate (η) can represent the energy dissipation effect of the TIED. The higher the energy dissipation rate, the better the energy dissipation effect. It can be expressed as:

$$\eta = \frac{h_w}{H} = \xi \frac{v^2}{2gH},\tag{5}$$

where h_w is the head loss between the front and back of TIED, H is the total test head, ξ is the head loss coefficient, and v is the average velocity of test pipe.

Figure 6 shows the variation trend of η with ε in the condition of different Re. η increases with Re for the same ε and decreases exponentially with ε for the same Re.

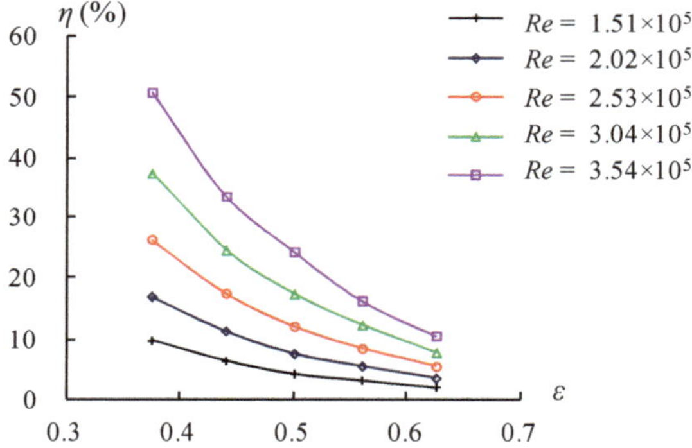

Figure 6. Relationship of the energy dissipation rate (η) and area contraction ratio (ε).

Because of the constant total head in the test, Equation (6) can be transformed as follows, aiming to analyze the change of η:

$$\eta == \xi \frac{Q^2}{2gA^2H},\tag{6}$$

In Equation (6), when ε is constant; ξ, g, A, and H are constant and η is proportional to the square of Q; when Q is constant, η is proportional to ε. Therefore, the value of η can be calculated by ξ in the case of a known Q.

In order to eliminate the influence of Re on ξ, the average value ($\overline{\xi}$) is obtained by taking an average of ξ in the testing flow range, and the change of $\overline{\xi}$ with the increase of ε is shown in Figure 7. $\overline{\xi}$ decreases from 5.9 to 1.2 when ε is between 0.375 and 0.625. The empirical formula of ξ can be expressed as follows:

$$\xi = 59.766e^{-6.198\varepsilon} \ (0.375 \leq \varepsilon \leq 0.625).\tag{7}$$

The calculated values of ξ are obtained by Equation (7), and the measured values of ξ are obtained with the help of the model test. Both of them are presented in Figure 7. Comparing the calculated values of ξ with the measured values, the errors are less than 5%. Thus, this formula can calculate ξ for the TIED in the testing flow range. Moreover, it can obtain η to substitute ξ into Equation (6) in a certain flow.

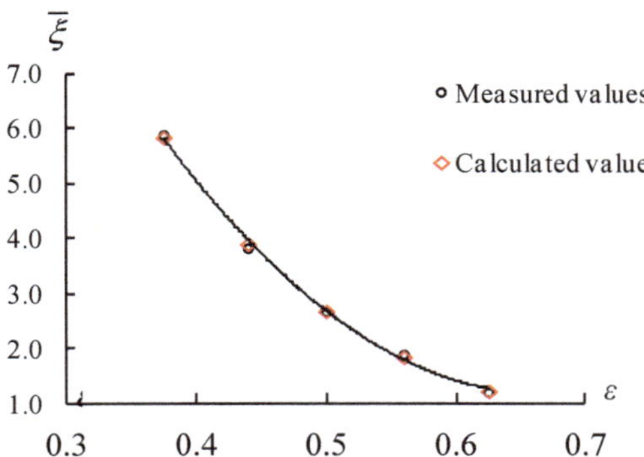

Figure 7. Relationship between the measured or calculated averaged value of the head loss coefficient ($\overline{\xi}$) and the area contraction ratio (ε).

3.3. The Variation of the Time-Averaged Pressure along the Test Pipe

When Q increases in the pipeline, Re becomes larger, the frictional head loss and local head loss also become greater, and the reduced amplitude of the time-averaged pressure ($\overline{p_i}$) decreases sharply along the test pipe. The time-averaged pressure coefficient (α) is introduced to express $\overline{p_i}$ better, established as:

$$\alpha = \frac{\overline{p_i} - \overline{p_{\min}}}{\rho v^2 / 2},\qquad(8)$$

where $\overline{p_i}$ is the time-averaged pressure for each measuring point; $\overline{p_{\min}}$ is the smallest value of the time average pressure among the measuring points along the test pipe; and its position pressure in the test is at $0.2D$. Figure 8a shows the change of α along the test pipe. It can be seen that the time-averaged pressure coefficients are less affected by Q and have similar change trends along the test pipe. The reason is that the difference of α between the two measured points is equal to the head loss coefficient, and the effect of Re on the value of α can be neglected, and it was only affected by the energy dissipater. In order to reduce the error caused by the changes of Re, the average value ($\overline{\alpha}$) is obtained to study α better in the range of Q. The relationship between $\overline{\alpha}$ and ε is shown in Figure 8b, and the changes of $\overline{\alpha}$ with the increase of ε for the inlet or outlet of the TIED are presented in Figure 9.

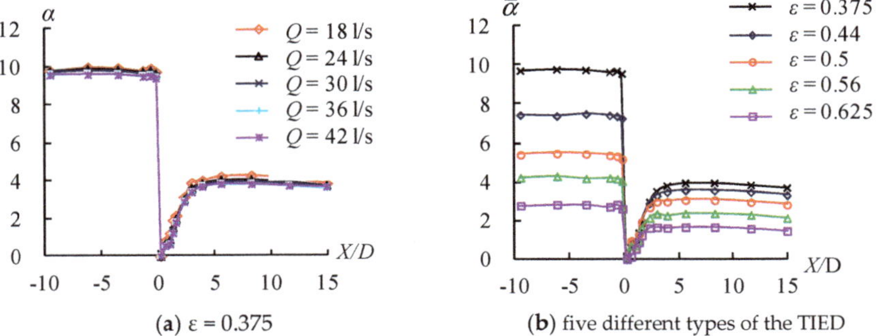

(a) $\varepsilon = 0.375$ **(b)** five different types of the TIED

Figure 8. Distribution of time-averaged pressure coefficient (α) and the mean of time-averaged pressure coefficient ($\overline{\alpha}$) along the test pipe.

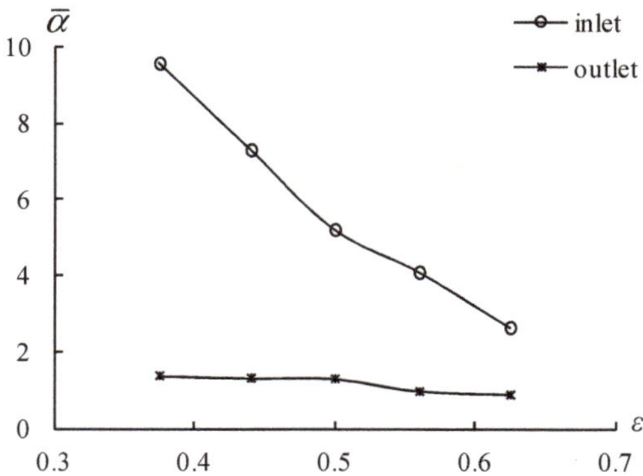

Figure 9. Relationship between the mean of time-averaged pressure coefficient ($\overline{\alpha}$) and the area contraction ratio (ε).

As shown in Figures 8 and 9, α drop sharply at the inlet of the TIED, then gradually increases and tends to remain nearly constant in the place of 4D after the inlet of the TIED. The reason is that the sudden changes of the flow velocity, caused by the sudden change of the flow cross-section through the TIED, lead to changes of α. For different types of TIED, α decreases from 9.8 to 2.8 before the inlet of the TIED and falls from 3.9 to 1.6 after the outlet of the TIED with the increase of ε from 0.375 to 0.46, and the minimum of α along the test pipe increases with the increase of ε. For different types of the TIED, $\overline{\alpha}$ in the outlet and inlet of the TIED both decrease with the increase of ε, and their differences drop from 8.2 to 1.8 when ε grows from 0.375 to 0.625. The change of α within and near the TIED is mainly due to the increase of the head loss coefficient, which is caused by the change of cross-sectional area.

When Q is constant, the smaller ε is, the larger $\overline{\alpha}$ in the front of the TIED is. Thus, the larger enough value of $\overline{p_6} - \overline{p_{min}}$ should be provided. Additionally, the value of $\overline{p_6}$ is constant when Q is constant. Therefore, the value of $\overline{p_{min}}$ will be negative with the decrease of ε. When ε is constant, $\overline{\alpha}$ is constant. The larger Q is, the smaller $\overline{p_i}$ becomes, so the value of $\overline{p_{min}}$ will be negative with the increase of Q. In the testing range flow, when the flow is about 42 l/s and ε is equal to 0.375 and 0.46, the value of $\overline{p_{min}}$ is negative. The air will enter the pipe when the value of time-averaged pressure is negative, and then cavitations are likely where the flow velocity becomes small.

3.4. Variation of Pulsating Pressure

The size of the pulsating pressure can be represented by the root mean square of the pulsating pressure (σ) in the Equation (9):

$$\sigma = \sqrt{\frac{1}{N}\Sigma_{i=1}^{N}(p_i - \overline{p})^2},\tag{9}$$

where N is the measuring times of the pressure and p_i is the instant pressure at the measuring points. The pulsation of pressure is caused by the mixing of particles in each layer of turbulence. If air enters the pipe, the stronger the pressure pulsation, the more likely it is for cavitations to occur.

The pulsating pressure coefficient (C_p) can be acquired to express σ, written as:

$$C_p = \frac{\sigma/\gamma}{v^2/2g} = \frac{\sigma/\gamma}{Q^2/2gA}.\tag{10}$$

Figure 10 presents the change of C_p along the test pipe. The variation trend of C_p is similar for different ε and Q, and the peaks of C_p appear at 1.3D after the outlet of the TIED, because the TIED increases the turbulence of water flow, resulting in the enhancement of pulsating intensity near the outlet of the TIED.

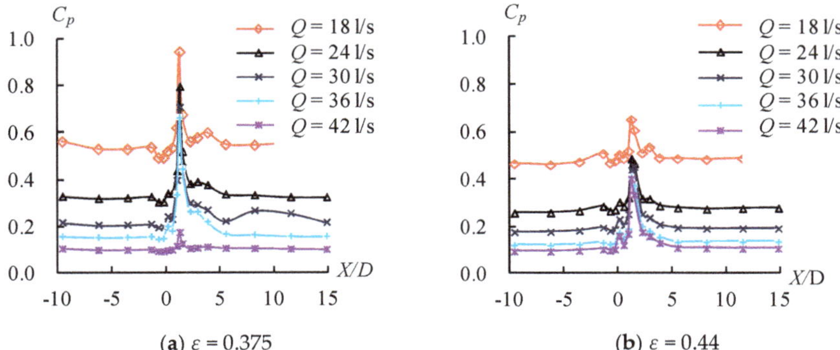

(a) $\varepsilon = 0.375$ (b) $\varepsilon = 0.44$

Figure 10. Distribution of pulsating pressure coefficient along for the typical body type of the TIED.

Table 1 shows the greatest peak of C_p for each test group. It is clear that the peaks of C_p decrease with the increase of Q and ε, because C_p is inversely proportional to the square of Q from Equation (10) and the changing rate of pressure decreases with the increase of ε. When ε increases from 0.375 to 0.625, the averaged peak of C_p $(\overline{C_{pmax}})$ decreases from 0.747 to 0.306 in the range of the testing flow. When Q is constant, the larger ε is, the smaller C_p is, and the larger σ is, indicating that the smaller ε is, the larger the pressure pulsation is.

Table 1. Peaks and average values of pulsating pressure coefficient for each test group.

ε C_{pmax} Q	0.375	0.44	0.5	0.56	0.625
18	0.942	0.654	0.652	0.644	0.64
24	0.794	0.489	0.355	0.355	0.316
30	0.707	0.442	0.282	0.272	0.233
36	0.661	0.413	0.268	0.216	0.184
42	0.633	0.405	0.247	0.203	0.16
$\overline{C_{pmax}}$	0.747	0.48	0.361	0.338	0.306

3.5. Change of the Time-Averaged Velocity

The water flow in the test pipe is a turbulent flow within the test flow range, and the transient velocity of the water flow (u_i) can be divided into two parts: the time-averaged velocity ($\overline{u}$) and the pulsating velocity (u'). The time-averaged velocity coefficient (β) is introduced to describe the characteristic of $\overline{u}$, written as:

$$\beta = \overline{u}/v. \tag{11}$$

$\overline{u}$ is logarithmic distribution along the radial direction in the pipe. The value of $\overline{u}$ at the center is greater than that at the side wall, and $\overline{u}$ in a different position of the pipe is affected by its position and sectional geometry parameters when Q is constant.

$Z/D = 0$ is at the central axis of the test pipe and $Z/D = 0.42$ is 1.2 cm away from the upper side of the test pipe. When Q is 18 l/s, the variation of β along the line of $Z/D = 0$ and $Z/D = 0.42$ for different ε is shown in Figure 11. It is clear that the change trends of β along the pipe are similar for each TIED. The values of β have little changes in the inlet and outlet sections of the test pipe, suddenly increasing at the entrance of the TIED, and then dropping sharply at the outlet of the TIED, because of the same flow cross section in the inlet and outlet sections in the test pipe and sudden changes near the TIED causing a sharp change of v. Further, the value of β on the line of $Z/D = 0.42$ reduced slightly before the inlet of the TIED. The value of β on the line of $Z/D = 0$ is larger than that on the line of $Z/D = 0.42$ for each measuring point. The values of β at the inlet and outlet sections of the test pipe are little affected by ε at the inlet and outlet sections of the test pipe, less than 1 on the line of $Z/D = 0.42$ and greater than 1 on the line of $Z/D = 0$. The main reason is that the inlet and outlet sections of the test pipe are far away from the TIED and $\overline{u}$ is the logarithmic distribution along the radial direction in these sections. The maximum value of β is located inside the TIED. When ε increases from 0.375 to 0.625, the maximum of β near the side wall reduces from 2.53 to 1.17 on the line of $Z/D = 0.42$ and decreases from 2.99 to 1.74 on the line of $Z/D = 0.42$.

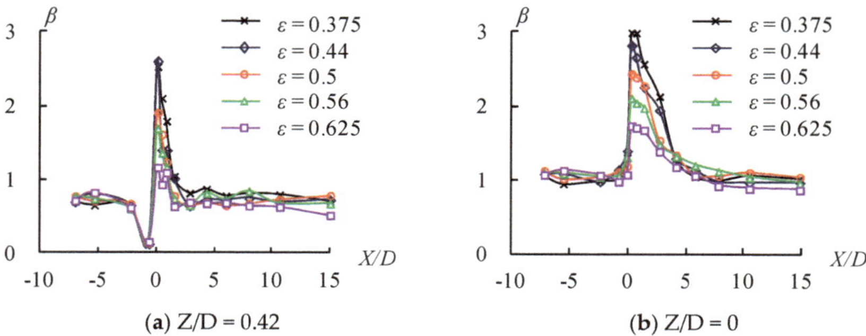

(a) Z/D = 0.42   (b) Z/D = 0

Figure 11. Changes of the time-averaged velocity along the test pipe.

According to the comprehensive analysis, the maximum value of $\overline{u}$ appears on the central axis of the pipe when X is constant. When Z is constant, the changed region of $\overline{u}$ is located inside and near the TIED, and its maximum value appears inside the TIED. That is caused by the sudden decrease of the cross-sectional area for the TIED. The larger ε is, the smaller the reduced amplitude of the cross-sectional area, and the smaller β is, the smaller the maximum of $\overline{u}$.

3.6. Change of the Pulsating Velocity

The pulsating strength (σ_u) is used to represent the fluctuating strength of the velocity for different measuring points and is denoted by the root mean square of the pulsating velocity (σ_u):

$$\sigma_u = \sqrt{\frac{1}{N_2}\Sigma_{i=1}^{N}(u_i - \overline{u})^2}, \tag{12}$$

where N_2 is the measured times of the velocity.

Introducing the turbulent strength (T_u), and defined as:

$$T_u = \sigma_u/v. \tag{13}$$

The change of T_u along the test pipe for different ε is illustrated in Figure 12 when Q is 18 l/s. It is shown that the variation of T_u is similar for the different ε, and the maximum value of T_u on the line of $Z/D = 0.42$ and $Z/D = 0$ appears at 1.57D and 2.72D away from the inlet of the TIED, respectively. ε has little effect on T_u in the constant section of $\overline{u}$ and has a great effect on the abrupt section of $\overline{u}$. When ε increases from 0.375 to 0.625, the maximum value of T_u on the line of $Z/D = 0.42$ reduces from 0.68 to 0.21 and decreases from 0.56 to 0.13 on the line of $Z/D = 0$.

As shown in Figure 12, T_u in the side wall is larger than that in the central axis when X is constant, which is mainly due to the diffusion of turbulent energy and the interference of the edge wall roughness causing the larger turbulence intensity at this position. The maximum value of T_u appears after the outlet of the TIED and decreases with the increase of ε when Z is constant because of the convection of turbulent energy, resulting in downstream movement of the interference wave.

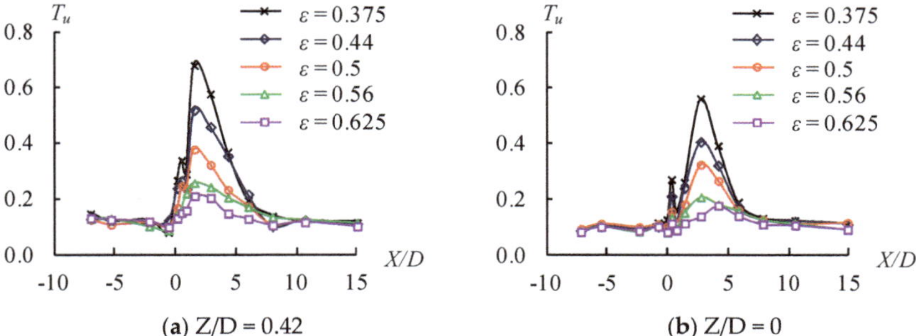

(a) Z/D = 0.42   (b) Z/D = 0

Figure 12. Distribution of turbulent strength along the test pipe.

4. Conclusions

In order to gain insight into the optimized body type parameters of the TIED, the effects of the area contraction ratio (ε) on the hydraulic characteristics of the TIED were discussed using the methods of a physical model test and theoretical analysis in this paper. In the testing flow range, the main conclusions are as follows.

During the test, the flow was basically in the square area of turbulent resistance when Re changed from 1.5×10^5 to 3.5×10^5, and the Re had little effect on the flow characteristics. The energy dissipation rate (η) was proportional to the head loss coefficient (ξ). The flow characteristics were mainly affected by the body type of the TIED. The over-current capability (μ_c) and the energy dissipation rate (η) can be characterized by μ_c and ξ, respectively. They mainly depended on ε. With the increase of ε, μ_c increased exponentially ($\mu_c = 0.1213e^{3.1948\varepsilon}$ ($0.375 \le \varepsilon \le 0.625$)) and ξ decreased exponentially ($\xi = 59.766e^{-6.198\varepsilon}$ ($0.375 \le \varepsilon \le 0.625$)).

The transient pressure of turbulent flow was composed of time-averaged pressure and pulsating pressure. The change trends of time-averaged pressure coefficients (α) only depended on ε; the differences of the averaged α between the inlet and outlet of the TIED decreased from 8.2 to 1.8 when ε increased from 0.375 to 0.625 in the range of the testing flow; when the flow was about 42 l/s and ε was equal to 0.375 or 0.46, the minimum of time-averaged pressure along the pipe was negative. The pulsating pressure coefficient (Cp) was determined by Re and ε, and its peaks appeared at 1.3D after the outlet of the TIED; the averaged peaks in the range of the testing flow decreased from 0.747 to 0.306 when ε increased from 0.375 to 0.625. Negative pressure and larger peaks of pulsating pressure coefficient were more prone to cavitations behind the outlet of TIED.

The transient velocity of turbulent flow was divided into time-averaged velocity and pulsating velocity, which can be represented by time-averaged flow velocity coefficients (β) and turbulent strength (T_u), respectively. When X was constant, the maximum of β appeared on the line of $Z/D = 0$ in pipe, and β near the side wall was larger than that on the line of $Z/D = 0$ in pipe due to the diffusion of turbulent energy and the interference of the wall roughness. When Z was constant, the maximum value of β appeared inside the TIED for different ε, decreasing with the increase of ε. Moreover, the maximum value of T_u appeared after the outlet of the TIED and decreased with the increase of ε because of the convection of turbulent energy. Therefore, the smaller ε is, the more likely it is that cavitations occur near the pipe wall behind the outlet of the TIED.

Comprehensive analysis of over-current capability, energy dissipation rate, and the distribution of flow velocity and pressure showed that the optimized body type parameter (ε) of the TIED in the test was 0.5 when flow in the pipe was about 42 l/s.

Author Contributions: T.Z., designed and performed the model test, and wrote the preliminary manuscript paper; Z.Z., conducted preparation of experiment and preliminary analysis the data of model test; R.H. and X.Z., provided guidance for model test and also further improved the concept, structure, contents and writing of the preliminary manuscript paper.

Funding: This work was supported by the Natural Science Foundation of Shanxi Province (Grand No: 2013011037-4) and National Natural Science Foundation of China (Grant No.51109155).

Conflicts of Interest: The authors declare no conflict of interest.

References

1. Fossa, M.; Guglielmini, G. Pressure drop and void fraction profiles during horizontal flow through thin and thick orifices. *Exp. Therm. Fluid Sci.* **2002**, *26*, 513–523. [CrossRef]
2. Wu, J.H.; Ai, W.Z. Flows through energy dissipaters with sudden reduction and sudden enlargement forms. *J. Hydrodyn.* **2010**, *22*, 360–365. [CrossRef]
3. Ai, W.Z.; Hu, L. Research on energy dissipation in a discharge tunnel with a plug energy dissipater. *Trans. Famena* **2016**, *40*, 57–66.
4. Tian, Z.; Xu, W.L.; Wang, W.; Liu, S.J. Hydraulic characteristics of plug energy dissipater in flood discharge tunnel. *J. Hydrodyn.* **2009**, *21*, 799–806. [CrossRef]
5. Wu, J.; Ai, W.Z.; Zhou, Q. Head loss coefficient of orifice plate energy dissipator. *J. Hydraul. Res.* **2010**, *48*, 526–530.
6. Yu, T.; Tian, Z.; Wang, W.; Xu, W.L. Energy dissipation and cavitation characteristics of contracted plug in discharge tunnel. *J. Hydraul. Eng.* **2011**, *42*, 211–217.
7. Lian, L.; Wang, W.; Tian, Z. Characteristics of energy dissipation and cavitation for complex plug in discharge tunnel. *J. Hydroelectr. Eng.* **2012**, *2*, 62–70.
8. Rydlewicz, W.; Rydlewicz, M.; Pałczyński, T. Experimental investigation of the influence of an orifice plate on the pressure pulsation amplitude in the pulsating flow in a straight pipe. *Mech. Syst. Signal Process.* **2019**, *117*, 634–652. [CrossRef]
9. Roul, M.K.; Dash, S.K. Single-phase and two-phase flow through thin and thick orifices in horizontal pipes. *J. Fluids Eng. Trans. ASME* **2012**, *134*, 091301. [CrossRef]
10. Ai, W.Z.; Wang, J.H. Minimum wall pressure coefficient of orifice plate energy dissipater. *Water Sci. Eng.* **2015**, *8*, 85–88. [CrossRef]
11. Dong, J.; Xu, W.; Deng, J.; Liu, S.; Wang, W.; Qu, J. Numerical simulation of turbulent flow through throat-type energy-dissipators. *J. Hydrodyn. Ser. B* **2002**, *14*, 135–138.
12. Tian, Z.; Deng, J.; Feng, X. Investigation of flow field of plug dissipaters by LDV. *J. Sichuan Univ. (Eng. Sci. Ed.)* **2014**, *46*, 1–5.
13. Ai, W.Z.; Ding, T.M. Orifice plate cavitation mechanism and its influencing factors. *Water Sci. Eng.* **2010**, *3*, 321–330.
14. Zhou, H. Numerical analysis of the 3-D flow field of pressure atomizers with V-shaped cut at orifice. *J. Hydrodyn. Ser. B* **2011**, *23*, 187–192. [CrossRef]
15. Zhang, C.-B.; Yang, Y.-Q. 3-D numerical simulation of flow through an orifice spillway tunnel. *J. Hydrodyn. Ser. B* **2002**, *14*, 83–90.
16. Zhang, T.; Tian, C.; Li, Y.; You, Z.; Fan, P. Characteristics of energy dissipation and pressure within inner energy dissipaters of tooth block. *J. Drain. Irrig. Mach. Eng.* **2014**, *32*, 136–139. (In Chinese)
17. Xue, D.; Tian, C. Influence of energy dissipation efficiency for different area reduction ratio on tooth pier energy dissipation. *Water Resour. Power* **2014**, *4*, 84–87. (In Chinese)
18. Jiang, X.; Hao, R.; Li, Y. Study on pressure and flow field characteristics of tooth block inner energy dissipater. *Water Resour. Power* **2015**, *8*, 94–97. (In Chinese)
19. Su, D.; Hao, R. Numerical simulation of hydraulic characteristic of tooth-block inner energy dissipaters. *Water Resour. Power* **2015**, *11*, 79–81. (In Chinese)
20. Yin, Z.; Shi, B.; Zhao, L.; Sun, D. Numerical simulation of plug energy dissipater flow. *Adv. Water Sci.* **2008**, *1*, 89–93.
21. Li, X. Numerical Simulation on Influence of Flow Characteristics for the Dental Length on Tooth Block Inner Dissipater. Master's Thesis, Taiyuan University of Technology, Taiyuan, China, October 2015.

22. Wang, H. Experimental Research about Tooth Block Height Effect on Inner Energy Dissipaters of Tooth Block. Master's Thesis, Taiyuan University of Technology, Taiyuan, China, October 2015.

Article

Physical Modeling of Ski-Jump Spillway to Evaluate Dynamic Pressure

Mehdi Karami Moghadam [1], Ata Amini [2,*], Marlinda Abdul Malek [3], Thamer Mohammad [4] and Hasan Hoseini [5]

[1]   Department of Agriculture, Payame Noor University (PNU), Tehran 19395-4697, Iran
[2]   Kurdistan Agricultural and Natural Resources Research and Education Center, AREEO, Sanandaj 6616949688, Iran
[3]   Institute of Sustainable Energy, Universiti Tenaga Nasional, Selangor 43000, Malaysia
[4]   Department of Water Resources Engineering, College of Engineering, University of Baghdad, Baghdad 10011, Iraq
[5]   Faculty of Water Sciences Engineering, Shahid Chamran University of Ahvaz, Ahvaz 6135783151, Iran
*   Correspondence: a.amini@areeo.ac.ir or ata_amini@yahoo.com; Tel.: +98-91-8371-4538

Received: 18 June 2019; Accepted: 11 July 2019; Published: 15 August 2019

Abstract: The effects of changes in the angle of pool impact plate, plunging depth, and discharge upon the dynamic pressure caused by ski-jump buckets were investigated in the laboratory. Four impact plate angles and four plunging depths were used. Discharges of 67, 86, 161, and 184 L/s were chosen. For any discharge, plunging depth and impact plate angle were regulated, and dynamic pressures were measured by a transducer. The results showed that with the increase in the ratio of drop length of the jet to its break-up length (H/L_b), and with an increase in the impact plate angle, the mean dynamic pressure coefficient decreased. An inspection of the plunging depth (Y) ratio to the initial thickness of the jet (B_j) revealed that when $Y/B_j > 3$, the plunging depth of the downstream pool reduced dynamic pressure. At the angle of 60°, the dynamic pressure coefficient due to increasing in plunging depth varied from 34% to 95%.

Keywords: jet falling; energy dissipation; surface disturbances; pressure fluctuations; water jet; physical modeling

---

1. Introduction

Owing to dynamic pressures resulting from the flow in hydraulic structures, the river bed is frequently affected with scouring [1,2]. To dissipate the flow energy and to avoid this scouring, dissipater structures such as a spillway with a fillip bucket, which is applied at the end of chute spillway, are used [3]. The flow in the structure is thrown into the air by a ski-jump and goes down after dissipation of part of the energy. Steiner et al. [4] conducted experiments on triangular jets, and compared parameters such as dynamic pressure over the bucket, as well as energy dissipation between triangular and circular buckets. Their results indicated that the relative energy dissipation hinges on the deflection angle and jet falling height from the take-off lip to tail water level. Jorabloo et al. [5] simulated the ski-jump stream outlet through the Fluent model and compared the model output with the results of the physical model. They concluded that pressure distribution, as well as jet trajectory in the two models, are close to each other. Turbulent jet into the flow was numerically analyzed by Mahmoud et al. [6], along with studying the recirculation bubbles in the flow. Their results show that the size and power of the recirculation bubbles increase with the enlargement of the nozzle size. Furthermore, the bubbles disappeared as the Froude number was reduced. Chakravarti et al. [7] have investigated the static and dynamic scouring caused by submerged circular vertical jets. They verified that the depth of dynamic scouring is greater than that of static scouring. Artificial neural network (ANN) was employed by Noori and Hooshyripor [8] based on the major effects of the input parameters on the downstream scouring of the fillip bucket. Their results showed that the Log–Sigmaid model had good performance in the modeling of the depth of scouring. The smooth particle hydrodynamics technique was adopted to study pressure distribution on the steps of a stepped spillway by Husain et al. [9]. Their results showed good consistency with the laboratory observations. Distribution of hydrostatic and non-hydrostatic pressure in shallow waters was investigated by Arico and Re [10]. Dividing the whole pressure into dynamic and hydrostatic components, they solved a hydrostatic and a non-hydrostatic problem sequentially in a fractional time step procedure. Simpiger and Bhalera [11] developed an equation for measuring the jet length according to the jet trajectory. They compared the obtained length with the observation values in a physical model and used the equation to calculate the jet trajectory length in the prototype. Aminoroayaie Yamini et al. [12] investigated the pressure fluctuations and the effect of the entering flow on the fillip bucket bed of Gotvand Dam in Iran. The results show that when the depth and discharge of the entering flow increases and Froude number decreases, the mean dynamic pressure declines and pressure fluctuations grow. They observed that the average pressure was at a maximum at the bucket entry, and was at a minimum at the end part of it. Wu et al. [13] studied the energy dissipation in a fillip bucket of slot-type, both numerically and in the laboratory, and suggested equations to estimate the energy dissipation value. The results showed that as the flow drops in downstream submerged or unsubmerged pools, the resulting dynamical pressure was transformed to the bottom and sidewalls.

Understanding the jet features is crucial in designing the pool and determining the plunging rate. Notwithstanding the research carried out in this regard, it seems that realizing the precise mechanism of ski-jet impacting on the pool requires more studies. In this research, the effects of the main variables upon changes in dynamical pressure and jet break-up length were investigated using the laboratory model of the spillway and the fillip bucket of a dam constructed in Iran. Comprehensive data, as well as the analyses presented in this study, can be used by researchers and engineers.

2. Materials and Methods

2.1. Mechanism and Effective Parameters

The effective parameters in jet break-up include fluid properties, the environment features, and the jet outlet conditions. A schematic drawing of regions and parameters in a jet break-up is depicted in Figure 1.

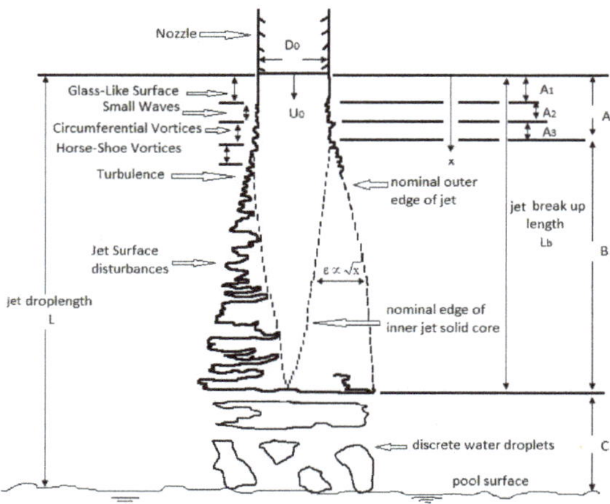

Figure 1. Features of falling jet and its development (adapted from Ervine et al. [14]).

As shown in Figure 1, there are three flow regimes to be considered before the vertical jet impacts the water surface. Region A is composed of three sub-regions: A_1, A_2, and A_3. In subregion A_1, when the flow exits the nozzle, the surface tension resists disturbance, meaning the jet surface remains flat and glass-like. In subregion A_2, roughness (waves) grows at the water surface. As for subregion A_3, surface waves turn into circumferential vortices. Region B is where surface disturbances (ε) increase with the square root of the fall distance ($\varepsilon \propto \sqrt{X}$), where X is the distance from the beginning of the jet. The air penetrates the jet perpendicular to its trajectory. The distance from the jet beginning up to the end of region B is called break-up length (L_b). Very intense surface disturbances enter the jet in region C, whereby the flow gets out of continuity mode. In region C, the flow is not of continuous mass and the flow masses are quite distinct. Surface tension and turbulence determine the distance of L_b, where the jet breaking-up occurs and causes the jet impacts with less energy (Ervine et al. [14]). The jet along its direction may be either core or non-core. In Figure 1, up till region B, the core jet exists, and in region C, there is the non-core situation. In case the plunging is sufficient, the non-core jet expands to the sides as vortices.

2.2. Dimensional Analysis

The parameters governing the dynamic pressure resulting from impacting the fillip bucket jet on the plunge pool floor are expressible as in Equation (1) [15]:

$$f(q, \rho_w, \Delta P_{max}, g, B_j, Y, H, R, \alpha, \varphi, \mu, H, L_b) = 0. \tag{1}$$

In Equation (1), q stands for discharge per unit width, ρ_w is water density, ΔP_{max} represents the maximum pressure head, g is the gravitational acceleration, B_j designates impingement jet thickness, Y is tail water depth, H is the drop height, R is radius of fillip bucket, α signifies the angle of impact plate with horizon, φ is the take-off angle of jet with horizon, μ stands for water dynamic viscosity, H is the jet length, and L_b is the jet break-up length. The dimensionless parameters may be represented, through dimensional analysis, as Equation (2) [15]:

$$f(Re, Fr, \frac{\Delta P_{max}}{Y}, \frac{H}{Y}, \frac{R}{Y}, \frac{B_j}{Y}, \alpha, \varphi, \frac{H}{Y}, \frac{L_b}{Y}) = 0, \tag{2}$$

where Re stands for Reynold's number and Fr is Froude number. By removing the fixed parameters, the average dynamic pressure coefficient (C_p) is obtained as follows:

$$C_P = \frac{H_m - Y}{\frac{U_j^2}{2g}} = f(\alpha, \frac{H}{L_b}, \frac{Y}{B_j}), \tag{3}$$

where H_m is the average of observed dynamic pressures, and U_j designates the jet velocity at the impingement moment. The break-up length is calculated from Equations (4) and (5) [16]:

$$\frac{L_b}{B_i Fr_i^2} = \frac{0.85}{(1.07 T_u Fr_i^2)^{0.82}}, \tag{4}$$

$$T_u = \frac{RMS(u')}{\bar{u}}, \tag{5}$$

where B_i, T_u, and Fri are initial thickness, turbulence intensity, and initial Froude number of the jet, respectively. RMS (u') is the root mean square of the values of the velocity fluctuations in the cross section and in the direction of the falling jet axis, and $\bar{u}$ is the average jet velocity.

2.3. The Experiments and Models

The experiments of the present research were conducted in the hydraulic laboratory of Chamran University, Ahwaz, Iran. The laboratory models of the spillway and fillip bucket were scaled down of

Balarood Dam spillway, located 27 km north of Andimeshk, Khouzestan Province. Balarood dam is constructed on the Balarood River and is of the earthy type with clay core. It is 75.5 m in height and 1070 m crest length, and has a reservoir volume of 131 MCM. The laboratory model of the spillway was scaled down with a scale of 1:40 based on the principle of dynamic simulation, and with due regard to the flume and discharge conditions in the laboratory. The dimensions of the flume were 0.5 m width, 9 m length, and 2 m height. Four discharges of 67, 86, 161, and 184 L/s were taken corresponding with the real discharges of 674.85, 870, 1622.2, and 1857.2 m^3/s in the prototype, respectively. The latter discharges correspond to those with the return periods of 2, 100, 1000, and 5000 years. The four downstream water depths used in the plunge pool were 0, 15, 30, and 45 cm. Also, angles of 0°, 30°, 60°, and 90° were chosen for the impact plate in the plunge pool. Taking these variables into account, a total of 64 experiments were carried out.

2.4. Experiments' Setup

During the experiments, the flow was established by a pump using a circulation system. To minimize turbulence, stilling reservoir and honeycomb were utilized before the flow reaches the experiment area. At the beginning of each experiment, the required discharge was regulated via a gate valve and a rectangular weir installed at the end of the flume. Figure 2 shows a schema of the ski-jump jet (Figure 2a) plus parts of the physical model (Figure 2b). A Plexiglass 0.5 m × 0.5 m square plate was employed in designing the impact plate on which the jet arising from the fillip bucket was to be impinged. A total of 37 pores, each with a diameter of 2 mm, was set on the impact plate in order for connection of the piezometer tubes. The impact plate was placed on a metal system movable in a vertical direction in order to set particular water depth at the position of jet impingement. Also, the impact plate was able to rotate around the horizontal axis to create various angles with the horizon. By moving and rotating, the impact plate angle (α) and the plunging depth (Y) were adjusted as in Figure 2a. Finally, the flow was discharged through the return channel system into the primary reservoir.

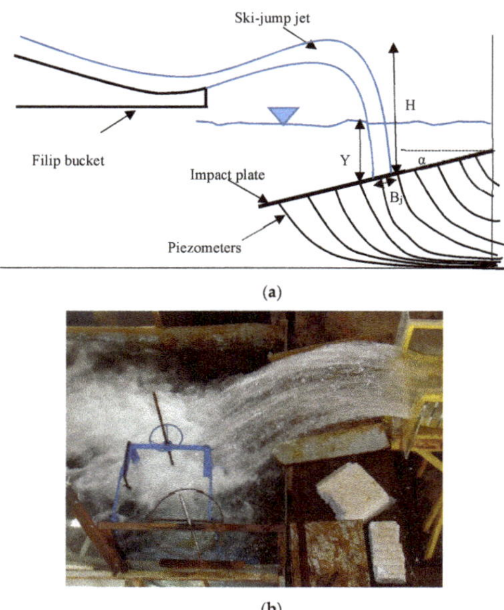

(a)

(b)

Figure 2. (a) Ski-jump, plunging depth, and angle of impact plate; (b) jet trajectory of ski-jump and impinge on impact plate.

2.5. Measurement of Dynamic Pressure

Piezometers were used to observe the fluctuations of the dynamic pressures on the impingement place. Those connected to the impact plate were stretched out from the bottom of the reservoir wall and moved to the dial board, as shown in Figure 3. After the flow exhausted from the bucket and the jet impinged on the impact plate, the dynamic pressures were measured. To demonstrate the dynamic pressures, two of the 37 piezometers, showing the highest pressures, were linked to a transducer. The transducer converts the dynamic pressure of the piezometers into electrical signals and transmitted to the amplifiers by the particular cables. For 10 min, 50 data of dynamic pressures per second were taken. The accuracy of the transducer was ±1 mm for the laboratory model corresponding to 0.04 m for the prototype (Balarood dam). The data obtained by the transducer were translated into the computer via Data Translation Scope, DT9800. This software recorded the sent information and presented them as graphs of pressure fluctuations versus time. For further analysis of the data, the output file of the software was provided to be used in MS Excel spreadsheet programs.

Figure 3. The 37 piezometers connected to impact plate for observation of dynamic pressure fluctuations.

3. Results and Discussion

3.1. Mean Dynamic Pressure Coefficient

The mean dynamic pressure coefficient (C_p) was used in the quantitative study of dynamic pressures (Equation (3)). Table 1 contains the mean values of dynamic pressure coefficients at various impact plate angles, plunging depths, and flow discharges.

Table 1. Values of mean dynamic pressure coefficients obtained in this research

α (deg.)	Y (cm)	67	86	161	184
			C_P		
0	0	0.360	0.386	0.812	0.858
	15	0.353	0.357	0.794	0.832
	30	0.109	0.105	0.461	0.544
	45	0.057	0.054	0.318	0.438
30	0	0.345	0.352	0.619	0.628
	15	0.321	0.328	0.574	0.605
	30	0.078	0.135	0.459	0.429
	45	0.023	0.052	0.328	0.365
60	0	0.154	0.215	0.553	0.455
	15	0.167	0.205	0.514	0.432
	30	0.082	0.015	0.503	0.345
	45	0.008	0.011	0.365	0.125
90	0	0.114	0.191	0.379	0.168
	15	0.099	0.171	0.356	0.168
	30	0.013	0.118	0.343	0.147
	45	0.012	0.077	0.136	0.044

The header spans: Q (L/s) over the four columns 67, 86, 161, 184.

3.1.1. Effect of Plunging Depth

Figure 4 shows the mean values of dynamic pressure coefficients measured at the place of jet impingement on the impact plate for different discharges and angles versus the ratio of plunging depth to the jet width (Y/B_j).

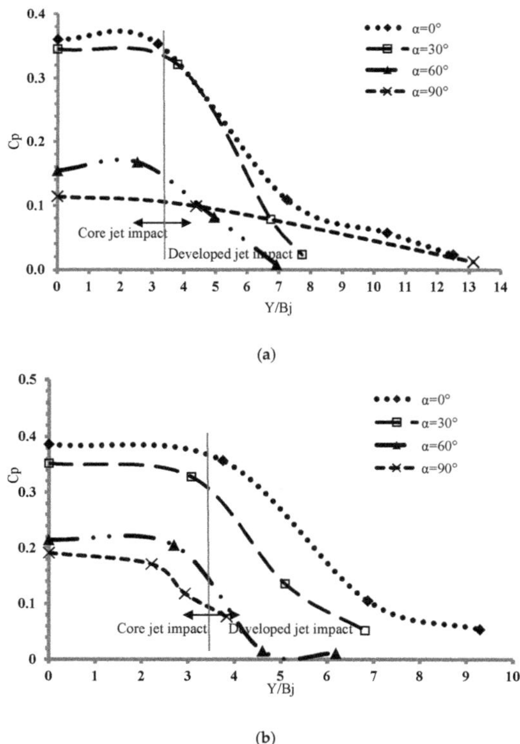

Figure 4. *Cont.*

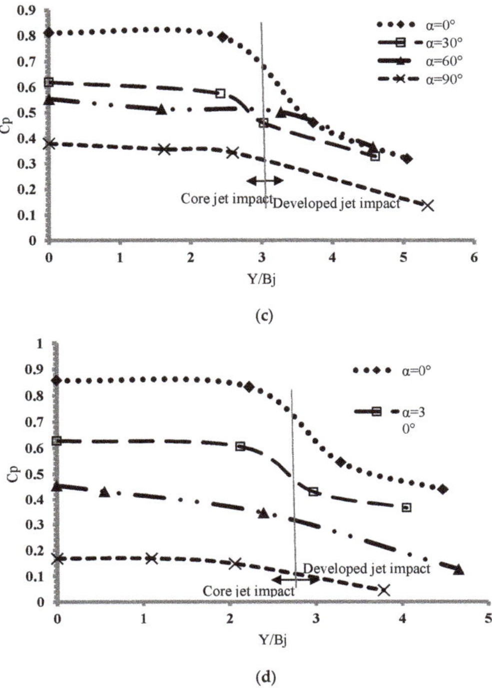

Figure 4. Variations of mean dynamic pressure coefficient versus the plunging depth ratio in discharges of (**a**) 67; (**b**) 86; (**c**) 161; and (**d**) 184 L/s.

Figure 4 reveals that for all discharges, when the plunging depth of the pool is zero ($Y/B_j = 0$), the mean dynamic pressure coefficient attains its maximum value at 0°, and its minimum value corresponds to 90°. The reason is that in the case of $\alpha = 0$, the jet impingement is centralized at just one point. By increasing the jet angle, part of the jet is tangent with the impact plate, so the coefficient reduces. Figure 4 also shows that once the plunging depth increases (increment of Y/B_j), the value of C_p is at first constant or has no significant reduction, whereas from a certain depth, this coefficient decreases. This occurrence may be argued on the basis of the jet features so that with an increase in the plunging depth, the jet would be of developed type. As regards to the developed jets, they generate more spectral energy at moderate frequencies (100–200 Hz) and low frequencies (less than 20 Hz) compared with the core jets. This is owing to the formation of greater vortices with fewer frequencies in the situation of developed jets [17,18]. The percentage reduction of the mean dynamic pressure coefficient due to the increase in plunging depth is variable from 34% for the 60° and discharge of 161 L/s to 95% for the same angle and with a discharge of 67 L/s.

The impingement of the core jet on the bottom of the downstream pool is a result of the plunge pool being shallow or ineffectiveness of its depth. When the plunge depth is sufficient, the jet impacts the bottom in the non-core or developed nature. In the case of non-core, the dynamic pressure on the bottom decreases, and consequently so does C_p. As seen in Figure 4, the value of C_p falls in the decline mode for the range of $2 < Y/B_j < 4$. Therefore, the boundary between the two jet areas, with the core and developed, is within this range. From the region of $Y/B_j > 3$, the plunge value of the downstream pool would be in effect as a result of the jet not impacting in the core state on the impact plate. In accordance with Figure 4a,c, increasing the water depth on the impact plate along with formation of the developed jet brings the diagrams close to each other, which is an indication of the reduction in the effect of impact plate angle. Furthermore, with the increases in discharge, the values of mean dynamic pressure coefficients increase. Figure 5 illustrates the relation between mean dynamic pressure coefficients and submerge ratio in previous studies compared with the data of this research.

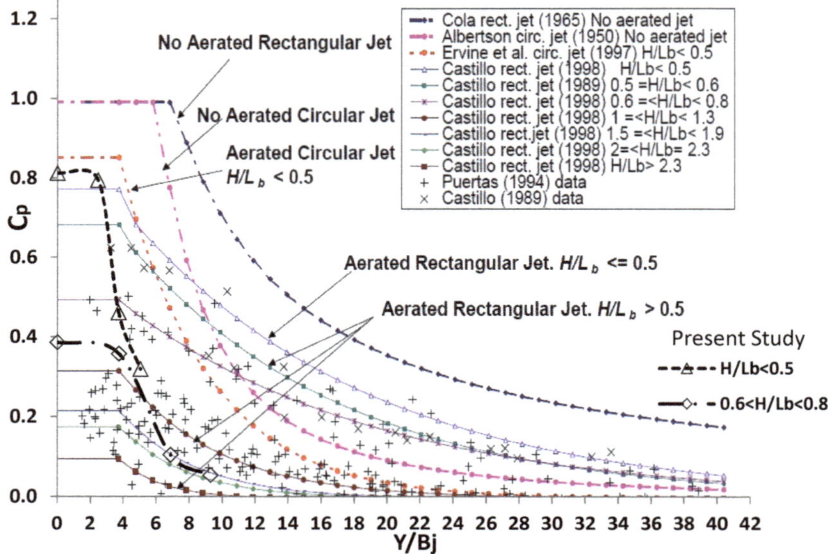

Figure 5. Comparison between correlation of Y/B_j and C_p in present research and previous works (reported by Castillo [19]).

Figure 5 shows that in previous research, in agreement with current research, the C_p value decreases from a certain plunge depth onwards. Converting core jet to developed jet, in most investigations, occurs when the ratio of Y/B_j is close to 4. This bound is commonly obtained in this research and some previous works such as Ervine et al. [14] and Castillo [20], and differs, to a degree, from the results of Cola [21] and Albertson et al. [22]. This dissimilarity arises from differences in drop height, jet thickness, and jet shape (aerated and non-aerated, rectangular and circular).

3.1.2. Effect of Jet Impact Angle

Figure 6 shows the diagram of C_p versus the impact plate angles in non-plunging depth.

Figure 6 confirms that as the impact plate angle increases, the C_p coefficient decreases. This decrement for the range between 0° and 60° occurs sharper relative to the range between 60° and 90°. An increase in the impact plate angle contributes to a decrease in discharges of 67, 86, 161, and 184 L/s, as almost 74%, 60%, 53%, and 51%, respectively. So, the effect of impact plate angle on the decrease of C_p is more in the lower discharges, and this coefficient increases along with increases in discharge. Hence, it is suggested that in executive works, the maximum discharges of the project design probability maximum flood (PMF) be used for the most critical situations being considered in applying dynamic pressures.

Figure 6 shows the diagram of C_p versus the impact plate angles in non-plunging depth.

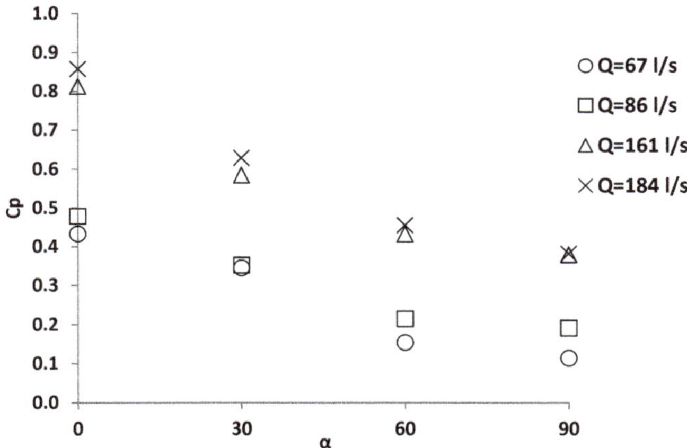

Figure 6. Mean dynamic pressure coefficient versus angle of impact plate.

3.2. Distribution of Dynamic Pressure on Impact Plate

To study dynamic pressure distribution on 0.5 × 0.5 m impact plate in different angles, the changes in C_p values at different radial distances from the impact axis are shown in Figure 7. The figure concerns the discharge of 184 L/s and 0° (with the greatest C_p value) for plunging depths of 0, 15, 30, and 45 cm.

Figure 7a shows the mean dynamic pressure coefficient in the non-plunging depths state. The coefficient value at the center of impact is 0.85, and it decreases away from the center. As seen in Figure 7b, compared with the non-depth case, an increase in the water depth of 15 cm does not have a significant effect on the reduction of C_p values. In fact, the plunging is still of low influence in this stage. At 30 cm depth (Figure 7c), the mean dynamic pressure coefficient is affected by a perceptible decrease. This result indicated that in this situation, the plunging depth was effective in decreasing dynamic pressure. Also, in Figure 7d, in which the plunging depth attained 45 cm, the decline in C_p values was negligible. Thus, the favorable thickness of the plunging depth for the decrease of C_p value happened at the 30 cm depth. An overall consideration of Figure 7 indicated that most dynamic pressures were made in longitudinal distances between 20 and 25 cm, and in transversal distances between 20 and 30 cm. In the cases in which the plunge depth was not so effective, the highest pressure occurred at the center of the impact plate. Any increase in plunging depth causes the center of the effective pressure on the bottom is inclined toward the sides. This may be related to the vortices and turbulent flows created at a slight distance from the point of impact in the direction of the jet central axis [19].

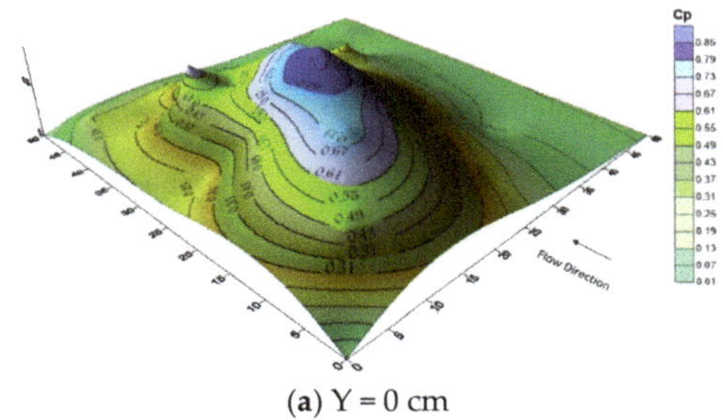

(a) Y = 0 cm

Figure 7. *Cont.*

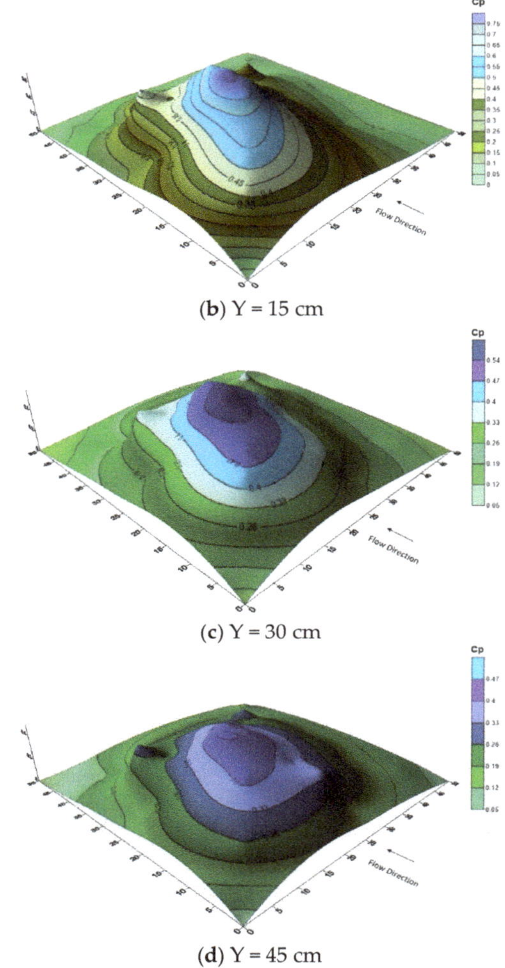

(**b**) Y = 15 cm

(**c**) Y = 30 cm

(**d**) Y = 45 cm

Figure 7. Variations of mean dynamic pressure coefficient at radial distances from falling jet axis for discharge 184 L/s at 0° in plunging depths of (**a**) 0; (**b**) 15 cm; (**c**) 30; and (**d**) 45 cm.

3.3. Mean Dynamic Pressure Coefficient and Break-Up Length

Experiments were carried out at various depths in the plunge pool to study alterations in the hydrodynamic pressure attributable to variations of the ratio of jet drop length to its break-up length (H/L_b). The results are given in Figure 8.

Figure 8 shows that as the impact plate angle relative to the horizontal increases, the value of C_p decreases in all depth situations in the plunge pool. The C_p values at different angles would come close to each other along with an increment in plunging depth. The reason behind this is that when the plunging depth increases, the energy dissipation caused by vortices grows too. Therefore, the effect of the jet impact angle upon dynamic pressure diminishes. Figure 8 confirms that as the H/L_b increases, the coefficient C_p decreases. If the plunging depth is zero or has a low value, the graphs are mostly linear, and the increase in plunging depths causes the graphs to be drawn exponentially. With decreasing plunging depth, the jet drop length increases and motivates the air penetration into the jet, which in turn brings about energy dissipation. As the plunging depth increases, the jet length, H, would shrink and vortices are produced. Consequently, the energy dissipation of the jet would be affected mainly by the vortices, as a result of which the decreasing trend of the coefficient C_p lessens. Enlargement of H means the extension of the jet drop length and the increase of air entry into the jet subsequently. Thus, it causes more dissipation of energy and decrease in C_p values. Moreover, with a decrease in the break-up length, L_b, the jet would be converted from the core to non-core occasion faster, whence a lower dynamic pressure would be exerted [19].

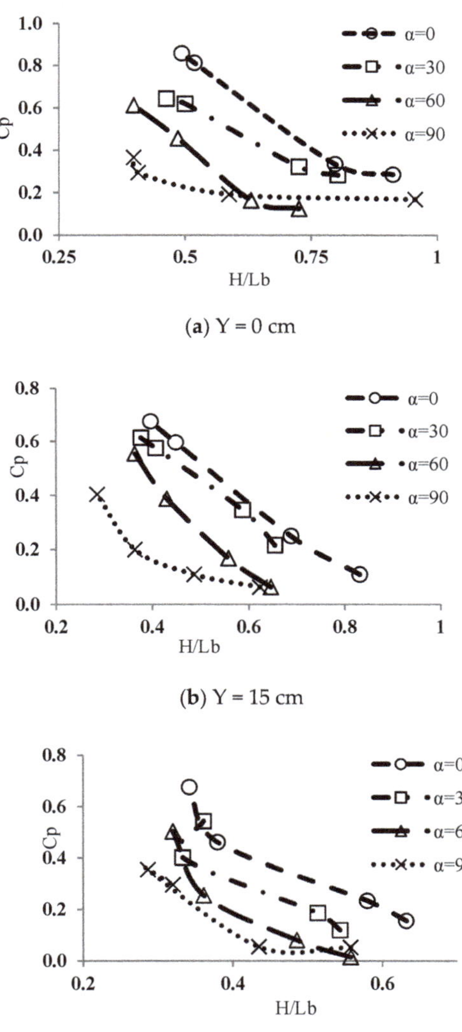

(a) Y = 0 cm

(b) Y = 15 cm

(c) Y = 30 cm

Figure 8. *Cont.*

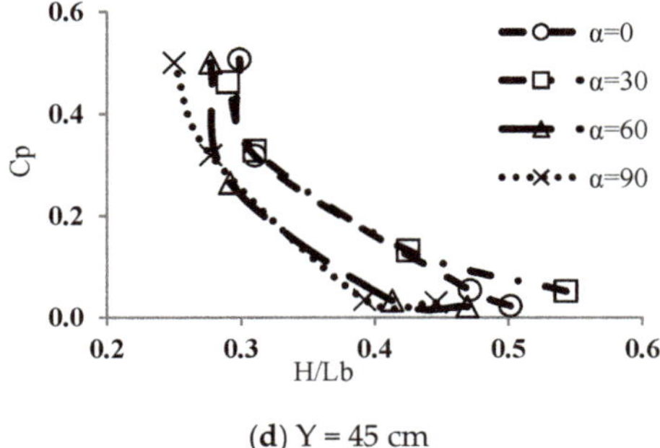

(**d**) Y = 45 cm

Figure 8. Mean dynamic pressure coefficient versus break-up length at different impact plate angles and plunging depths: (**a**) Y = 0 cm; (**b**) Y = 15 cm; (**c**) Y = 30 cm; and (**d**) Y = 45 cm.

4. Conclusions

In this research, a laboratory model was used to investigate the effect of impingement of a jet after take-off from the fillip buckets. The variables of take-off discharge, jet impact angle, and the downstream plunging depth were simulated. The experiments were carried out both without and with plunging depths of 15, 30, and 45 cm. The most important results obtained from this research are as follows:

- At different discharges, with an increase in plunging depth, the mean dynamic coefficient of the pressure on the impact plate was primarily constant and then decreased.
- As the impact plate angle increased the mean dynamic pressure reduced, this reduction occurred more often from the angles of 0 to 60°.
- The greater the plunging depth, the less the effect of impact plate angle on the mean dynamic pressure was observed.
- A low plunging depth did not give a reduction of dynamic pressure on the walls. This depth was in effect at the time when it reached a certain limit. The bound was obtained as $Y/B_j > 3$ in this research.

Author Contributions: M.K.M. and H.H. collected the data and revised original draft preparation, A.A.; provided critical comments in planning this paper, writing and editing the paper, T.M.; M.A.M. read and approved the final manuscript and M.A.M. funded the APC.

Funding: This research received no external funding.

Acknowledgments: The authors acknowledge Shahid Chamran University of Ahvaz, Iran, for their technical supports.

References

1. Karami Moghadam, M.; Amini, A.; Keshavarzi, A. Intake design attributes and submerged vanes effects on sedimentation and shear stress. *Water Environ. J.* **2019**, 1–7. [CrossRef]
2. Eghbalzadeh, A.; Javan, M.; Hayati, M.; Amini, A. Discharge prediction of circular and rectangular side orifices using artificial neural networks. *KSCE J. Civ. Eng.* **2016**, *20*, 990–996. [CrossRef]
3. Eklund, S. CFD modeling of ski-jump spillway in Stornnforsen. Master's Thesis, Royal Institute of Technology, Stockholm, Sweden, 2017.
4. Steiner, R.; Heller, V.; Hager, W.H.; Minor, H.E. Deflector ski jump hydraulics. *J. Hydraul. Eng.* **2008**, *134*, 562–571. [CrossRef]
5. Jorabloo, M.; Maghsoodi, R.; Sarkardeh, H. 3D simulation of flow over flip buckets at dams. *J. Am. Sci.* **2011**, *7*, 931–936.

6.   Mahmoud, H.; Kriaa, W.; Mhiri, H.; Le Palec, G.; Bournot, P. Numerical analysis of recirculation bubble sizes of turbulent co-flowing jet. *Eng. Appl. Comput. Fluid.* **2012**, *6*, 58–73. [CrossRef]

7.   Chakravarti, A.; Jain, R.K.; Kothyari, U.C. Scour under submerged circular vertical jets in cohesionless sediments. *ISH J. Hydraul. Eng.* **2014**, *20*, 32–37. [CrossRef]

8.   Noori, R.; Hooshyaripor, F. Effective prediction of scour downstream of ski-jump buckets using artificial neural networks. *Water Resour.* **2014**, *41*, 8–18. [CrossRef]

9.   Husain, S.M.; Muhammed, J.R.; Karunarathna, H.U.; Reeve, D.E. Investigation of pressure variations over stepped spillways using smooth particle hydrodynamics. *Adv. Water Resour.* **2014**, *66*, 52–69. [CrossRef]

10.   Aricò, C.; Re, C.L. A non-hydrostatic pressure distribution solver for the nonlinear shallow water equations over irregular topography. *Adv. Water Resour.* **2016**, *98*, 47–69. [CrossRef]

11.   Simpiger, B.M.; Bhalerao, A.R. Estimation of throw distance in the design of ski-jump bucket. *Int. J. Sci. Eng. Res.* **2016**, *7*, 1214–1221.

12.   Aminoroayaie Yamini, O.; Kavianpour, M.R.; Mousavi, S.H.; Movahedi, A.; Bavandpour, M. Experimental investigation of pressure fluctuation on the bed of compound flip buckets. *ISH J. Hydraul. Eng.* **2018**, *24*, 45–52. [CrossRef]

13.   Wu, J.H.; Li, S.F.; Ma, F. Energy dissipation of slot-type flip buckets. *J. Hydrodynam.* **2018**, *30*, 365–368. [CrossRef]

14.   Ervine, D.A.; Flavey, H.R.; Withers, W. Pressure fluctuation on plunge pool floor. *J. Hydraul. Res. IAHR.* **1997**, *35*, 257–279. [CrossRef]

15.   Nazari, O.; Jabbari, E.; Sarkardeh, H. Dynamic pressure analysis at chute flip buckets of five dam model studies. *J. Civ. Eng. Trans. A Civ. Eng.* **2015**, *13*, 45–54. [CrossRef]

16.   Castillo, E.L.G. Aerated jets and pressure fluctuation in plunge pools. In Proceedings of the 7th ICHE, Philadelphia, PA, USA, 10–13 September 2006.

17.   Bollaert, E.; Schleiss, A. Scour of rock due to the impact of plunging high velocity jets. Part I: A state-of-the-art review. *J. Hydraul. Res.* **2003**, *41*, 451–464. [CrossRef]

18.   Bollaert, E.; Schleiss, A. Scour of rock due to the impact of plunging high velocity jets. Part II: Experimental results of dynamic pressures at pool bottoms and in one-and two-dimensional closed end rock joints. *J. Hydraul. Res.* **2003**, *41*, 465–480. [CrossRef]

19.   Castillo, E.L.G. Pressure characterization of undeveloped and developed jets in shallow and deep pool. In Proceedings of the 32nd Congress of IAHR, Venice, Italy, 1–6 July 2007.

20.   Castillo, E.L.G. Metodología experimental y numérica para la caracterización del campo de presiones en los disipadores de energía hidráulica. Aplicación al vertido en presas bóveda. Ph.D. Thesis, Universitat Politècnica de Catalunya, Barcelona, Spain, 1989.

21.   Cola, R. Energy dissipation of a high-velocity vertical jet entering a basin. In Proceedings of the 11th IAHR, Leningrad, Russia, Ex-USSR, 6–11 September 1965.

22.   Albertson, M.L.; Dai, Y.B.; Jenson, R.A.; Rouse, H. Diffusion of submerged jets. *Trans. Am. Soc. Civ. Eng.* **1950**, *115*, 639–664.

 *water*

Article

Effects of Spur Dikes on Water Flow Diversity and Fish Aggregation

Tingjie Huang [1,2], **Yan Lu** [1,*] **and Huaixiang Liu** [1]

[1] State key laboratory of Hydrology-Water Resource and Hydraulic Engineering, Nanjing Hydraulic Research Institute, Nanjing 210029, China

[2] College of Water Conservancy and Hydropower Engineering, Hohai University, Nanjing 210098, China

* Correspondence: ylu@nhri.cn; Tel.: +86-025-8582-9305

Received: 1 August 2019; Accepted: 29 August 2019; Published: 31 August 2019

Abstract: As a typical waterway modification, the spur dike narrows the water cross section, which increases the flow velocity and flushes the riverbed. Meanwhile, it also protects ecological diversity and improves river habitat. Different types of spur dikes could greatly impact the interaction between flow structure and local geomorphology, which in turn affects the evolution of river aquatic habitats. Four different types of spur dikes—including rock-fill, permeable, w-shaped rock-fill, and w-shaped permeable—were evaluated using flume experiments for spur dike hydrodynamics and fish aggregation effects. Based on Shannon's entropy, an index for calculating water flow diversity is proposed. Additionally, the impact of the different spur dikes on water flow diversity and the relationship between water flow diversity and fish aggregation effects were studied. The water flow diversity index around the spur dike varied from 1.13 to 2.96. The average aggregation rate of test fish around the spur dike was 5% to 28%, and the attraction effect increased with increasing water flow diversity. Furthermore, we plotted the relationship between water flow diversity index and average fish aggregation rate. A fish hydroacoustic study conducted on the Laohutan fish-bone dike in the Dongliu reach of downstream Yangtze River showed that the fish aggregation effect of the permeable spur dike was greater than the rock-fill spur dike. These research results could provide theoretical support for habitat heterogeneity research and ecologically optimal design of spur dikes.

Keywords: water flow diversity; permeable spur dike; fish aggregation effect; channel regulation

1. Introduction

Shipping is one of the important functions of rivers. Meanwhile, waterway remediation is an important means to improve waterway conditions. As a typical waterway modification, the spur dike narrows the water cross section, which increases flow velocity and flushes the riverbed. However, it affects the local river habitat as well by changing the interaction between water flow and sediment, and local geomorphology within the range of influence of the spur dike.

Recently, many researchers and engineers have studied the ecological effects of spur dikes through model tests, numerical simulations and on-site observations. The results show that spur dikes can enhance river habitat heterogeneity and improve the suitability of fish habitats under low and medium water flows. The tributary area behind the spur dike provides a habitat for plankton and benthic animals, and it also provides a good living environment and shelter for juveniles with a particular fish-aggregation effect [1–7]. The improvement of habitat quality can provide more living space for aquatic organisms [8,9]. Concurrently, a low-speed zone, or even still water zone, is formed behind the spur dike due to its blocking effect. Consequently, it is easy for fine sand and silt to accumulate, making the environment unstable and not conducive to benthic organism survival [10]. Furthermore, the still water area is not suitable fish habitat [11]. To increase quantity and quality of the habitat around the spur dike and strengthen its ecological effect, new layouts of dikes or permeable ecological spur dike are proposed. While retaining the effects of narrowing the water cross section to increase the flow velocity and flush the riverbed, the new spur dike has interrelated and complementary diverse flows, refuges and bait effects. In particular, the diverse flows encourage various plankton and benthic animals to remain as food for fish. The shelter formed by the proposed structure could be used by small animals such as juveniles to avoid predation and should become an excellent micro-habitat. The spur dike of new layouts such as notched spur dikes, J-Hook, cross vane and double cross vane were proposed and applied in the Upper Mississippi River restoration and the Austrian Danube River [12–15]. Presently, there are a variety of ecological spur dikes such as the fish nest ecological, the deflector ecological, the four-side six-edge permeable, the square ecological, the open-hole trapezoidal ecological, and the open-hole semi-circular ecological spur dike in China. Great parts of them have already been applied in channel regulation projects along the Yangtze River and Xijiang River [3,16–19].

The impacts of spur dikes on habitats has mainly been assessed by calculating and comparing the weighted available area (WUA) or the suitable habitat area, under different combinations of water flow and dike type. Additionally, the weighted available area (WUA) or the suitable habitat area is simulated using a habitat model (e.g., PHABSIM- the Physical Habitat Simulation System, [20,21]), which is based on the preference of aquatic organisms for habitat factors such as water depth, flow velocity, and sediment quality. However, fish use of habitats is not only determined by flow velocity, water depth, or sediment quality, but also closely related to water flow turbulence. Turbulence is common in rivers, rough terrain, plants in the river, stone blocks and artificial structures such as spur dikes, which can cause turbulence in the water flow. However, turbulence is not always correlated with velocity—particularly in pools and around roughness elements [22,23]. Fish will adjust their swimming mode and frequency of fishtail swing to conserve energy from the turbulent water flow. Furthermore, the turbulent water flow can reduce the probability of fish being preyed upon. Moreover, the turbulent kinetic energy can better reflect on the fish aggregation status than the flow velocity. Therefore, some scholars have suggested incorporating turbulence into fish habitat assessment management [11,24–32]. Turbulence and its relationship with fish swimming and shoaling are complex, so the effects of turbulence are generally not considered in the habitat models. Some scholars have suggested that turbulence could be described with vorticity or vortex scale to analyze the complexity of water flow. However, the vortex is three-dimensional, time-dependent, and the scale varies greatly. Even when using a two-dimensional vortex, it is difficult to describe the complexity of water flow. Furthermore, it is difficult to extrapolate laboratory results to the field [4,30,33,34]. Generally, there is a lack of a clear calculation method for water flow diversity using physical mechanisms that accurately reflect the impact of spur dikes on water flow diversity and its relationship to fish aggregation.

W-Weir is a habitat enhancement structure applied in middle and small rivers. It is a submerged closure dam and its symmetry plane coincides with the centerline of the river [35]. It is w-shaped along the direction of the water flow. The angle between the first section of the dike and the flume is 20–30°. W-Weir can lower the water surface gradient along the river and decreases flow velocity nearshore, which improves the stability of the riverbank. Two back flow areas and two pools are formed downstream of W-Weir, where rapid flow and slow alternates, and step-pool morphology is generated. It provides habitat and spawning ground for fish and other dwellers. Based on the ecological mechanism of W-Weir, a w-shaped spur dike was proposed by Lu Y. et al. (2018) [36]. Its layout is similar to W-Weir, but not cover the whole cross section river. The w-shaped permeable spur dike is stacked with permeable frames of four sides and six edges, each with a length of 1.0 m and a section diameter of 0.1 m × 0.1 m, which is widely used in the Yangtze river waterway training as revetment. The w-shaped permeable spur dike is applied in the ecological conservation area in Lianhuazhougang within Dongliu reach, which is part of the waterway Project 645—the process of dredging the Yangtze to increase the depth of the waterway between Wuhan to Anqing to 6 m, and to 4.5 m from Wuhan to Yichang.

To determine the influence of different types of spur dikes on water flow diversity and the relationship between water flow diversity and fish aggregation, we therefore used flume experiments in rock-fill, permeable, w-shaped rock-fill, and w-shaped permeable spur dikes. The flow velocity, turbulent kinetic energy and fish distribution around the spur dike were measured. Subsequently, a water flow diversity index was proposed, and its relationship with the average fish aggregation rate was determined. A fish hydroacoustic study was carried out on Laohutan fish-bone dike in the Dongliu reach of the Yangtze River to investigate the fish aggregation effects of rock-fill and permeable spur dikes in the field.

2. Experiments

2.1. Experimental Design

The flume experiments were carried out at the Basic Theory Sediment Laboratory of the Tiexinqiao Test Base, Nanjing Hydraulic Research Institute. The test setup is shown in Figure 1. The flume is 42 m long, 2 m wide and 1 m deep, and flow velocity is controlled by a rectangular thin-walled weir with a maximum flow velocity of 0.60 m^3/s. There are two 1 m energy-dissipating walls at the flume water inlet used to smooth the water flow. Nets are placed at the second wall position and at 32 m from the beginning of the flume inlet. The test section is 10 m long and located in the middle of the flume.

The test was divided into two groups. The first group consisted of a rock-fill spur dike and a permeable spur dike (Figure 1a,c). The rock-fill spur dike was made with 80 mm particle sized gravel with a porosity of about 10% (proportion of void volume with spur dike to the volume of spur dike, the same below). The permeable spur dike was stacked with permeable frames of four sides and six edges each with a length of 20 cm and a section diameter of 2 cm × 2 cm with porosity of about 80%. The permeable frame spur dike was placed 5 m upstream of the rock-fill spur dike on the left side, both of them were perpendicular to the side wall of the flume. The spur dike is 0.55 m long and 0.50 m high. Its relative length is 0.28 (ratio of length of spur dike to width of flume). The second group consisted of a w-shaped rock-fill spur dike and a w-shaped permeable spur dike (Figure 1b,c). It is w-shaped along the direction of the water flow, and the spur dike root to the spur dike head is composed of four straight-line dikes, each with an 80 cm axis length. The w-shaped permeable spur dike is located 5 m upstream of the w-shaped rock-fill spur dike and on the left side of the flume. The angle between the first section of the dike and the flume is 30°, and the angle between adjacent dikes is 60°. The straight-line distance from the starting point of the spur dike root to the spur dike head is 1.2 m (relative length is 0.60, ratio of the straight-line distance from the starting point of the spur dike root to the spur dike head to width of flume), and the height of the spur dike is 0.35 m.

The flow velocity was measured with an acoustic doppler velocimetry system (Sontek ADV, Sontek, Inc., San Diego California, CA, USA). The data acquisition frequency was 25 Hz and processed by winADV (a post-processing software, developed by the USA Department of the Interior). After removing data with a correlation coefficient lower than 70 and a signal-to-noise ratio less than 20, the average flow velocity and the three-dimensional pulsation flow square root at each measurement point were obtained.

The crucian, which are sensitive to changes in water flow, were selected as the test fish. The fish had a body length of about 15 cm and were caught from a natural reservoir. Initially, they were kept in a 2 m × 1.2 m × 0.9 m (length × width × depth) pool for three days, with regular feeding and the water changed regularly. Feeding was curtailed one day before the test. The fish distribution during the test was recorded by a high-definition camera mounted about 3 m above the spur dike. The camera resolution was 1920 × 1280 pixels, 25 frames per second.

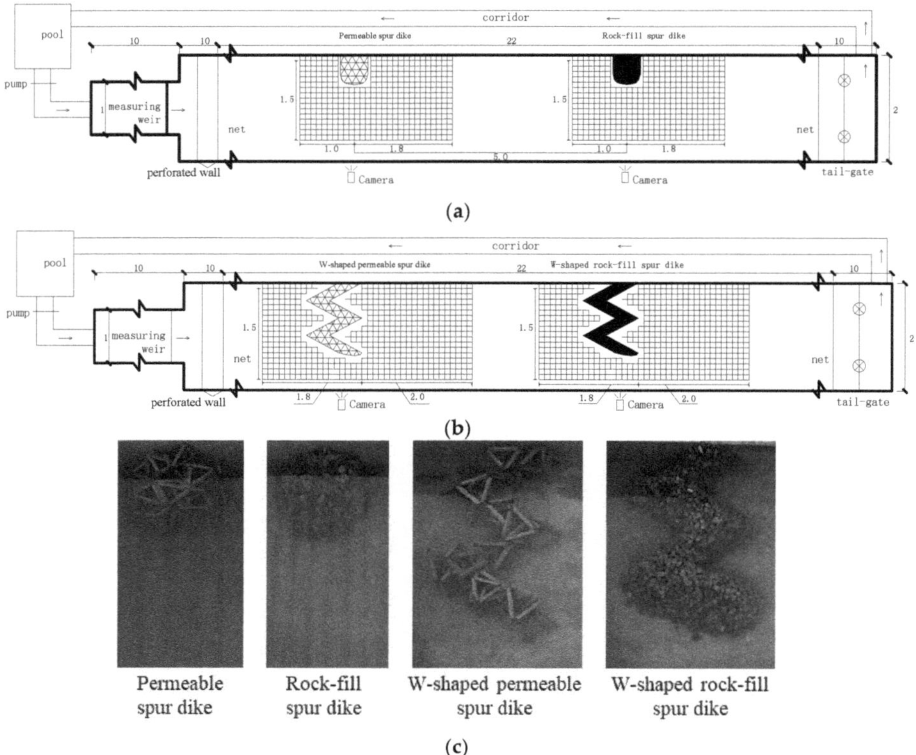

Figure 1. Test layout: (**a**) experimental layout of the rock-fill spur dike and the permeable spur dike, (**b**) test layout of the w-shaped rock-fill spur dike and the w-shaped permeable spur dike and (**c**) pictures of spur dikes. The unit in the figure is m. The grid points around the spur dike are the measurement points.

2.2. Experimental Procedure

For all experiments, the test flow discharge of the rock-fill and permeable spur dikes was 0.12, 0.18 and 0.24 m^3/s, and the water depth was 0.60 m. The test flow discharge of the w-shaped rock-fill and w-shaped permeable spur dikes was 0.18 and 0.20 m^3/s, and the water depths were 0.35 and 0.40 m, respectively. Flow conditions of all experiment were listed in Table 1. The flow velocity of the water layer 5 cm from the bottom was measured as the representative flow velocity of the corresponding point. A total of 900 data measurements were made at each point. The measurement range of the rock-fill and permeable spur dikes is as follows: 1.0 m upstream from the dike axis and 1.80 m downstream of the dike in the direction of water flow. The streamwise direction starts from the side of the dike root, from 0.1 to 1.6 m, and the measurement interval is 0.1 m. The measurement range of the w-shaped rock-fill and w-shaped permeable spur dikes is from 1.8 m upstream to 2.0 m downstream from the dike along the direction of water flow. The streamwise direction is from 0.1 to 1.6 m, and the measurement interval is 0.1 m, as shown in Figure 1.

The water flow conditions for the fish aggregation test and the spur dike setup were the same as the hydrodynamic test. For each test, after the water level and flow velocity were adjusted to working conditions, 25 fish were placed near the net at the tail of the flume. The camera was used to record the aggregation status of the test fish around the spur dike. After the test, the fish were removed and separated. It should be noted that different fish groupings were used in each test. Each test started at

9:30 a.m. and ended at 15:30 p.m. on the same day. During the test, no one was close to the flume, and the test fish distribution was observed on the video recorder through the camera.

Table 1. Test conditions and water flow diversity.

Spur Dikes	Q (L/s)	H (m)	U(m/s) Min–Average–Max	k (m²/s² (×10⁻⁴)) Min–Average–Max	H_F	AAR (%)
rock-fill spur dike	0.12	0.60	0–0.09–0.14	0.2–2.1–6.6	1.25	20
	0.18	0.60	0–0.13–0.18	0.0–2.9–12.5	1.13	5
	0.24	0.60	0–0.18–0.25	0.0–5.2–24.0	1.19	16
permeable spur dike	0.12	0.60	0–0.08–0.11	1.0–1.8–4.9	2.21	20
	0.18	0.60	0–0.12–0.16	0.8–3.0–8.8	2.05	13
	0.24	0.60	0–0.17–0.24	0.9–5.2–13.9	2.89	28
w-shaped rock-fill spur dike	0.18	0.35	0–0.21–0.46	0–19.4–84.5	2.86	23
	0.20	0.40	0–0.19–0.48	0–12.1–70.3	2.17	27
w-shaped permeable spur dike	0.18	0.35	0.01–0.18–0.31	1.6–8.5–58.1	3.07	29
	0.20	0.40	0.04–0.17–0.30	1.4–7.0–39.5	2.68	27

2.3. Data Processing

(1) Flow data process:

The flow velocity of each point was calculated using the recorded average flow velocity. The calculation formula is as follows

$$U = \sqrt{\bar{u}^2 + \bar{v}^2 + \bar{w}^2},\tag{1}$$

where $\bar{u}$, $\bar{v}$, and $\bar{w}$ are the average flow velocity in the streamwise, transverse, and vertical directions of the measurement point, respectively.

The turbulent kinetic energy can be calculated by the following formula:

$$k = \frac{1}{2}\left(\overline{u'u'} + \overline{v'v'} + \overline{w'w'}\right),\tag{2}$$

where u', v', and w' are streamwise, transverse, and vertical pulsating flow velocity. $u' = u - \bar{u}$, $v' = v - \bar{v}$, $w' = w - \bar{w}$ are the difference between the instantaneous flow velocity and the average flow velocity. u, v, and w are instantaneous flow velocity in the streamwise, transverse, and vertical directions.

(2) Flow data standardized:

When the test group is different, the numerical values of the hydrodynamic parameters are quite different, and the flow velocity and the turbulent kinetic energy are inconsistent, making comparisons difficult. Therefore, the data was standardized. The standardization of dispersion (min–max standardization) preserves the data trend and "compresses" the data to 0–1, facilitating the comparison of multiple datasets, so this method was used to normalize the flow velocity and turbulent kinetic energy data. The calculation for standardization of dispersion can be expressed by:

$$y_i = \frac{X_i - X_{min}}{X_{max} - X_{min}},\tag{3}$$

where y_i is the standardized data, X_{min}, X_{max}, X_i are the minimum, maximum and *i-th* data respectively.

(3) Fish aggregative data process:

At the beginning of the test, the fish were adapting to the conditions, so their positions changed greatly. Therefore, only the fish data from 10:00 a.m. to 15:00 p.m. were used in the actual analysis. A sample was taken every 30 min, and each sample was 5 min long. A picture was taken at every 10 s to record the frequency and position of the test fish. To eliminate influence of flow conditions and spur dike size on laboratory test data and on-site observation data, the average aggregation rate (*AAR*) was used to express the attraction effect of the spur dike on test fish [37]. This index indicates the ratio of the amount of gathered fish around the spur dike to the total number of test fish after setting each spur dike and is calculated by the following formula:

$$AAR(\%) = \frac{\sum_{i=1}^{n} N_i}{nN} \times 100\%, \tag{4}$$

where N_i is the number of fish gathered around the spur dike of the *i-th* observation, N is the total number of test fish, and n is the number of observations.

3. Results and Discussion

3.1. Mean Flow Characteristic

Table 1 shows the range of average flow velocity and turbulent kinetic energy around different types of spur dikes in the two different test groups. As listed in Table 1, mean flow velocity of different spur dikes at the same statistical area within a test is almost the same, but the max-min range is different, as that of the rock-fill spur dike is larger than the permeable spur dike. Both mean turbulent kinetic energy and the max-min range of the rock-fill spur dike are larger than that of the permeable spur dikes of the same statistical area within a test, especially for the w-shaped spur dikes. Generally speaking, spur dikes extend max-min range of velocity and turbulent kinetic energy, especially the rock-fill and the w-shaped rock-fill spur dikes.

Figure 2 shows the flow velocity and turbulent kinetic energy distribution around the rock-fill spur dike and permeable spur dike at a discharge of 0.18 m^3/s and average depth of 0.60 m. Figure 3 shows the flow velocity and turbulent kinetic energy distribution around the w-shaped rock-fill spur dike and w-shaped permeable spur dike at a discharge of 0.20 m^3/s and average depth of 0.40 m. Four types of spur dike are submerged in water.

When water flows through the rock-fill spur dike (Figure 2b), the flow velocity on the water-facing side of the dike body gradually decreases due to water-blocking by the dike body. The upstream is restricted and moves towards another side of the flume. After passing the head of the spur dike, the narrowest main flow cross-section will be formed due to inertia (A-A in Figure 2b). Then, the main flow diffuses gradually along the flume. Meanwhile, a very small flow velocity region is formed on the backwater side. The flow boundary lay behind the spur dike sperate from the groin when flow is passing the groin head. Eddies are formed downstream of the separated point and move forward one by one. The moving eddies collide, break and merge with each other, and the corresponding turbulent kinetic energy is large. However, the turbulent kinetic energy in other areas is low, and in the vicinity of the dike body it is almost zero.

The permeable spur dike can divert flow and narrow flow area as well as rock-fill spur dike. At the same time, as the permeable spur dike had good water permeability, the narrowest main flow cross-section moves down (B-B in Figure 2a), the flow mixing area with high turbulent kinetic energy move towards the main flow. Flow velocity and turbulent kinetic energy in the vicinity of the dike body was low—but not zero. The zone with low flow velocity and low turbulence was diminished compared with the rock-fill spur dike. The gradient of flow velocity and turbulent kinetic energy is lower than that of the rock-fill spur dike, as the disturbance on flow of permeable spur dike is reduced.

The w-shaped spur dike preserves the water flow characteristics of the spur dike, but there are differences as well—mainly in the vicinity of the dike. Oblique dikes (1, 2, 3, 4 in Figure 2b) of the w-shaped rock-fill spur dike are inclined, which causes large flow velocity at the two junctions on the lower side of the dike (area around A and B in Figure 3b). After flow passes over the first two straight-line dike (1 and 2 in Figure 3b), flow from two directions come across and mix in the downstream area of the spur dike (area around D in Figure 3b), where flow velocity and turbulent kinetic energy are very high (area around D in Figure 3d). A similar area around E (in Figure 3b) is affected by flow around the groin head as well as the mix effect of flow from dike 3 and dike 4 (in Figure 3b), which makes flow velocity and turbulent kinetic energy small.

The w-shaped permeable spur dike has a similar diversion function as the w-shaped rock-fill spur dike, which causes flow to come across and mix around the two junctions on the lower side of the dike (the area around A and B in Figure 3a). However, as the w-shaped permeable spur dike has large porosity, flow velocity of the main flow (the area around F in Figure 3a) is smaller than that of the w-shaped rock-fill spur dike, and opposite in the vicinity of the dike (the area around A, B, D, F in Figure 3a). Frames of w-shaped permeable spur dike have excellent energy dissipation capacity. Flow run through those frames would generating high level turbulent kinetic energy in the vicinity areas downstream of the w-shaped dike (D, E, and C in Figure 3c).

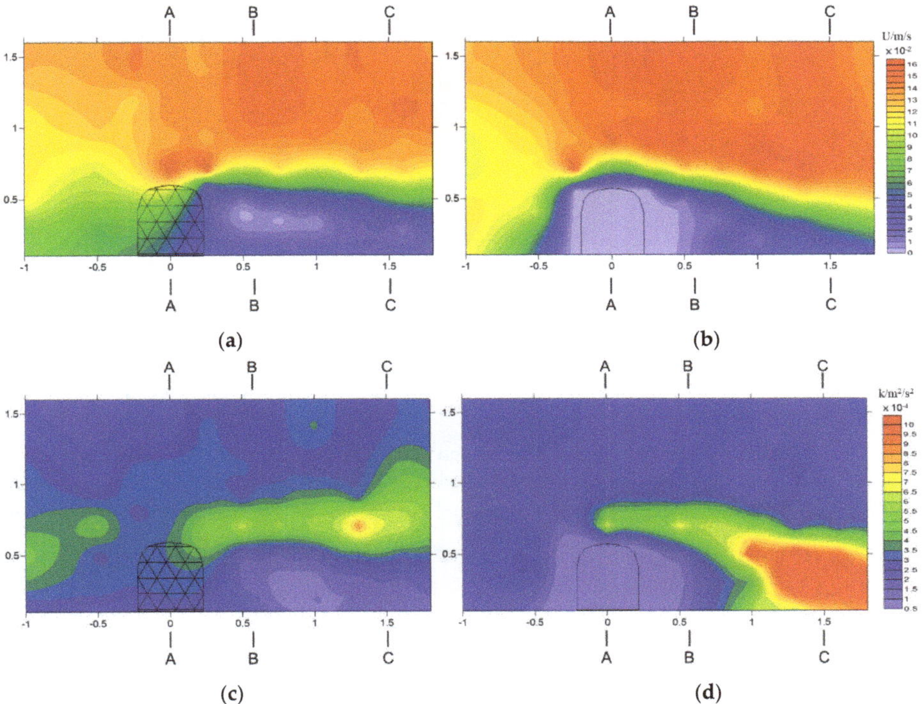

Figure 2. Hydrodynamic distribution of permeable spur dike and rock-fill spur dike (flow velocity is 0.18 m^3/s, water depth is 0.60 m; flow direction is from left to right, distance unit in the figure is m, same as below); (**a**) Average flow velocity distribution of permeable spur dike; (**b**) Average flow velocity distribution of rock-fill spur dike; (**c**) The turbulent kinetic energy distribution of permeable spur dike; (**d**) The turbulent kinetic energy distribution of rock-fill spur dike .

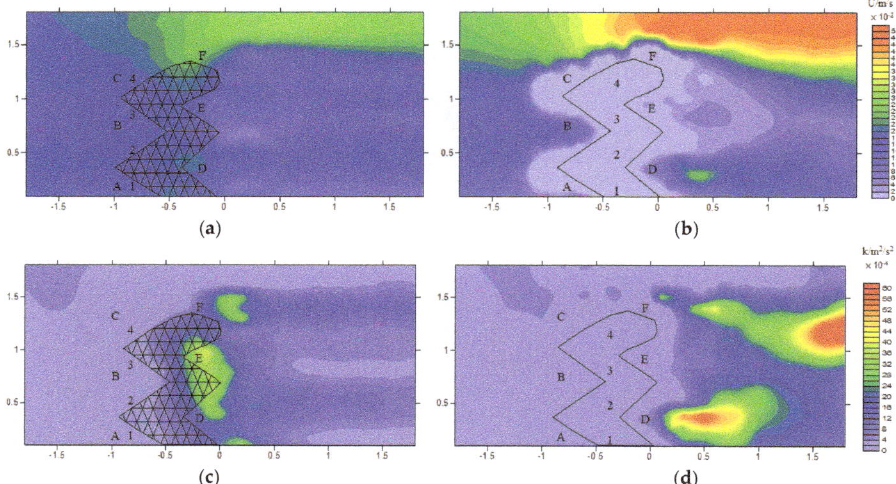

Figure 3. Hydrodynamic distribution of w-shaped permeable spur dike and rock-fill spur dike and aggregation of test fish (flow velocity is 0.20 m^3/s, water depth is 0.40 m). (**a**) Average flow velocity distribution of w-shaped permeable spur dike; (**b**) Average flow velocity distribution of w-shaped rock-fill spur dike; (**c**) The turbulent kinetic energy distribution of w-shaped permeable spur dike; (**d**) The turbulent kinetic energy distribution of w-shaped rock-fill spur dike.

3.2. Water Flow Diversity

Entropy is a measure of system disorder, which is an extensive property of a thermodynamics system. In 1948, Shannon introduced the notion of entropy to information theory in order to measure the quantity of information contained in a message [38]. Entropy is a fundamental concept linking together information theory and statistical physics [39]. For a set of possible events $\{x_1, x_2, \dots, x_n\}$ whose probabilities of occurrence are $p_1, p_2, \dots, p_n$. The information entropy or Shannon's entropy is expressed by:

$$H = -\sum_{i=1}^{n} P_i ln P_i. \tag{5}$$

The more uncertain and heterogeneous the possible event, the greater Shannon's entropy; the more certain and homogeneous the possible event, the smaller Shannon's entropy. For only one type in the data set, Shannon's entropy equals zero. Therefore, high Shannon's entropy stands for high diversity, low Shannon's entropy for low. The Shannon's entropy equation provides a way to quantifies the randomness of probability laws and is a measure of heterogeneity commonly applied in the fields of statistical physics, information theory, ecology [39,40].

The spur dike is a permanent disturbance for a river flow system at reach scale. In order to eliminate the disturbance, the river flow regime is changed and spatial distribution, range of mean velocity and turbulent kinetic energy are modified (analyzed in Section 3.1), which can be used to describe flow diversity around the spur dike. Mean velocity and turbulent kinetic energy are important parameters reflecting the engineer effect, structure stability as well as the habitat quality. Based on Shannon's entropy, we devised a water flow diversity index H_F, to measure the diversity of water flow around the spur dike. It is defined as the product of the regional flow velocity diversity index H_V and the turbulent kinetic energy diversity index H_k, which can be expressed by:

$$H_F = H_V \times H_k, \tag{6}$$

$$H_V = -\sum\nolimits_{i=1}^{n} P_{Vi} ln P_{Vi}, \tag{7}$$

$$H_k = -\sum\nolimits_{i=1}^{n} P_{ki} ln P_{ki}, \tag{8}$$

where n is the level of flow velocity and turbulent kinetic energy. n is 10, meaning that the measured flow velocity and turbulent kinetic energy are divided into 10 levels after standardization, which is from 0.1 to 1.0. P_{Vi} and P_{ki} are the ratios of i-th flow velocity and turbulent kinetic energy to the total area of the studied area, respectively. According to Equation (6), for the present study, H_F reaches a maximum when $P_{Vi} = P_{ki} = 0.1$, and $H_{Fmax} = ln(10)^2 = 5.30$. Namely, the more uniform the area ratio of flow velocity and turbulent kinetic energy, the larger the H_F, indicating a higher water flow diversity around the spur dike. The area ratio of some level(s) of the flow velocity (turbulent kinetic energy) is not zero—however, it is extremely small, making it impossible to provide an effective habitat for fish. Therefore, the flow velocity and turbulent kinetic energy with an area ratio of less than 5% was excluded in actual calculations. Additionally, gaps among gravels of flume experiments are too small to use for fish, so this area was excluded in the statistics when calculating the water flow diversity index.

According to Equations (6)–(8), the water flow diversity index H_F around the spur dikes for each working condition were calculated and the results are shown in Table 1. Under the different waterflow conditions, the values and distribution of flow velocity and turbulent kinetic energy are different. However, the water flow diversity index calculated after the nondimensionalization fluctuates within a certain range, which means the water flow diversity around a particular spur dike structure can be represented by a constant, such as the rock-fill spur dike, where the mean H_F is 1.19 ± 0.05. According to the analysis above, the larger the constant—the higher the diversity of the water flow. Based on our calculations, the water flow diversity index around the four spur dikes ranged from 1.13 to 2.896, in the order (from high to low): w-shaped permeable > w-shaped rock-fill > permeable > rock-fill spur dike. The regional ratio of standardized velocity and standardized turbulent kinetic energy of different spur dikes is shown in Figure 4. Flow conditions are the same as discussed in Section 3.1. Under the same water flow conditions, the flow velocity and turbulent kinetic energy around the rock-fill spur dike was higher than that of the permeable spur dike, but the area of particularly large flow velocity and turbulent kinetic energy is small, and the area of slow flow zone is large, which makes small parts of standardized velocity and turbulent kinetic energy account for most of the area. The rock-fill spur dike, with a standardized velocity of 0.8, 0.9 and 1.0 accounts for 70% of the area, and the standardized turbulent kinetic energy of 0.2 takes up 67% of the area. The flow velocity and turbulent kinetic energy area ratios around the rock-fill spur dike are quite different at each level, while those around the permeable spur dike are relatively uniform. The corresponding water flow diversity index value is relatively high as well. For the w-shaped spur dike, its structural design increases the complexity of the water flow in the adjacent area of the dike (refer to the analysis above). As shown in Figure 4, subsequently, the corresponding water flow diversity index increases as well.

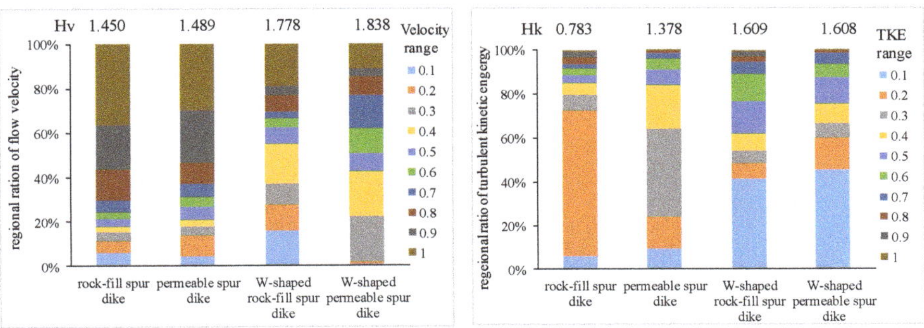

Figure 4. Regional ratio of velocity and turbulent kinetic energy (flow condition is the same as Figures 2 and 3).

3.3. Relationships between Fish Aggregation and Water Flow Diversity Index

The aggregation of the test fish around the rock-fill spur dike and permeable spur dike at discharge of 0.18 m^3/s are shown in Figure 5, and the gathering of the test fish around the w-shaped rock-fill spur dike and the w-shaped permeable spur dike at discharge of 0.20 m^3/s are shown in Figure 6. Overall, the test fish mainly gathered in the slow-flow area on the back side of the spur dike where flow velocity are nearly the lowest (Figures 2, 3, 5 and 6). For spur dikes of the same shape, the average aggregation rate of the test fish around the permeable spur dike was higher, and the aggregation range larger than that of the rock-fill spur dike. The test fish in the w-shaped spur dike would gather in the slow-flow area on the back side of the dike, but gathered in the slow-flow area on the water-facing side of the dike as well—especially for the W-shaped permeable spur dike. This can be explained by the following: the water-blocking effect of the rock-fill spur dike reduces flow velocity on the back side of the dike to extremely low levels, and the flow velocity and turbulent kinetic energy in the water flow mixing zone vary greatly. Additionally, the test fish need to use more energy if they want to stay, and the drastic change of the water flow environment makes it difficult for them to maintain their stability as well. However, the volume velocity of the slow-flow zone on the backwater side of the permeable spur dike is not zero—and there is a certain pulsation. In the water flow mixing zone behind the spur dike, flow velocity and turbulent kinetic energy vary only slightly so fish can gain power from the turbulence to reduce their own energy loss. Additionally, compared with the rock-fill spur dike, the structural design of the w-shaped rock-fill spur dike makes water flow around the spur dike more complex. The flow velocity and the turbulent kinetic energy gradient around the w-shaped permeable spur dike are lower than that of the w-shaped rock-fill spur dike as permeability of the permeable spur dike, which is easier for fish to maintain their stability. Besides, compared with the rock-fill spur dikes, cavity within the permeable spur dikes provides refuge for fish, which is an important factor for fish gathering and is widely applied in artificial reef design.

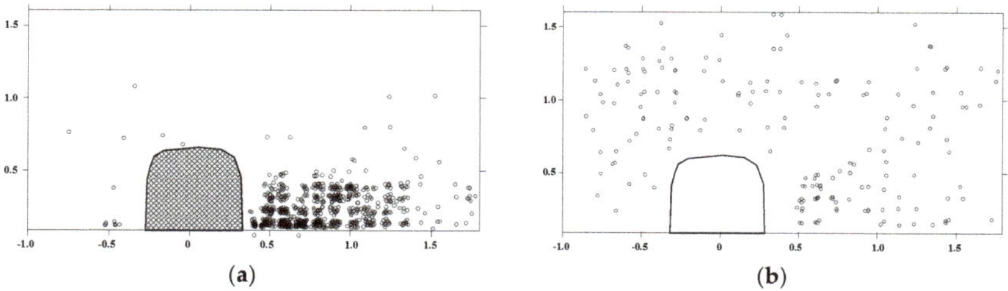

Figure 5. Aggregation of test fish of permeable spur dike and rock-fill spur dike (flow velocity is 0.18 m^3/s, water depth is 0.60 m; "o" represents fish location, flow direction is from left to right, distance unit in the figure is m, same as below). (**a**) Test fish distribution of permeable spur dike; (**b**) Test fish distribution of rock-fill spur dike.

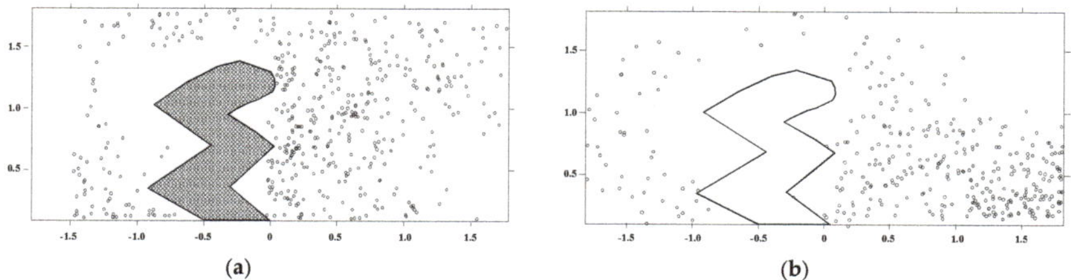

Figure 6. Aggregation of test fish of w-shaped permeable and rock-fill spur dikes (flow velocity is 0.20 m^3/s, water depth is 0.40 m). (**a**) Test fish distribution of w-shaped permeable spur dike; (**b**) Test fish distribution of w-shaped rock-fill spur dike.

Test fish average aggregation rate around the spur dike of each test was listed in Table 1. The average aggregation rate of the test fish around the w-shaped permeable spur dike was the highest (about 28%), while around the rock-fill spur dike it was the lowest (about 14%). Figure 7 shows the relationship between the test fish average aggregation rate and the water flow diversity index. It is not difficult to conclude that for a certain type of spur dike under different flow conditions, fish average aggregation rate increases with water flow diversity index—except for one group. For different types of spur dikes, average aggregation rate of all test increases with average water flow diversity index. In other way, the aggregation effect of the spur dike is enhanced with an increase in water flow diversity. A relation curve between average aggregation rate and water flow diversity index was plotted with a linear fitting method using experiment data. However, fish movement is random, turbulent flow and its relationship with complex correlations between water flow diversity index and average the fish aggregation is not very high.

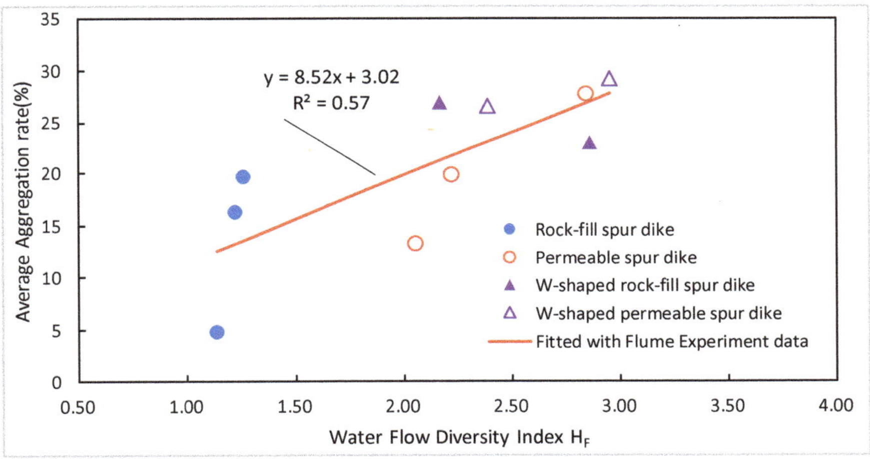

Figure 7. Relationship between water flow diversity and average aggregation rate.

At present, before a newly designed structure applied in channel regulation or river restoration, a series of flume experiments and numerical studies have to be conducted to research flow characteristics (mean and turbulent), geomorphologic changes, habitat suitability, which would cost a lot of time and money. Although new layouts or structures of the spur dike were proposed and applied, it's still difficult to quantified parameters of flow diversity around spur dikes. Studies have demonstrated the importance of turbulence on fish swimming [30,41] and habitat selection [11,26,27,32]. However, fish movement is random, different fish have their own habits, and the complexity of turbulence and mechanisms of fish assemblage in a turbulent environment is still not clear, all these factors increase the difficulty of quantified research on fish responses to turbulent flow. The new relationship between structure and water flow complexity, average aggregation rate of test fish and water flow complexity in our research can provide a tool and a new method to determine which type of structure to be chosen and how to optimize the structure.

However, flow characteristic of spur dikes is affected not only by layouts and made-up elements, but also by parameters of spur dikes (such as length, height, porosity), their relative relations with flow (like ratio of height of spur dike to depth), which may change water flow diversity. Besides, as mention above, different fish (including different species and a specie of different age) have different habits. More research needs to be conducted—especially quantitative studies on flow characteristics and flow diversity of different spur dikes and studies on the relationship between more fish (different species, a specie of different life stage) aggregation and turbulent flow characteristics.

4. Hydroacoustic Investigation of Laohutan Fish-Bone Dike, the Dongliu Reach and the Yangtze River

4.1. Project Background and Investigation Techniques

The Dongliu reach is located between Jiujiang and Anqing in the downstream part of the Yangtze River (Figure 8). The waterway is complex and variable in the dry season with poor navigation conditions. It has always been one of the key shallow waterways of the Yangtze River. To limit the development of Donggang, control the width of the left-hand rectification line of Laohutan, increase the flow velocity of the left-hand of Laohutan, and improve the water depth of the left-hand waterway, the Yangtze River Waterway Bureau built a fish-bone dike at the head of Laohutan. The project began in November, 2012 and was completed in April, 2014. The fish-bone dike consists of a 1050 m ridge dike and two thorn dikes (342 and 521 m, respectively). The first 100 m in front of the right side of the thorn dike is a four-side six-edge permeable dike, and the other part is a rock-fill structure. Two investigations were carried out for the water flow during the season, using the flow of the Datong Station which can be used to represent the flow of the Dongliu reach as it is located 120 km downstream, with no large tributaries in between. During the dry season in 7 December 2016 the flow at the Dantong station was 13,022 m^3/s and the flood flow there on 8 August 2017 was 46,190 m^3/s.

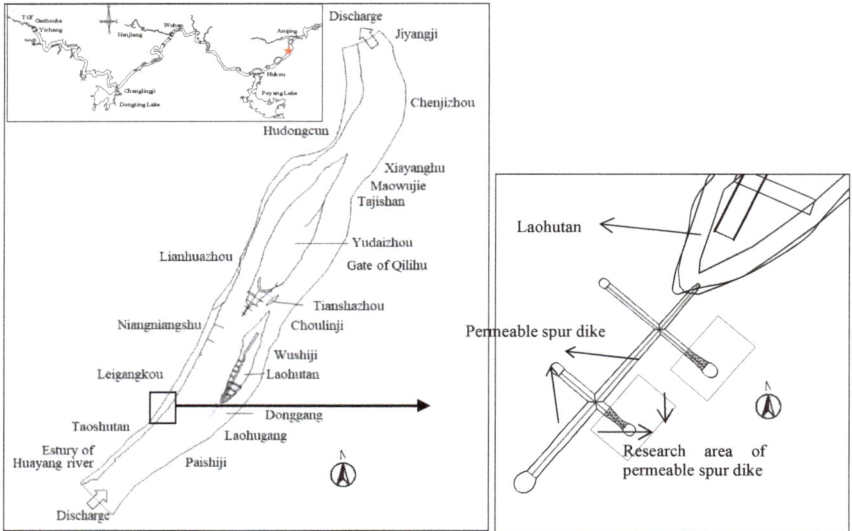

Figure 8. Fish-bone dike Project at Laohutan of the Dongliu reach Waterway.

A BioSonics-DT-X-Digital Scientific Echosounder (BioSonics, Inc., Seattle, WA, USA) was used to investigate the distribution of fish around the fish-bone dike. Its transducer has an operational frequency of 208 khz and split-beam of 6.5°. The investigation time was from 9:00 a.m. to 16:00 p.m. The transducer of the Biosonics DT-X echo detector (BioSonics, Inc., Seattle, WA, USA) was fixed to the trimaran and the trimaran was tied to the fishing boat's head by a rope. The transducer was about 0.4 m into the water and inclined at 45° to the water surface. Multiple navigation measurements were conducted around the fish-bone dike to cover as much of the dike as possible. Hydroacoustic data acquisition was performed using BioSonics Acquisition 6.0 software (BioSonics, Inc., Seattle, WA, USA). The pulse frequency of the transducer was 8 pps, the pulse width was 0.5 ms, the data acquisition threshold was −130 dB, and the data acquisition distance was 1–30 m. Using a Garmin GPS 17x HVS, GPS data was synchronously collected and stored as well. The instrument was calibrated in the field using a 36 mm tungsten carbide standard ball prior to measurement. Hydroacoustic data collected from on-site observations were analyzed using BioSonics Visual Analyzer 4.3 (BioSonics, Inc., Seattle, WA, USA) to identify fish location and to record fish frequency of occurrence. The data analysis started from the 1 m position of the beam and synchronously outputted the monomer echo recognition result. The single echo recognition parameters were as follows: the echo threshold was −60 dB, the correlation coefficient was 0.90, the minimum pulse width coefficient was 0.75, the maximum pulse width coefficient was 3, the termination pulse width was −12 dB and the time-varying gain (TVG) was 40 lgR.

Field velocity in large river are usually measured with ADCP (Acoustic Doppler Current Profilers, produced by Teledyne RD Instruments, Poway California, CA, USA), and Vector Current Meter (point measure) (Nortek AS, Rud, Norway). Flow turbulent characteristics can be derived from the measured data of ADV (Nortek AS, Rud, Norway), or ADCP (Teledyne RD Instruments, Poway California, CA, USA) measure data [42,43], but the velocity measured devices need to be held still for a certain period. It means the devices must be fastened to a holder that can be fixed on the riverbed. According to the physical model research results, velocity around heads of the fish-bone is about 1.5 m/s, and D-shaped mattress and four-side six-edge permeable frames were placed around the dike to protect the structure [44]—which makes it difficult and dangerous to set a holder to fix ADCP and ADV. Besides, flow characteristic of rock-fill spur dike and permeable were measured and analyzed using experiment data. Flow around the rock-fill dikes and the permeable in Dongliu reach were not measured.

4.2. Results

During on-site observation, it was difficult to accurately determine the total fish nN due to the ship's disturbance and limited measurement coverage. Therefore, the total number of fish observed was taken as the total fish nN. The permeable spur dike area was 200 m × 300 m and surrounded by the 100 m of water-permeable dike facing the water side, the 100 m in front of the dike head, and the 200 m backwater side of the dike. The counting range of the rock-fill dike was the left-hand area of the fish-bone dike. The habitat selection of fish is affected by many factors, including hydrological, hydraulic, biological and the life stage of the fish. In our research, permeable spur dike and rock-fill spur dike are adjacent, and flow discharge, sediment size, temperature, fish composition are almost the same. The difference is the range and spatial distribution of velocity and turbulence, mainly caused by makeup of the spur dike, which affect fish aggregation of the spur dike. In other words—the fish aggregation rate in the Dongliu reach is comparable.

During water flow in the dry season on 7 December 2016, there were 65 fish observed in the fish-bone dike, including eight fish in the rock-fill spur dike and 17 fish in the permeable spur dike. The corresponding average aggregation rates were 12% and 26%, respectively. During flood flow on 8 August 2017, there were 69 fish observed in the fish-bone dike, including 11 fish in the rock-fill spur dike and 19 fish in the permeable spur dike. The corresponding average aggregation rates were 16% and 28%. Thus, it can be concluded that fish aggregation was better in the permeable spur dike than the rock-fill spur dike.

The permeable spur provides more heterogeneous flow environment than rock-fill spur dike (analyzed above), which is attractive for different fish (different species and a species of different life stage). Flow velocity in downstream of Yangtze River is very fast, which make it difficult for plankton and macroinvertebrate survive as their weak swimming ability. The flow velocity behind the permeable spur dike is slow but not almost still as where of the rock-fill spur dike, which provides a stable habitat for plankton and macroinvertebrate, increases feeding potential for fish. Besides, cavities with the permeable spur dike provide refuge for fish larvae and other small dwellers from rapid flow and predators [3,45,46]. In some degree, the permeable spur dikes work as an artificial reef applied to increase reef fish habitat quantity and quality, protect marine life, reduce user conflicts, and so on [47].

The artificial reef-like permeable spur dike makes it more attractive for fish than the rock-filled spur dike and has a greater fish aggregative rate.

In this section, hydraulic parameters were not measured for the complex flow environment. But Hydraulic characteristics are a key factor affecting habitat use of fish, and a link between results of flume experiments and field. Systematically monitors and analysis on hydraulic distribution, geomorphic change, dwellers (plankton, macroinvertebrate and fish) diversity around the permeable and rock-fill spur dikes in Laohutan, Dongliu are necessary in the future.

5. Conclusions

Rock-fill, permeable, w-shaped rock-fill, and w-shaped permeable spur dikes were selected for flume experiments to investigate their influence on water flow diversity and the relationship between water flow diversity and fish aggregation. Additionally, a fish hydroacoustic study was conducted on the Laohutan fish-bone dike in the Dongliu reach of the Yangtze River to investigate fish aggregation effects around rock-fill and permeable spur dikes in the field.

(1) Based on Shannon's entropy theory, a water flow diversity index, H_F was proposed to calculate water flow diversity.

(2) The water flow diversity index values of different types of spur dikes under different flow condition were calculated, reflecting the degree of water flow diversity around the different spur dikes. The water flow diversity index values of the four different types of spur dikes are 1.13–2.96, and from high to low are: w-shaped permeable > w-shaped rock-fill > permeable > rock-fill spur dike.

(3) The test fish gathered in the backwater side slow-flow area of the spur dike, and the average aggregation rate was 5% to 28%. The attraction effect of the spur dike to fish increases with an increasing water flow diversity. The relationship between water flow diversity index and average fish aggregation rate was plotted, which provides a new method to assess the environmental effects of spur dikes.

(4) The permeable spur dike had a better fish aggregation effect than the rock-fill spur dike in the field at Laohutan fish-bone dike of the Dongliu reach Waterway in dry and flood seasons.

(5) The permeable spur dike works as a river training structure as well as an artificial reef: it alters hydrodynamic conditions, the sediment transport regime, and at the same time generates more heterogeneous flow than rock-fill spur dike, providing shelter for fish and other small organism, increasing fish habitat quantity and quality and reducing conflicts of river training and river protection.

The research results of this paper could provide theoretical support for the evaluation and analysis of habitat water flow diversity, as well as ecologically optimal design of spur dikes.

These results derive from some certain types of spur dikes among different river training structures and a limited number of individuals—all of the same species. Future works should expand this work with more river training structures and fish across different species and life stages.

Author Contributions: Flume experiments, T.H.; hydroacoustic investigation, Y.L., T.H., H.L.; writing-original draft preparation, T.H.; writing-review and editing, Y.L., H.L.; supervision, Y.L.

Funding: This research was funded by the National Key Research and Development Program of China (Grant No. 2016YFC0402108) and Hydraulic Science and Technology Program of Jiangsu Province, China (No. 2017044 and 2018038).

Acknowledgments: We are particularly grateful to Dingan Zhang, Yi Liu and Weixu Wang for assisting on experiments and the hydroacoustic investigation.

References

1. Walker, J. Geomorphological considerations for the introduction of boulders and groynes for fisheries enhancement: Assessment of River Eden at Carhead. *J. Phycol.* **1998**, *38*, 862–871.
2. Pinn, E.H.; Mitchell, K.; Corkill, J. The assemblages of groynes in relation to substratum age, aspect and microhabitat. *Estuar. Coast. Shelf Sci.* **2005**, *62*, 271–282. [CrossRef]

3. Lu, Y.; Lu, Y.; Li, S.; Ding, J.; Liu, H. *Study on Process of Flow and Sediment and Ecological Effects on River with Waterway Regulation*; Nanjing Hydraulic Research Institute: Nanjing, China, 2011. (In Chinese)

4. Lacey, R.W.J.; Neary, V.S.; Liao, J.C.; Enders, E.C.; Tritico, H.M. The IPOS Framework: Linking Fish Swimming Performance in Altered Flows from Laboratory Experiments to Rivers. *River Res. Appl.* **2012**, *28*, 429–443. [CrossRef]

5. Kang, J.G.; Yeo, H.K.; Jung, S.H. Flow characteristic variations on groyne types for aquatic habitat. *Engineering* **2012**, *4*, 811–812. [CrossRef]

6. Erick, D.; Thiel, R. Key environmental variables affecting the ichthyofaunal composition of groyne fields in the middle Elbe River, Germany. *Limnol. Ecol. Manag. Inland Waters* **2013**, *43*, 297–307.

7. Pan, B.; Wang, Z.; Li, Z.; Yong-Jun, L.; Wen-Jun, Y.; Yi-Ping, L. Macroinvertebrate assemblages in relation to environments in the West River, with implications for management of rivers affected by channel regulation projects. *Quat. Int.* **2015**, *384*, 180–185. [CrossRef]

8. Wang, W.; Yin, C.; Lu, J.; Wang, H. Application of submerged groin systems in the ecological restoration of littoral zones. *Chin. J. Environ. Eng.* **2007**, *1*, 135–138. (In Chinese)

9. Wang, Z.; Wu, B.; Wang, G. Fluvial processes and morphological response in the Yellow and Weihe Rivers to closure and operation of Sanmenxia Dam. *Geomorphology* **2007**, *91*, 65–79. [CrossRef]

10. Allan, J.D.; Castillo, M.M. *Stream Ecology: Structure and Function of Running Waters*, 2nd Ed. ed; Springer: Dordrecht, The Netherlands, 2007.

11. Enders, E.C.; Roy, M.L.; Ovidio, M.; Hallot, E.J.; Boyer, C.; Pettit, F. Habitat Choice by Atlantic Salmon Parr in Relation to Turbulence at a Reach Scale. *North Am. J. Fish. Manag.* **2009**, *29*, 1819–1830. [CrossRef]

12. Burch, C.W.; Abell, P.R.; Stevens, M.A.; Dolan, R.; Dawson, B.; Shields, F.D., Jr. Environmental Guidelines for Dike Fields. In Proceedings of the Conference River 83, ASCE, New Orleans, LA, USA, 24–26 October 1983.

13. Rosgen, D. *Applied River Morphology*; Wildland Hydrology: Pagosa Springs, CO, USA, 1996.

14. Joseph, W.J. River Training structures and Secondary Channel Modifications. In *U.S. Army Corps of Engineers, Rock Island District*; Upper Mississippi River Restoration—Environmental Management Program's (UMRR-EMP) Environmental Design Handbook: Rock Island, IL, USA, 2012.

15. Martin, G.; Kurt, G.; Michael, T.; Liederman, M.; Habersack, H. Hydrodynamic and morphodynamic sensitivity of a river's main channel to groyne geometry. *J. Hydraul. Res.* **2018**, *56*, 714–726.

16. Zheng, J.; Lei, G.; Zhu, W.; Gu, Z.; Yu, Z. Comparative analysis of layout and structure of filter dam at Zhangjiawan of Zhougongdi waterway. *Port Waterw. Eng.* **2014**, *3*, 130–133. (In Chinese)

17. Li, Y.; Ma, Y. *Research on the Development of New Spur Dike Applied in the Taicang–Nantong Reach [R]*; CCCC First Harbor Consultants Co Ltd.: Tianjin, China, 2014. (In Chinese)

18. Liu, H. Ecological measures and technology prospect for trunk waterway management in the Yangtze River. *Port Waterw Eng.* **2016**, *1*, 114–118. (In Chinese)

19. Chang, L.; Xu, B.; Xiao, Z.; Zhang, P. Experimental study on permeability characteristics of hollow trapezoidal block spur dike. *Chin. J. Hydrodyn.* **2019**, *34*, 103–109. (In Chinese)

20. Milhous, R.T.; Wegner, D.L.; Waddle, T. *User Guide to the Physical Habitat Simulation System (PHABSIM)[R]*; Fish and Wildlife Service: Washington, DC, USA, 1984.

21. Bovee, K.D. *Development and Evaluation of Habitat Suitability Criteria for Use in the Instream Flow Incremental Methodology*; Fish and Wildlife Service: Washington, DC, USA, 1986.

22. McVicar, B.J.; Roy, A.G. Hydrodynamics of a forced riffle pool in a gravel bed river: 1. Mean velocity and turbulence intensity. *Water Resour. Res.* **2007**, *43*, W12401. [CrossRef]

23. McVicar, B.J.; Roy, A.G. Hydrodynamics of a forced riffle pool in a gravel bed river: 2. Scale and structure of coherent turbulent events. *Water Resour. Res.* **2007**, *43*, W12402. [CrossRef]

24. Liao, J.C.; Beal, D.N.; Lauder, G.V.; Triantafyllou, S.M. Fish Exploiting Vortices Decrease Muscle Activity. *Science* **2003**, *302*, 1566–1569. [CrossRef] [PubMed]

25. Liao, J.C.; Beal, D.N.; Lauder, G.V.; Triantafyllou, M.S. The Karman gait: Novel body kinematics of rainbow trout swimming in a vortex street. *J. Exp. Biol.* **2003**, *206*, 1059–1073. [CrossRef] [PubMed]

26. Smith, D.L.; Brannon, E.L.; Shafii, B.; Odeh, M. Use of the Average and Fluctuating Velocity Components for Estimation of Volitional Rainbow Trout Density. *Trans. Am. Fish. Soc.* **2006**, *135*, 431–441. [CrossRef]

27. Smith, D.L.; Goodwin, R.A.; Nestler, J.M. Relating Turbulence and Fish Habitat: A New Approach for Management and Research. *Rev. Fish. Sci. Aquac.* **2014**, *22*, 123–130. [CrossRef]

28. Enders, E.C.; Buffin, B.T.; Boisclair, D.; Roy, A.G. The feeding behavior of juvenile Atlantic salmon in relation to turbulent flow. *J. Fish Biol.* **2005**, *66*, 242–253. [CrossRef]

29. Cotel, A.J.; Webb, P.W.; Tritico, H. Do brown trout choose locations with reduced turbulence? *Trans. Am. Fish. Soc.* **2006**, *135*, 610–619. [CrossRef]

30. Tritico, H.M.; Cotel, A.J. The effects of turbulent eddies on the stability and critical swimming speed of creek chub (Semotilus atromaculatus). *J. Exp. Biol.* **2010**, *213*, 2284–2293. [CrossRef] [PubMed]

31. Webb, P.W.; Cotel, A.J. Assessing possible effects of fish–culture systems on fish swimming: The role of stability in turbulent flows. *Fish Physiol. Biochem.* **2011**, *37*, 297–305. [CrossRef] [PubMed]

32. Tullos, D.; Walter, C. Fish use of turbulence around wood in winter: Physical experiments on hydraulic variability and habitat selection by juvenile coho salmon, Oncorhynchus kisutch. *Env. Biol. Fishes* **2015**, *98*, 1339–1353. [CrossRef]

33. Crowder, D.W.; Diplas, P. Using two-dimensional hydrodynamic models at scales of ecological importance. *J. Hydrol.* **2000**, *230*, 172–191. [CrossRef]

34. Crowder, D.W.; Diplas, P. Vorticity and circulation: Spatial metrics for evaluating flow complexity in stream habitats. *Can. J. Fish. Aquat. Sci.* **2002**, *59*, 633–645. [CrossRef]

35. Dong, Z.; Sun, Y. *Principals and Technologies of Eco-Hydraulic Engineering*; China Water & Power Press: Beijing, China, 2007; pp. 285–287. (In Chinese)

36. Lu, Y.; Lu, Y.; Huang, T. *Interim Report of Waterway Regulation Technology and Demonstration in Typical Ecological Protection reach of the Yangtze River*; Nanjing Hydraulic Research Institute: Nanjing, China, 2018. (In Chinese)

37. Chen, Y.; Liu, X.-D.; Wu, X.-Y.; Shi, G.-F. Distribution of schlegel's rockfish (Sebastes schlegeli Hilgendorf) in different artificial reef models. *J. Dalian Fish. Univ.* **2006**, *21*, 153–157.

38. Shannon, C.E. A Mathematical Theory of Communication. *Bell Syst. Tech. J.* **1948**, *27*, 379–423. [CrossRef]

39. Jaynes, E.T. Information Theory and Statistical Mechanics. *Phys. Rev.* **1957**, *106*, 620–630. [CrossRef]

40. Margalef, R. La Teoría de la Información en Ecología. *Memorias de la Real Academia de Cienciasy Artes de Barcelona* **1957**, *32*, 373–449. (In Spanish)

41. Enders, E.C.; Boisclair, D.; Roy, A.G. The effect of turbulence on the cost of swimming for juvenile Atlantic salmon (Salmo salar). *Can. J. Fish. Aquat. Sci.* **2003**, *60*, 1149–1160. [CrossRef]

42. Lohrmann, A.; Hackett, B.; Røed, L.P. High Resolution Measurements of Turbulence, Velocity and Stress Using a Pulse-to-Pulse Coherent Sonar. *J. Atmos. Ocean. Technol.* **1990**, *7*, 19–37. [CrossRef]

43. Lu, Y.; Lueck, R.G. Using a Broadband ADCP in a Tidal Channel. Part II: Turbulence. *J. Atmos. Ocean. Technol.* **1999**, *16*, 1568–1579. [CrossRef]

44. Li, G.; Xu, H.; Gao, Y.; Song, D.; Zhu, Z.; Zhao, J. *Flood Control Evaluation of the Second Phase of Waterway Regulation of Dongliu Reach Downstream of the Yangtze River*; Nanjing Hydraulic Research Institute: Nanjing, China, 2011. (In Chinese)

45. Li, S.; Xiong, F.; Wang, G.; Duan, X.; Liu, S.; Chen, D. Effect of Tetrahedron Permeable Frames on the Community Structure of Benthic Macroinvertebrates in the Middle Yangtze River. *J. Hydroecol.* **2015**, *36*, 72–79. (In Chinese)

46. Guo, J.; Wang, K.; Duan, X.; Chen, D.; Fang, D.; Liu, S. Effect of Tetrapod Clusters on Fish Assemblage in Channel Improvement Projects. *J. Hydroecology* **2015**, *36*, 29–35. (In Chinese)

47. Seaman, J.W. *Artificial Reef Evaluation with Application to Natural Marine Habitats*; CRC Press: New York, NY, 2000.

Article

Simulation of a Hydrostatic Pressure Machine with Caffa3d Solver: Numerical Model Characterization and Evaluation

Rodolfo Pienika [1,2,*], Gabriel Usera [1] and Helena M. Ramos [2]

[1] IMFIA, Facultad de Ingeniería, Universidad de la República, Montevideo 11300, Uruguay;
gusera@fing.edu.uy

[2] Civil Engineering, Architecture and Georesources Departament, CERIS, Instituto Superior Técnico,
Universidade de Lisboa, 1049-001 Lisbon, Portugal; helena.ramos@tecnico.ulisboa.pt

* Correspondence: rpienika@fing.edu.uy; Tel.: +598-27113386

Received: 21 July 2020; Accepted: 25 August 2020; Published: 28 August 2020

Abstract: The Hydrostatic Pressure Machine (HPM) is a novel energy converter for micro and pico hydropower that becomes very suitable for installation in channels with very low head, where conventional hydraulic turbines are inadequate or too expensive. Although this technology has been studied through several experimental tests and also by numerical simulations, open source flow solvers have not been used yet. The research team on Computational Fluid Mechanics of IMFIA-Universidad de la República (Uruguay) has been developing a CFD open source solver named caffa3d, which has obtained great results in a few international challenges, although it has not been used yet for free surface flows or turbomachinery simulations. The present work shows the contributions made within caffa3d in order to enable its use for simulating a HPM. The Large Eddy Simulation (LES) method is used to model the turbulence structures of the flow. Sliding Mesh (SM) and Volume of Fluid (VOF) methods were chosen respectively to resolve the rotation of the wheel and the position of the free surface. The SM module was already validated in the past, but the VOF module needed to be validated in the present work through the simulation of free surface over a semicylindrical dam. Finally, the performance of a small 12-straight-blade HPM was simulated with caffa3d, with quite satisfactory results. Some issues of the solver yet need to be solved before other HPM with more complex designs could be studied.

Keywords: hydrostatic pressure machine; micro hydropower; CFD; open source; sliding mesh; volume of fluid; caffa3d

1. Introduction

Hydropower, being one of the oldest energy sources developed by humankind, continues to also be one of the most prominent ones, in terms of costs, social, and environmental impacts. A renovated interest in micro and pico hydropower has been developing in recent years to accommodate the increasing electricity demand avoiding the environmental impacts of large power stations [1]. Very few non exploited sites along the entire world are suitable for medium and large hydropower, considering the social and environmental negative impacts. In the opposite side, micro to pico hydropower generates minimum impact on the environment in addition to when they are installed in already existing infrastructures such as water distribution systems, irrigation dams or water channels. Conventional hydraulic turbines become very expensive (especially in relation to the total cost of the project) when it comes to very low heads and low power ratings [2], and most of the suitable micro hydropower converters present poor efficiencies or are still in a development stage such as the ones presented in [3–5].

The Hydrostatic Pressure Machine (HPM) is a novel energy converter first introduced by [6] inspired in the ancient water wheels used within mills [7], but with the main difference of extracting energy from the pressure of the water flow instead of the kinetic or potential energy. Among the recent publications, there are several studies on waterwheels' performances and improvements, but a great part of these refer to stream or gravity wheels [8–12] and only a few refer to HPM [13–18] that operate in a different way and should be distinguished from the formers. The HPM becomes very suitable for installation in water channels with very low head, where conventional hydraulic turbines or hydrokinetic turbines are inadequate or just too expensive because of the low head or power ratings. The installation of a HPM requires very little modifications to the existing infrastructure, and has proven to work with acceptable efficiencies (up to 80%), which make them the best choice for very low head waterways like irrigation or waste water channels. The principle of operation of a generic HPM is explained in Figure 1. The presence of the HPM covering almost the entire water passage of a rectangular channel section (leaving only the required gap between it and the channel walls and bed) generates a water level difference between upstream and downstream (just like a sluice gate). Then, like the sluice gate, the blades of the HPM receive a force due to this level or pressure difference, which generates a torque at the rotating centre and consequently a shaft mechanical power.

The rotational speed in this machines is very low, so, on one hand, there is a technological challenge to convert the mechanical to electrical power, but on the other hand and in combination with the narrow bottom opening, enables sediment transport and fish passage through it.

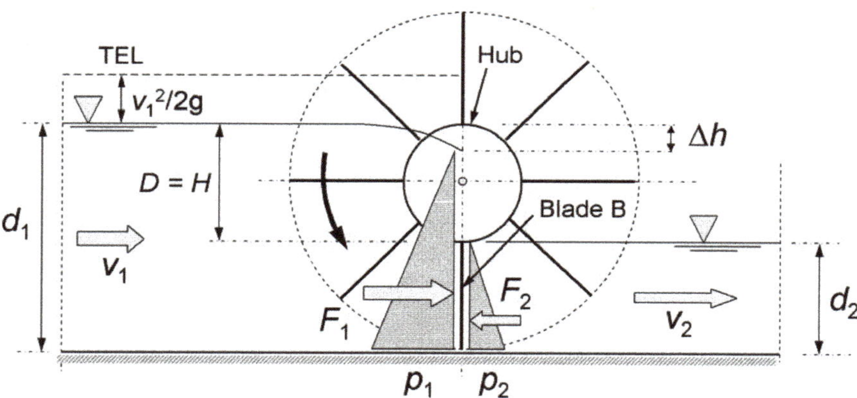

Figure 1. Explanation of the operation of a generic HPM, taken from HYLOW Internal Task Report 2.3 [19].

The operation of some HPM designs has been assessed during the last years, mostly within the HYLOW project (of which CERIS, from IST of Lisbon, was part of), by means of experimental tests in reduced scale models [13] and field tests [14]. In addition, some numerical assessment with commercial software has been done in order to better understand the optimum configuration for the performance of the HPM [15]. The highlighted modifications introduced to the design shown in Figure 1 are the alteration of the bed of the channel below the HPM in order to reduce the leakage flow, and the inclination of the blades remaining straight or radially twisted (see Figure 2), in order to facilitate the filling and emptying of the buckets. This latter improvement would also reduce the impact when the blades enter the water and would make the torque more continuous along the rotation. However, some doubts regarding the shape optimization of the blades to extract the most of the energy still remain. In fact, there are currently several ongoing projects that try to search for the best configuration, not only of the geometry of the wheel [16], but also of the geometry of the channel, water levels and type of generator [17,18].

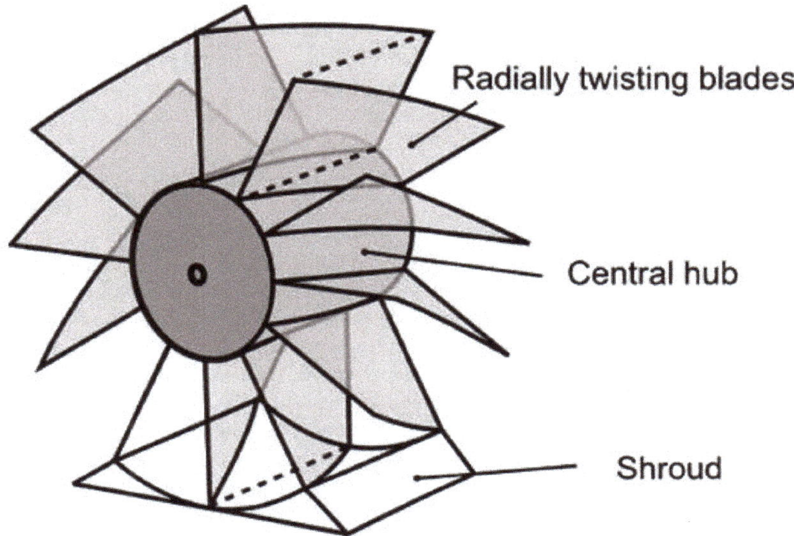

Figure 2. HPM with radially twisted blades and modified bottom shroud, taken from HYLOW Internal Task Report 2.3 [19]. In the present work, only the modification of the bottom shroud was modeled, maintaining the radially straight blades as in Figure 1.

Regarding the use of numerical tools for performance assessment of turbomachinery, the majority of academic researchers use commercial 3D flow solvers, while only few are using open source and freely available solvers. Within the latter, OpenFOAM is by far the most known and used around the world. In addition, a small but growing group of researchers from Uruguay have been developing another CFD solver named caffa3d [20]. It has evolved from a family of 2D flow solvers introduced by Ferziger and Peric [21], it is also open source, freely available, and used by researchers all around the world. The flow solver caffa3d is an implementation of the finite volume method in Fortran 90; it uses block structured body fitted grids, it can be combined with the immerse boundary condition method, it can be parallelized through the domain decomposition under a distributed memory model using the MPI library, and include a few methods for modeling turbulence like RANS (Reynolds Average Navier–Stokes) and LES (Large Eddy Simulation). The research team on Computational Fluid Mechanics of School of Engineering from Universidad de la República (Uruguay) has successfully participated with caffa3d in few CFD international challenges, working with an immerse boundary condition in hemodynamic simulation to predict the rupture of an intracranial aneurysm [22,23] and with the actuator line model in the simulation of wakes behind wind turbines [24]. Other uses of the solver include simulation of wind around buildings, pollutant dispersion, numeric wind tunnel, wind farm evaluation, and more [25]. However, it has not been successfully used yet for the simulation of rotating machinery and free surface flows.

Although the original version of the caffa3d solver from 2008 already included approaches for the solution of this kind of flows, they have not been used since then, and therefore became out of date within the last updates of the solver. Then, a big effort was done in updating the specific part of the code to make it work properly. This paper presents the most important parts of the updating and the results of preliminary simulations of straight-blades HPM in order to validate the solver. The next step will be to modify the geometry of the wheel and assess its optimum shape.

2. Materials and Methods

The numerical simulation of the operation of a HPM is obviously transient, and the main challenges are: the rotation of the wheel, the presence of a free surface and the generation of an adequate mesh. Further description of the solver caffa3d can be found in [20,25]. Other configuration parameters like boundary conditions and time step are also covered within this section. Finally, the procedure for the calculation of power generated by the HPM is stated.

2.1. Rotation of the Wheel

The rotation of the wheel can be simulated by means of different approaches, of which the multiple reference frames (MRF) and sliding meshes (SM) are highlighted.

In the MRF approach, the solver runs in a relative reference frame for the rotating part of the domain (working with relative velocities and adding the transport and Coriolis terms to the momentum equations) and in an absolute reference frame for the stationary part of the domain (working with absolute velocities and no addition of terms to the momentum equations). In MRF, all the blocks in the domain remain fixed and there has to be a steady flow condition at the interface.

In the SM approach [21], the entire flow is solved in the absolute reference frame for two separate parts of the domain (the moving and the stationary), with no addition of terms to the momentum equations. Because it is solved in the absolute frame, the mesh of the rotating part of the domain must also rotate, and some fluxes must be corrected because of mesh movement and some connectivity information through the interface must be updated. It is an inherently transient simulation, and captures the actual position of the rotor at any time, making it the best approach to solve the simulation of a HPM.

In the interface between the rotating and the stationary meshes, there is a perfect match between cells only for null relative displacement angle α (Figure 3). For other values of α, every cell in one side of the interface relates to two cells in the other side.

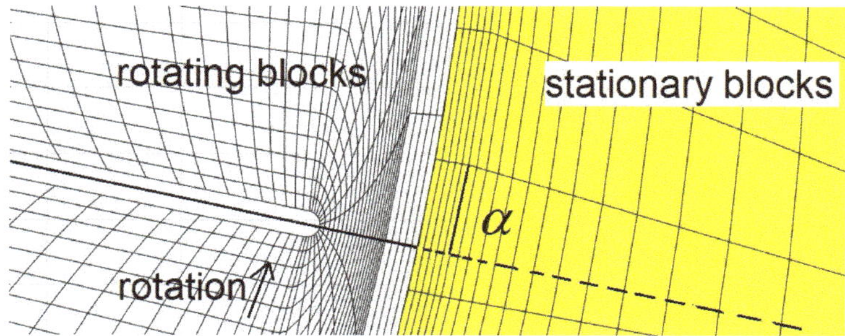

Figure 3. Sketch of the sliding interface, showing the relative displacement angle α between stationary blocks (painted) and rotating blocks (white).

The momentum balance equation for each moving block must consider the local time derivative of the absolute velocity field ($\vec{v}_A$), and the relative velocities ($\vec{v}_R$) for the mass fluxes, in a way as [20]:

$$\int_\Omega \rho \frac{\partial_R \vec{v}_A}{\partial t} d\Omega + \int_S \rho \vec{v}_A (\vec{v}_R \times \hat{n}_S) dS = \int_\Omega \rho \vec{g} d\Omega + \int_S -p\hat{n}_S dS + \int_S \mu (\nabla \vec{v}_A + \nabla \vec{v}_A{}^T)\hat{n}_S dS \quad (1)$$

where subindex A refers to the inertial reference system and subindex R refers to the reference system that rotates with the blocks, ρ is the density, μ is the dynamic viscosity, $\vec{g}$ is the gravity force, and p is the pressure.

2.2. Free Surface Flow

For the solution of the free surface flow, there also exists several methods, some of which are explained by Ferzger and Peric [21]. Free surface simulation methods can be classified as Interface-Tracking where each fluid occupies a different part of the grid, and the grid changes each time step to accommodate to the free surface geometry, as Interface-Capturing where the grid is fixed and the free surface geometry is found by different approaches, or as an hybrid method. Among the Interface-Capturing methods, the most popular one is the Volume of Fluid (VOF), developed by Hirt and Nichols [26] with latter updates and modifications [27]. In the VOF method, a scalar (V_f) is used to assess the fraction of volume of each fluid that occupies each cell volume at each time. For V_f=1, the cell is filled with water, for V_f=0, the cell is filled with air, and, in the middle, there is an interface between both fluids. Fluid properties such as density and viscosity are derived from the value of V_f for each cell at a given time, and the densities and viscosities of water and air:

$$\rho = V_f \rho_w + (1 - V_f) \rho_a \tag{2}$$

$$\mu = V_f \mu_w + (1 - V_f) \mu_a \tag{3}$$

Then, the corresponding transport Equation (4) is solved together with conservation equations for mass (5) and momentum (6), to find the geometry of the free surface at each time step:

$$\frac{\partial V_f}{\partial t} + \nabla(V_f \vec{v}) = 0 \tag{4}$$

$$\frac{\partial \rho}{\partial t} + \nabla(\rho \vec{v}) = 0 \tag{5}$$

$$\frac{\partial \rho \vec{v}}{\partial t} + \nabla(\rho \vec{v} \vec{v}) + \nabla p = \rho \vec{g} + \vec{F}_s + \nabla[\mu(\nabla \vec{v_A} + \nabla \vec{v_A}^T)] \tag{6}$$

where u_i is the component of the velocity vector $\vec{v}$ in the direction of x_i, and $\vec{F}_s$ is the force due to surface tension.

The solver caffa3d uses the VOF method together with a CICSAM approach (Compressive Interface Capturing Scheme for Arbitrary Meshes) for flux calculation [28,29], and the inclusion of the surface tension force [30,31]. In addition, in order to avoid non physical values of V_f, it is implemented a predictor-corrector procedure that resets these values to zero or one [28].

2.3. Model Geometry and Meshing

The analysed HPM (see Figure 4) is a 12-straight-blade-wheel with 150 mm inner diameter, 450 mm outer diameter, and 235 mm width. The ratio between the channel width and the wheel width is 1.26, meaning that the distance from the wheel to the channel's lateral walls is 30 mm. The thickness of the blades is 2 mm and the tips have a semi-spherical shape to reduce the risk of boundary layer separation. The shroud has a modified arc-shaped bottom that is attached to at least one full bucket every time, looking for a reduction of leakage flow. The gap between the blades and the bottom shroud is 10 mm (smaller gap could not be modeled accurately because it affected the simulation stability), while the gap with the lateral wall is 30 mm. Due to the straight blades, the geometry and thus also the flow, have a symmetry plane in the midsection of the channel. This enables the definition of a symmetry boundary condition at the midsection and thus only half of the entire domain needed to be modeled, reducing the computational time.

Figure 4. Computational model of the analysed HPM.

The full mesh, generated within caffa3d solver, is composed of three parallel bands of blocks (seventy-five in total): two bands for the half width of the HPM and the remaining for the width of the lateral gap. The only blocks that cover the entire width of the domain (half of the channel) are the two ring-shaped blocks corresponding to the sliding interface. Each of the bands including parts of the HPM are composed of sixteen structured grid blocks of H-type, O-type and C-type [21]: twelve body-fitted blocks for the wheel (one for each sector around blades) of the C-type, one H-type block surrounding the ring blocks, and three orthogonal blocks for the inlet, outlet and upper regions of the domain. These latter blocks are long enough to avoid interference between the wheel and the inlet and outlet sections. Moreover, both sections are located 5 times the outer diameter away from the periphery of the wheel. The band of blocks covering the lateral gap is similar to the others, but with more blocks to model the space adjacent to the hub and blades of the HPM, totaling forty-one blocks for this band. Currently, the band of blocks containing the lateral gap is modeled as a narrow passage of constant width along the entire length of the channel. However, in the real HPM, the lateral gap is of course much smaller than the difference between the channel and wheel width, and this is realized by positioning a wall normal to the flow at the plane of the rotation axis. This was not done in the present simulation, just for simplifying the geometry modeling and the simulation itself, but, in the future, this wall will be included in the model with help of the immerse boundary conditions module implemented in caffa3d.

The solver is domain parallelised within regions which are integrated by one or more grid blocks [20,25]. For the implementation of SM approach in caffa3d, the pair of blocks related by the sliding interface must be within the same region. The rest of the blocks are arranged in fifteen regions searching for a rough equality of element count among them all.

The grid is refined near the walls (to correctly model the viscous sublayer) and in a vertical direction around the expected position of the free surface, in order to capture its shape accurately. Because the free surface will occupy variable parts of the interior of the wheel as it rotates, the grid in this region also needs to be refined.

The final version of the mesh has 1.2 million volume cells,strongly concentrated around the HPM, as can be appreciated in Figures 5 and 6. The smallest cells are 0.3 mm in length and correspond to cells close to the tips of the blades.

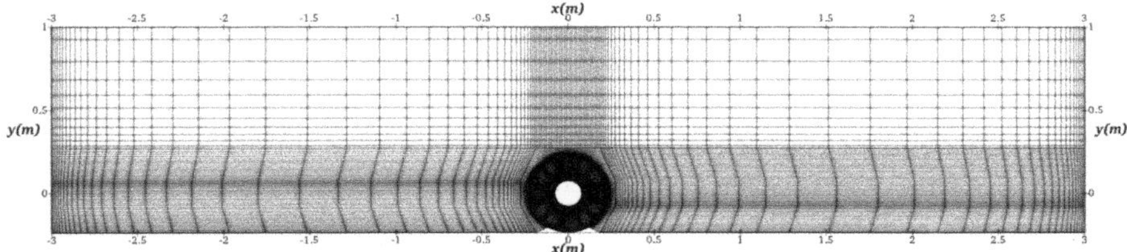

Figure 5. Computational mesh of the entire domain with dimensions in meters.

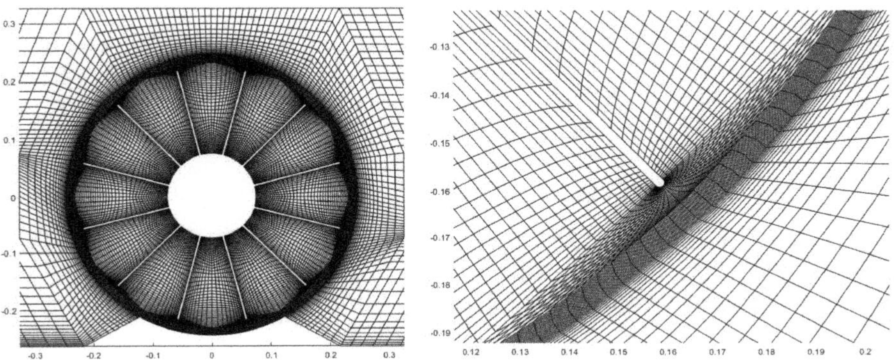

Figure 6. Computational mesh of the region surrounding the HPM (**left**) and around a blade (**right**) with dimensions in meters.

2.4. Boundary Conditions

Symmetry boundary condition is applied to one of the lateral walls of the domain (that corresponds to the midsection of the actual channel), while the other lateral wall is treated as a non-slip wall, as also are the surfaces of the wheel (blades and hub) and the bottom of the channel.

For the inlet section, it is assigned uniform velocity for the water region and zero velocity for the air region, thus imposing the inlet water flow (discharge).

Zero manometric pressure was assigned to the top of the domain, to simulate the atmosphere, and also to the outlet section of the channel, to simulate a free discharge.

2.5. Other Configuration Parameters

The rotational speed of the HPM at the start of the simulation is increased smoothly from zero to the nominal value during the first 50 steps, in order to prevent instabilities during the start-up of the simulation. In this simulation, the rotational speed is fixed in its nominal value, but the solver has the possibility to calculate it from the interaction with the fluid if such would be the case.

Regarding the time step, for the SM method, it would be sufficient to be less than the time that takes a cell to rotate one position. However, the VOF method imposes higher restrictions in order to satisfy the Courant–Friedrichs–Lewy condition. This restriction implies that, for any given cell, Equation (7) must verify, where u is the velocity of the fluid, Δt is the time step, and Δx is the smallest dimension of the cell.

$$\frac{u\Delta t}{\Delta x} \leq 1 \tag{7}$$

The solver is able to adapt the time step to accomplish the latter condition. The simulation is initiated with a time step of 5×10^{-5} s, but it is reduced whenever necessary. Actually, the ideal time step would have been close to 1.5×10^{-5} s.

The large eddy simulation (LES) technique, with the Smagorinsky model [32], is attached to the solver to model turbulence. The size of the cells adjacent to wall surfaces in normal direction (y) needs to be controlled in order to be included inside the viscous sublayer. According to Kirkgoz and Ardiclioglu (1997) cited in [33], the dimensionless distance must satisfy $y^+ = \rho u_* \frac{y}{\mu} \leq 10$, where $u_* = \sqrt{\frac{\tau_o}{\rho}}$ and τ_o is the boundary shear stress.

In order to increase stability of the simulation, strong under relaxation in the time scheme, and an advance for momentum equations is introduced, although with the risk of obtaining results with too many errors.

Diffusive fluxes are discretized in space using CDS (Central Differencing Scheme), while the contribution of CDS vs. UDS (Upwind Differencing Scheme) in convective fluxes is blended with a factor of 0.5.

The convergence of the iterations at each time step was checked when the residuals of each of the equations solved attained a value lower than 1×10^{-6}, while the maximum number of iterations for each time step was set to 5.

2.6. Power Calculation

The available (maximum) power is the theoretical hydraulic power of the flow through the HPM:

$$P_h = \rho g Q \Delta H = \rho g Q (h_1 - h_2 + \frac{v_1^2 - v_2^2}{2g}) \tag{8}$$

where Q is the discharge, ΔH the head difference upstream and downstream of the HPM calculated from the water levels (h_1 and h_2) and the free surface velocities (v_1 and v_2). The dynamic head (last term in Equation (8)) can be neglected as the velocities are very low. Due to intrinsic losses during the energy transfer, the hydraulic power exerted on the HPM is lower than the available hydraulic power.

The total hydraulic moment ($\vec{M}_h$) acting over the HPM should be calculated from the simulation results, through the contribution of pressure and viscous tensions acting on every cell face of the rotating grid blocks corresponding to the wall type boundary condition (those adjacent to the wheel), multiplied by the distance between the center of the HPM and the center of the cell face ($\vec{d}$):

$$\vec{M}_h = \int_S \vec{d} \times [-p\hat{n}_S + \mu(\nabla \vec{v}_A + \nabla \vec{v}_A{}^T)] dS \tag{9}$$

Because of the particular shape of the analysed HPM, the contribution of viscous tensions could be neglected, as its direction is parallel to the distance vector for almost every cell face. This was confirmed by the results of the simulations made by [15], even for the HPM with radially twisted blades.

Then, the hydraulic power (P_h) exerted on the HPM could be obtained by multiplying the total moment and the rotating speed ($\vec{\omega}$):

$$P_h = \vec{M}_h \times \vec{\omega} \tag{10}$$

3. Results and Discussion

As the solver has not been used extensively for free surface flows or turbomachinery simulations, it was advisable to assess each approach individually. The SM method has been previously used and verified within caffa3d [20]. First, a simulation of a free surface flow over a rigid and stationary object using the VOF module was performed. Then, both modules were incorporated into the solver, and a simulation of the HPM performance was conducted with the configuration mentioned in Section 2. Different meshes and time steps were used for each part of the assessment process, since each approach has distinct simulation demands. The post-processing was made in Paraview.

3.1. VOF Assessment

For the assessment of the VOF module, it was performed a simulation of a free surface flow over a semicylindrical object of 0.196 m diameter lying down roughly on the middle of the bed of a 2.4 m long and 0.2 m width channel. The case was chosen equal to the one analysed in [33] by means of numerical simulations in 2D (in ANSYS-Fluent v.12.1 with RANS method and several turbulence closure models) and experimental tests, in order to validate the simulations run with caffa3d. The mesh used here (see Figure 7) was different than the one in Section 2.3 (mainly because the HPM is not part of the simulation). The maximum value of the normal dimensionless distance to the wall is $max(y^+) = 7.0$.

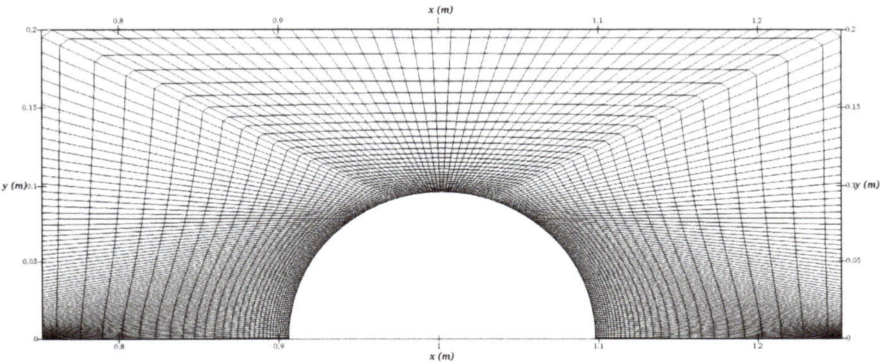

Figure 7. Computational mesh around the semicylindrical dam.

Lateral sides of the channel were treated as non-slip walls, while the boundary conditions for the inlet, top and outlet were the same as in Section 2.4. Upstream of the object the flow is subcritical with a depth $h_o = 0.147$ m (roughly 1.5 times the radius of the cylinder), and an inlet uniform velocity $v_o = 0.187$ m/s (corresponding to a discharge $Q = 0.055$ m^3/s as in [33]), resulting in a Froude number $Fr_1 = v_1/\sqrt{g \times h_1} = 0.156$. At $t = 0$ the water was at rest occupying the entire channel up to a depth $h_1 = 0.147$ m.

Downstream of the object the flow becomes supercritical with $Fr_2 = 2.26$. This change from subcritical to supercritical also occurs during the operation of the HPM, thus this validation becomes more attractive.

After a few seconds, the flow becomes stable, so only the results corresponding to the last time ($t = 20$ s) are shown. In addition, although the simulation was done in 3D, the flow is quite bi-dimensional, and thus the results are shown only for the midsection of the channel. The comparison of the free surface profile over the semicylindrical dam is shown in Figure 8. The profile in Figure 8b was obtained through experiments and numerical simulations performed with the Reynolds Stress Model (RSM) as turbulence closure by [33]. Similar comparison for the computed streamlines over the dam are shown in Figure 9.

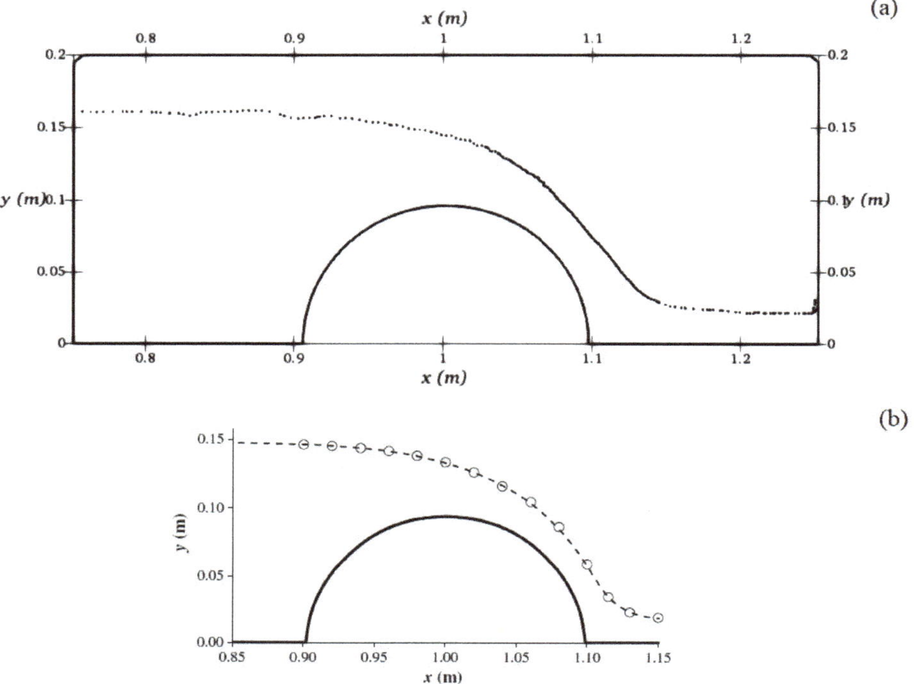

Figure 8. Comparison of free surface profile in the channel over the semicylindrical dam: (**a**) obtained with caffa3d and (**b**) presented in [33] through exprimental (o) and numerical (-).

These results show a quite well performance of the simulation, and are very similar to the results presented for similar studies [33,34].

The free surface level upstream of the dam obtained with caffa3d is a little higher than the one obtained by [33], where it remained in the initial value of $h_1 = 0.147$ m. This might be related with the fact that the streamlines in Figure 9a show a separation region bigger than in Figure 9b, thus raising the level of water. However, the free surface level and streamlines downstream of the dam are very similar in both figures, even with the separation region obtained with caffa3d being a little bigger. In [33], the authors also found differences in the size and shapes of the separation regions among the several turbulence models analysed, so the difference encountered here may be due to an error in the implementation of LES and/or the Smagorinsky model.

There is a small recirculation of air from the top surface to the outlet section above the free surface, but as this occurs just near the outlet section far away from the dam, it does not appear any effect of it in Figure 9a.

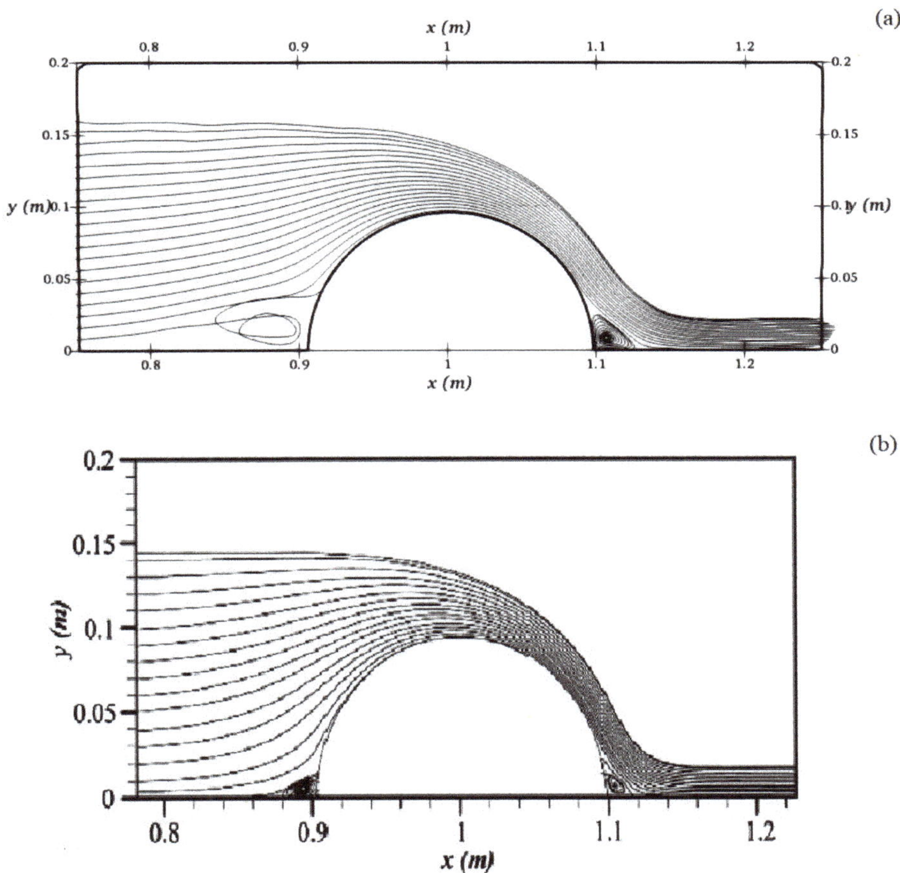

Figure 9. Comparison of streamlines in the channel over the semicylindrical dam: (**a**) obtained with caffa3d and (**b**) presented in [33].

The presence of spurious velocities in the air region has been previously reported by several authors to be caused by errors related to the calculation of the surface tension force [27,35], and several methods to solve it are also reported in [31,36–38]. However, when performing simulations without surface tension force, the recirculation persisted.

Another cause of spurious velocities could be the use of a predictor–corrector procedure [28] to reset values of V_f greater than 1 or lower than 0, which could also affect stability of the solution. Because non physical values of V_f are rare, the simulation without the corrector step was performed. Nevertheless, the recirculation did not disappear.

Afterwards, the recirculation was ignored, as it appears not to affect the flow near the object (this is where the HPM would be located).

Three horizontal velocity profiles at different sections of the channel, upstream of the dam ($x = 0.8$ m), over the dam ($x = 1.0$ m) and downstream of the dam ($x = 1.3$ m), are shown in Figure 10, enabling a direct comparison between the results with those presented in [33] for all the analysed turbulence models and the experimental test.

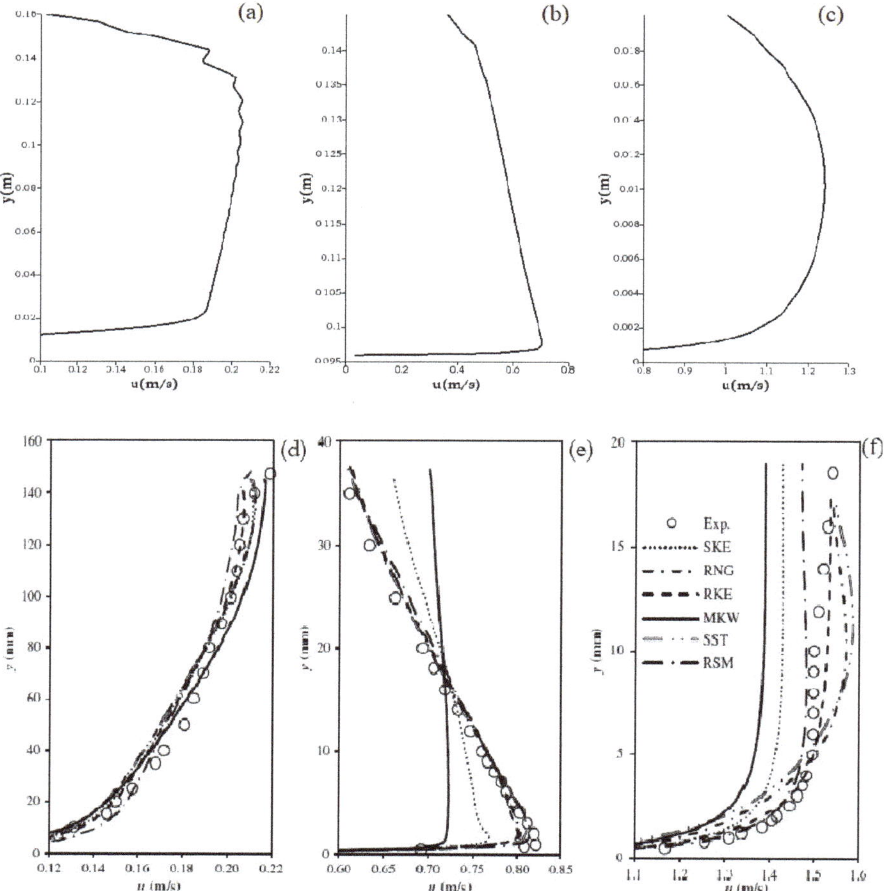

Figure 10. Horizontal velocity profiles obtained with caffa3d for sections: (**a**) $x = 0.8$ m; (**b**) $x = 1.0$ m; (**c**) $x = 1.3$ m, and by experiments and simulations [33] for sections: (**d**) $x = 0.8$ m; (**e**) $x = 1.0$ m; (**f**) $x = 1.3$ m.

One should compare the results obtained in caffa3d with the experimental and the numerical results obtained with the RSM, as it is the one that showed the best performance among the six analysed models in [33]. In that case, the shape of the profiles are similar, but the maximum velocity reached with caffa3d simulations in the three sections observed in Figure 10 are lower than those presented in [33].

3.2. Simulation of a HPM

The mesh, boundary conditions and general configuration of this simulation correspond to the ones explained in Section 2. For the initial condition of this simulation, the water level for the upstream zone of the channel corresponded to the height of the top of the HPM's hub, and to the height of the bottom of the HPM's hub for the downstream zone of the channel. All the fluid is at rest at the initial time. From the laboratory tests performed with the full width HPM, the actual flow rate corresponding to the best efficiency point at a rotational speed of 15 rpm was found to be $Q_{bep} = 11.5 \times 10^{-3}$ m^3/s.

The wheel rotates at a constant speed of 15 rpm and there is an inlet flowrate $Q = 5.75 \times 10^{-3}$ m^3/s (resulting from a uniform velocity $v_1 = 0.13$ m/s and a water depth $h_1 = 0.31$ m).

The predictor–corrector procedure was affecting the stability of the simulation, especially in the zone close to the inlet of the channel. This was suspected to happen according to [28], so only the predictor step was used. Despite disabling the corrector step, non-physical values of fraction of volume were not observed.

The maximum value of the normal dimensionless distance to the wall is $max(y^+) = 40$ in the vertical plane, but reaches much bigger values for the z direction in band of blocks containing the lateral gap (up to 200). Even though this will probably affect the numerical stability, it was a compromise between this and the computational extra time of dealing with a finer mesh.

3D images of the water flow and the streamlines at time $t = 4.0$ s (after one exact full revolution of the wheel has been performed) in the surroundings of the HPM are shown in Figures 11 and 12, respectively, where the water flows from left to right of the image. In addition, Figure 13 shows the magnitude of the velocity in a plane normal to the flow at the section of the rotation axis ($x = 0.0$ m), in order to better assessed the quantity of flow through the lateral gap.

Figure 11. 3D image of water flow near the HPM, at time $t = 4.0$ s.

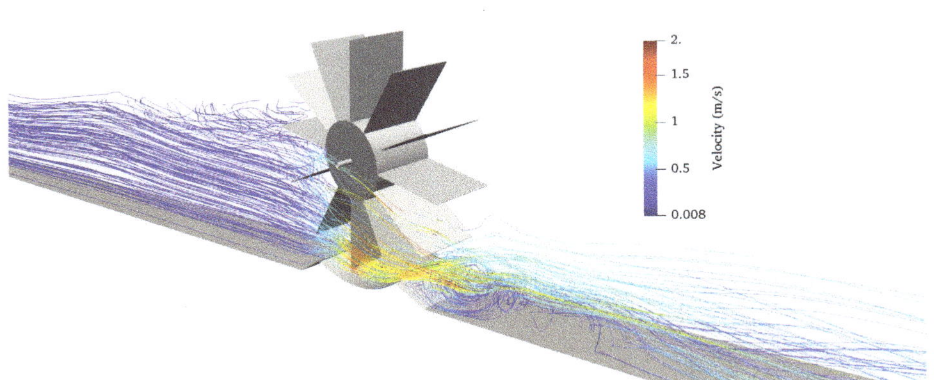

Figure 12. Streamlines of water flow near the HPM, at time $t = 4.0$ s.

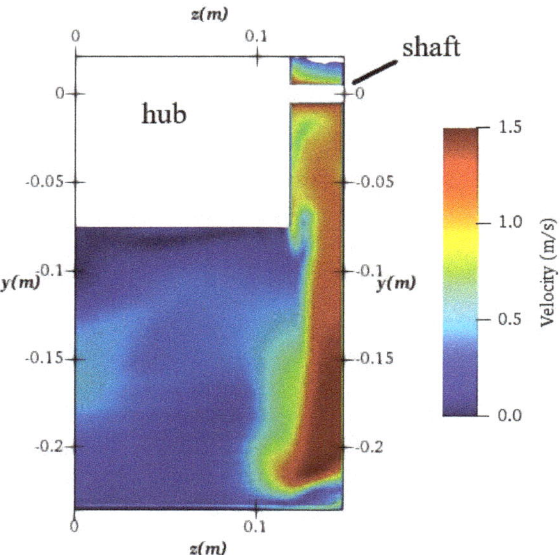

Figure 13. Magnitude of the velocity at the section of the rotation axis ($x = 0.0$ m), at time $t = 4.0$ s.

The importance of reducing the width of this gap at the section of the rotation axis becomes clear observing the high velocities of water passing through the lateral gaps in Figures 12 and 13. In the present simulation, the channel was modeled with constant width along its entire length, meaning that the lateral gap was equal to the distance between the wheel and the channel's wall (see Section 2.3). This distance needed to be big enough to enable the water entering laterally (besides longitudinally) facilitating the filling process, as well as to enable the water leaving the buckets laterally (besides longitudinally) to facilitate the emptying process. It was observed during the study that, without modeling this distance (as in a 2D simulation), the filling and emptying process were greatly hindered and, as a result, a poor performance of the HPM was obtained, as also observed by [15]. In fact, it has been pointed out by [18] that an important gain in power and efficiency is obtained by increasing the ratio of the channel width to the wheel width up to a value of 1.3 (for bigger ratios, both power and efficiency would remain almost constant). This was the reason for choosing the width of the gap as 0.3 m. In summary, the filling and emptying of the buckets of the wheel were well performed (as can be seen in Figure 11), but a great amount of water passed through the lateral gap reducing the power absorbed by the HPM and affecting its performance.

The pressure acting over the surfaces of the blades and hub of the HPM can be analysed in Figure 14, where the water flows from the left to right of the image.

Figure 14. Manometric Pressure acting on the surface of the HPM, at $t = 4.0$ s.

Negative values of pressure can be observed at the edges of some blades, the edge of the hub and over the surface of the shaft, due to flow separation in those zones.

Slice views at the midsection of the channel at $t = 4.0$ s of fraction of volume, manometric pressure and 2D velocity vectors are shown in Figures 15–17, respectively. The good performance of the filling and emptying of the wheel is also observed in Figure 15. The reduction of the pressure of the water while passing through the wheel also has a good and expected behaviour. A small zone of water with negative pressure is observed in the bucket leaving the bottom shroud at the downstream side, but the value does not represent any risk of cavitation. Lastly, the velocity vectors look quite well in almost every zone, noting a reverse flow near the free surface just upstream the wheel that could be due to the impact of the blade entering the water.

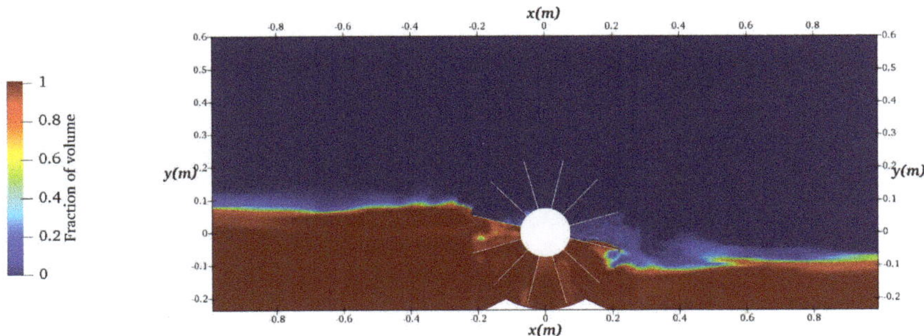

Figure 15. Slice view at the midsection of the channel of fraction of volume, at $t = 4.0$ s.

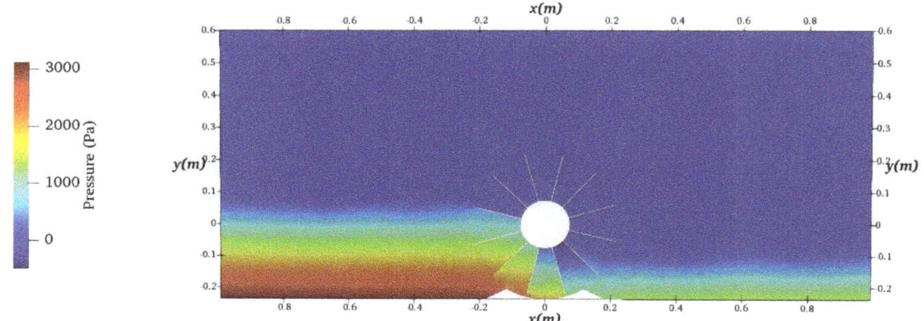

Figure 16. Slice view at the midsection of the channel of manometric Pressure, at $t = 4.0$ s.

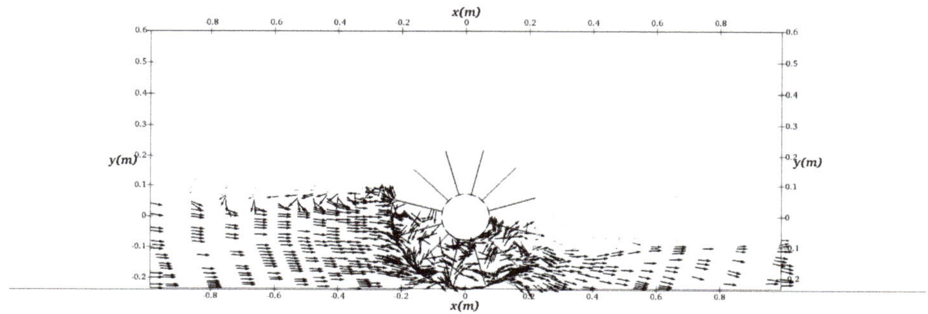

Figure 17. Velocity vectors of the slice view at the midsection of the channel, at $t = 4.0$ s.

In the following, a sequence of images for three instants of the rotation of the wheel, showing the fraction of volume and pressure fields around the HPM in a slice view at the midsection of the channel, is presented in Figure 18.

The three instants correspond to times $t = 2$ s (half revolution of the wheel), $t = 3$ s (three quarters of revolution of the wheel) and $t = 4$ s (full revolution of the wheel). The wheel in the images appears not to be moving, but it is just a visual effect because between each of the three instants the wheel rotates exactly a quarter of revolution.

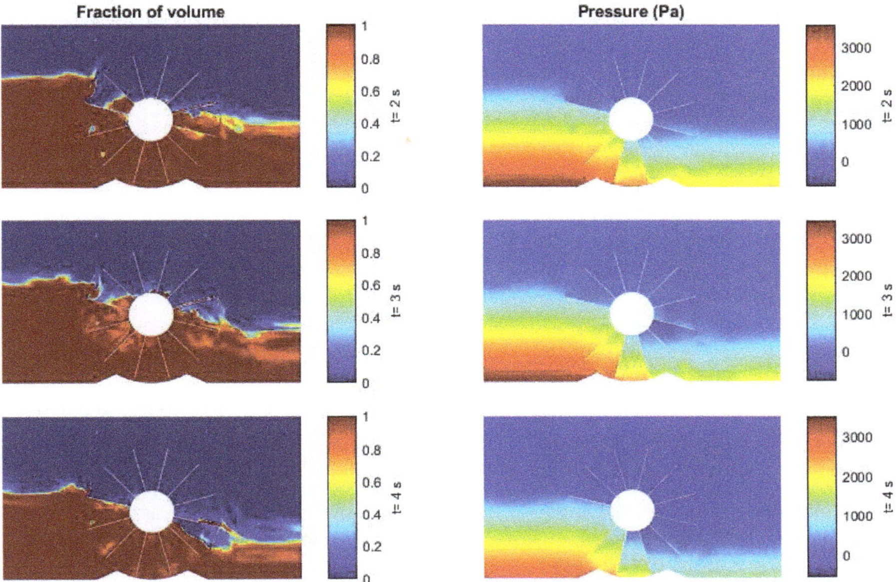

Figure 18. Sequence of fraction of volume (**left**) and pressure (**right**) around the HPM at times $t = 2$ s (**up**), 3 s (**middle**) and 4 s (**down**).

From the evolution of the free surface until the first full revolution of the wheel, it seems that the channel is emptying downstream of the wheel. This could be due to the zero manometric pressure acting on the outlet section of the channel, and perhaps it would be better to try a hydrostatic pressure profile or install a little dam to generate the water level. The evolution of the pressure field follows the behaviour of the evolution of the fraction of volume field.

Regarding the hydraulic power exerted on the HPM, at time $t = 5.0$ s, it reached a value of $P_h = 1.0$ W, and before that time it presented negative values. It is thought that this low value could be associated with the high leakage flow through the lateral gap. From previously conducted experiments ([19]), the shaft mechanical power generated by the HPM for the analysed discharge was $P_{exp} = 5.4$ W (corresponding to the best efficiency point of this machine). Then, if we would consider the mechanical losses, it could be concluded that the power exerted on the HPM obtained through the numerical simulations are of the same order of magnitude of the experimental value (if such would have been calculated).

Finally, it is worth noting that a mesh independence analysis was not carried out during the present study, but it would be interesting to do it, as it was found that a steeper refinement near the walls would be needed to improve the numerical accuracy.

4. Conclusions

The recent academic research on HPM has been reviewed and, despite all the advances in its knowledge, a lack of open source solvers usage in the numerical simulations was recognised. Then, the solver caffa3d, developed within an academic environment in Uruguay, was chosen for performing the simulations. This is a well proven solver; however, it had not been used until now for free surface flows or for turbomachinery simulations.

The Sliding Mesh approach for solving the rotation of the HPM was chosen among other methods available in the solver. It has been previously validated, but yet it required some adjustments because it was outdated.

For the solution of the free surface flow, the VOF method is implemented in the solver. Some issues regarding the use of VOF still exist, such as spurious velocities in the air zone, recirculation of air from the top surface to the outlet section and instabilities near the inlet section of the channel caused by the predictor–corrector procedure. It was tested through a simulation of a free surface flow over a semicylinder lying on the bottom of the channel, and the results were similar to the results obtained by other authors.

Preliminary quite acceptable results from the merging of both modules (SM and VOF) were obtained after a simulation of the performance of a small scale HPM with straight blades. Thanks to the significant space between the wheel and the lateral walls of the channel, the filling and emptying process of the buckets of the HPM were adequately performed. However, in contrast, a great amount of leakage flow passed through the lateral gap, hindering the performance of the HPM. The boundary condition applied to the outlet section of the channel should be changed to a hydrostatic profile (of the desired depth), or a small dam should be positioned, in order to assure the downstream level of the free surface, avoiding the emptying of the channel. The value of the power exerted on the HPM assessed by the pressure field obtained through the numerical simulations resulted of the same order of magnitude (and very close) of the shaft mechanical power obtained through previous experimental tests.

The next steps in this line of research will be:

- to add the vertical wall normal to the flow at the section of the rotation axis, in order to reduce the gap between the wheel and the lateral walls, while maintaining the channel wider than the wheel, in order to enable the filling and emptying process;
- to perform a mesh independence analysis while maintaining low values of y^+;
- to assess other turbulence techniques and associated models (particularly the $\kappa - \epsilon$ model for RANS already implemented in the solver);
- to incorporate the continuous calculation (at each time step) of torque and shaft power into the solver.

Once these aspects are covered, modifications of the geometry of the HPM pursuing better performance could be assessed with this open source flow solver.

Author Contributions: Conceptualization, H.M.R.; methodology, R.P.; software, R.P. and G.U.; validation, R.P.; formal analysis, R.P., G.U. and H.M.R.; writing–original draft preparation, R.P.; writing–review and editing, R.P., G.U. and H.M.R.; supervision, G.U. and H.M.R.; project administration, H.M.R.; funding acquisition, R.P. and H.M.R. All authors have read and agreed to the published version of the manuscript.

Funding: The authors wish to thank to the project REDAWN (Reducing Energy Dependency in Atlantic Area Water Networks) EAPA 198/2016 from INTERREG ATLANTIC AREA PROGRAMME 2014–2020 and the programme Young Professors and Researchers of Santander Universidades, for the corresponding grants.

Acknowledgments: The simulations reported in this article were performed in ClusterUY, a newly installed platform for high performance scientific computing at the National Supercomputing Center, Uruguay.

Conflicts of Interest: The authors declare no conflict of interest. The funders had no role in the design of the study; in the collection, analyses, or interpretation of data; in the writing of the manuscript, or in the decision to publish the results.

References

1. Paish, O. Small hydro power: technology and current status. *Renew. Sustain. Energy Rev.* **2002**, *6*, 537–556. [CrossRef]
2. Elbatran, A.H.; Yaakob, O.B.; Ahmed, Y.M.; Shabara, H.M. Operation, performance and economic analysis of low head micro-hydropower turbines for rural and remote areas: A review. *Renew. Sustain. Energy Rev.* **2015**, *43*, 40–50. [CrossRef]
3. Sari, M.A.; Badruzzaman, M.; Cherchi, C.; Swindle, M.; Ajami, N.; Jacangelo, J.G. Recent innovations and trends in in-conduit hydropower technologies and their applications in water distribution systems. *J. Environ. Manag.* **2018**, *228*, 416–428. [CrossRef]

4. Samora, I.; Hasmatuchi, V.; Münch-Alligné; C.; Franca, M.J.; Schleiss, A.J.; Ramos, H.M. Experimental characterization of a five blade tubular propeller turbine for pipe installation. *Renew. Energy* **2016**, *95*, 356–366. [CrossRef]

5. Vagnoni, E.; Andolfatto, L.; Richard, S.; Münch-Alligné, C.; Avellan, F. Hydraulic performance evaluation of a micro-turbine with counter rotating runners by experimental investigation and numerical simulation. *Renew. Energy* **2018**, *126*, 943–953. [CrossRef]

6. Senior, J.; Wiemann, P.; Müller, G. The Rotary Hydraulic Pressure Machine for Very Low HEAD Hydropower Sites. Available online: http://www.hylow.eu/knowledge/all-download-documents/ (accessed on 7 August 2020).

7. Viollet, P.-L. From the water wheel to turbines and hydroelectricity. Technological evolution and revolutions. *Comptes Rendus MéCanique* **2017**, *345*, 570–580. [CrossRef]

8. Quaranta, E.; Revelli, R. Performance characteristics, power losses and mechanical power estimation for a brestshot water wheel. *Energy* **2015**, *87*, 315–325. [CrossRef]

9. Quaranta, E.; Revelli, R. Optimization of breastshot water wheels performance using different inflow configurations. *Renew. Energy* **2016**, *97*, 243–251. [CrossRef]

10. Quaranta, E.; Revelli, R. Gravity water wheels as a micro hydropower energy source: A review based on historic data, design methods, efficiencies and modern optimizations. *Renew. Sustain. Energy Rev.* **2018**, *97*, 414–427. [CrossRef]

11. Cleynen, O.; Kerikous, E.; Hoerner, S.; Thévenin, D. Characterization of the performance of a free-stream water wheel using computational fluid dynamics. *Energy* **2018**, *165*, 1392–1400. [CrossRef]

12. Zhao, M.; Zheng, Y.; Yang, C.; Zhang, Y.; Tang, Q. Performance Investigation of the Immersed Depth Effects on a Water Wheel Using Experimental and Numerical Analyses. *Water* **2020**, *12*, 982. [CrossRef]

13. Senior, J.; Saenger, N.; Müller, G. New hydropower converters for very low head differences. *J. Hydraul. Res.* **2010**, *48*, 703–714. [CrossRef]

14. Linton, N.P. Trials and Development of a Hydrostatic Pressure Wheel. Ph.D. Thesis, Faculty of Engineering and The Environment, University of Southampton, Southampton, UK, December 2013.

15. Narrain, A.G. Low Head Hydropower for Local Energy Solutions. Ph.D. Thesis, Delft University of Technology and UNESCO-IHE Institute for Water Education, Delft, The Netherlands, October 2017.

16. Licari, M.; Benoit, M.; Anselmet, F.; Kocher, V.; Clément, S.; Le Faucheux, P. Study of low-head hydrostatic pressure water wheels for harnessing hydropower on small streams. In Proceedings of the 24 Congrès Français de Mécanique, Brest, France, 26–30 August 2019.

17. Paudel, S.; Linton, N.; Zanke, U.; Saenger, N. Experimental investigation on the effect on channel width on flexible rubber blade water wheel performance. *Renew. Energy* **2013**, *52*, 1–7. [CrossRef]

18. Butera, I.; Fontan, S.; Poggi, D.; Quaranta, E.; Revelli, R. Experimental Analysis of Effect of channel Geometry and Water Levels on Rotary Hydrostatic Pressure Machine. *J. Hydraul. Eng.* **2020**, *146*, 04019071. [CrossRef]

19. Schneider, S.; Müller, G.; Saenger, N. *HYLOW Project Report: Converter Technology Development—HPM and HPC*; Internal Task Report 2.3; Not published.

20. Usera, G.; Vernet, A.; Ferré, A. A parallel block-structured finite volume method for flows in complex geometry with sliding interfaces. *Flow Turbul. Combust* **2008**, *81*, 471–495. [CrossRef]

21. Ferziger, J.; Peric, M. *Computational Methods for Fluid Dynamics*; Springer: Berlin/Heidelberg, Germany, 2002.

22. Usera, G.; Mendina, M. CFD Challenge: Solutions using open source flow solver caffa3d.MBRi with immersed boundary condition. In Proceedings of the ASME 2012 Summer Bioengineering Conference, Farjardo, Puerto Rico, 20–23 June 2012.

23. Berg, P.; Roloff, C.; Beuing, O.; Voss, S.; Sugiyama, S.-I.; Aristokleous, N.; Anayiotos, A.S.; Ashton, N.; Revell, A.; Bressloff, N.W.; et al. The Computational Fluid Dynamics Rupture Challenge 2013-Phase II: Variability of Hemodynamics Simulations in Two Intracranial Aneurysms. *J. Biomech. Eng.* **2015**, *137*, 121008.

24. Mühle, F.; Schottler, J.; Bartl, J.; Futrzynski, R.; Evans, S.; Bernini, L.; Schito, P.; Draper, M.; Guggeri, A.; Kleusberg, E.; et al. Blind test comparison on the wake behind a yawed wind turbine. *Wind. Energy Sci.* **2018**, *3*, 883–903. [CrossRef]

25. Mendina, M.; Draper, M.; Kelm Soares, A.P.; Narancio, G.; Usera, G. A general purpose parallel block structured open source incompressible flow solver. *Clust. Comput.* **2014**, *17*, 231–241. [CrossRef]

26. Hirt, C.W.; Nichols, B.D. Volume of Fluid (VOF) Method for the Dynamics of Free Boundaries. *J. Comput. Phys.* **1981**, *39*, 201–225. [CrossRef]

27. Ketabdari, M.J. Chapter 15:Free Surface Flow Simulation Using VOF Method . In *Numerical Simulation—From Brain Imaging to Turbuloent Flows*; Lopez-Ruiz, R., Ed.; IntechOpen: London, UK, 2016.

28. Ubbink, O.; Issa, R.I. A method for computing sharp fluid interfaces on arbitrary meshes. *J. Comput. Phys.* **1999**, *153*, 26–50. [CrossRef]

29. Hogg, P.W.; Gu X.J.; Emerson, D.R. An implicit algorithm for capturing sharp fluid interfaces in the volume of fluid advection method. In Proceedings of the European Conference on Computational Fluid Dynamics, Egmond aan Zee, The Netherlands, 5–8 September 2006.

30. Brackbill, J.U.; Kothe, D.B.; Zemach, C. A continuum method for modeling surface tension. *J. Comput. Phys.* **1992**, *100*, 335–354. [CrossRef]

31. Vachaparambil, K.J.; Einarsud, K.E. Comparison of surface tension models for the Volume of Fluid method. *J. Eng. Math.* **1988**, *22*, 3–13. [CrossRef]

32. Smagorinsky, J. General circulation experiments with the primitive equations: I. The basic experiment. *Mon. Weather Rev.* **1963**, *91*, 99–-164. [CrossRef]

33. Akoz, M.S.; Gumus, V.; Kirkgoz, M.S. Numerical Simulation of Flow over a Semicylinder Weir. *J. Irrig. Drain. Eng.* **2014**, *140*. [CrossRef]

34. Forbes, L.K. Critical free-surface flow over a semi-circular obstruction. *J. Eng. Math.* **1988**, *22*, 3–13. [CrossRef]

35. Abadie, T.; Aubin, J.; Legendre, D. On the combined effects of surface tension force calculation and interface advection on spurious currents within Volume of Fluid and Level Set frameworks. *J. Comput. Phys.* **2015**, *297*, 611–636. [CrossRef]

36. Jamet, D.; Torres, D.; Brackbill, J.U. On the theory and computation of surface tension: the elimination of parasitic currents through energy conservation in the second-gradient method. *J. Comput. Phys.* **2002**, *182*, 262–276. [CrossRef]

37. Harvie, D.J.E.; Davidson, M.R.; Rudman, M. An analysis of parasitic current generation in Volume of Fluid simulations. *Appl. Math. Model.* **2006**, *30*, 1056–1066. [CrossRef]

38. Pan, Z.; Weibel, J.A.; Garimella, S.V. Spurious Current Suppression in VOF-CSF Simulation of Slug Flow through Small Channels. *Processes* **2019**, *7*, 542. [CrossRef]

Article

Hybrid Pumped Hydro Storage Energy Solutions towards Wind and PV Integration: Improvement on Flexibility, Reliability and Energy Costs

Mariana Simão * and Helena M. Ramos

Civil Engineering, Architecture and Georesources Department, CERIS, Instituto Superior Técnico, Universidade de Lisboa, 1049-001 Lisboa, Portugal; hramos.ist@gmail.com or helena.ramos@tecnico.ulisboa.pt
* Correspondence: m.c.madeira.simao@tecnico.ulisboa.pt

Received: 16 June 2020; Accepted: 29 August 2020; Published: 1 September 2020

Abstract: This study presents a technique based on a multi-criteria evaluation, for a sustainable technical solution based on renewable sources integration. It explores the combined production of hydro, solar and wind, for the best challenge of energy storage flexibility, reliability and sustainability. Mathematical simulations of hybrid solutions are developed together with different operating principles and restrictions. An electrical generating system composed primarily by wind and solar technologies, with pumped-storage hydropower schemes, is defined, predicting how much renewable power and storage capacity should be installed to satisfy renewables-only generation solutions. The three sources were combined considering different pump/turbine (P/T) capacities of 2, 4 and 6 MW, wind and PV solar powers of 4–5 MW and 0.54–1.60 MW, respectively and different reservoir volume capacities. The chosen hybrid hydro-wind and PV solar power solution, with installed capacities of 4, 5 and 0.54 MW, respectively, of integrated pumped storage and a reservoir volume of 378,000 m^3, ensures 72% annual consumption satisfaction offering the best technical alternative at the lowest cost, with less return on the investment. The results demonstrate that technically the pumped hydro storage with wind and PV is an ideal solution to achieve energy autonomy and to increase its flexibility and reliability.

Keywords: pumped hydro storage (PHS); hybrid hydro-wind-solar solutions; technical feasibility; new power generation

---

1. Introduction

Hydropower plays an important role today and will become even more important in the coming decades, since hydropower can be a catalyst for the energy transition in Europe [1]. The ambitious plan for energy transition in Europe seeks to achieve a low-carbon climate-resilient future in a safe and cost-effective way, serving as a worldwide example [2]. The key role of electricity will be strongly reinforced in this energy transition. In many European countries, the phase out of nuclear and coal generation has started with a transition to new renewable sources comprising mainly of solar and wind for electricity generation. However, solar and wind are variable energy sources and difficult to align with demand. Hydropower already supports integration of wind and solar energy into the supply grid through flexibility in generation as well as its potential for storage capacity. These services will be in much greater demand in order to achieve the energy transition in Europe, and worldwide [1,2].

Hydropower, with its untapped potential, has all the characteristics to serve as an excellent catalyst for a successful energy transition. However, this will require a more flexible, efficient, environmentally and socially acceptable approach to increasing hydropower production to complement wind and solar energy production. In particular, (a) increasing hydropower production through the implementation of new environmentally friendly, multipurpose hydropower schemes and by using the hidden potential in existing infrastructure; (b) increasing the flexibility of generation from existing hydropower plants by adaptation and optimization of infrastructure and equipment combined with innovative solutions for the mitigation of environmental impacts and (c) increasing storage by the heightening of existing dams and the construction of new reservoirs, which have to ensure not only flexible energy supply but also support food and water [2]. Therefore, a hybrid system between several renewable energy resources, a complementary nature among other sources in an integrated and flexible way is of utmost importance in the electric producer system and energy management. The most traditional and mature storage technology, pumped hydro storage, is adopted to support both the grid connection, as well as the standalone hybrid hydro-wind-solar grid system.

Energy and water are closely interlinked. These interconnections intensify as the demand for resources increases with population growth and changing consumption patterns [3]. Due to the growing awareness about environment, climate changes, pollution and waste footprint, clean and renewable sources of energy are being encouraged and used globally. With these growing trends of using intermittent renewable sources of energy, there will be a greater need to make flexible the modern energy production and distribution systems. The challenge of using renewable sources such as hydro, wind and solar energies is their variability, intermittency, non-predictability and under-weather dependency [4]. Kapsali and Kaldellis [5] assessed the techno-economic viability of a system incorporating simultaneous wind farms with pumped storage and hydro turbines for remote islands and determined that the contribution of renewable energy was increased by 15%. Castronuovo et al. [6] worked on optimum operation and hydro storage sizing of a wind and hydro hybrid power plant and calculated an annual profit of 11.91% by purchasing energy during periods of low demand and selling during periods of high demand. Vieira et al. [7] studied optimization and operational planning of wind and hydro hybrid water supply systems and concluded that up to 47% of energy costs can be saved in the optimization mode compared to normal operating mode. More detailed information can be found in [8–11] on feasibility, optimal design and operation of pumped-hydro hybrid systems.

Combining renewable energy resources is the best way to overcome energy shortcomings, which not only provides more reliable power systems increasing the storage capacity but also leads to the reduction in climate change effects [1,12]. In addition, renewable-based technologies can make water accessible for domestic, industry and agricultural purposes, improving supply security while decoupling growth in water from fossil fuels. Renewable energy technologies offer opportunities to address trade-offs and to leverage on synergies between sectors to enhance water and energy nexus. However, connecting renewable power plants to the grid can cause dynamic controlling problems if the electricity network is not prepared for handling such variations due to the intermittency of renewable sources availability [13]. Therefore, a continuous and reliable power supply is hardly possible without energy storage. Using an energy storage system, the surplus energy can be stored when the power generation exceeds the demand and then released to cover the periods when the net load exists, providing a robust flexible back-up for intermittent renewable energy sources [14,15]. This has the advantage in increasing the system flexibility and reliability, decreasing the variability of renewable sources availability, since the variable power output can be levelled out due to a complementary nature between renewable resources through their integration in the hydropower by a pumped storage solution.

Although an oversized hybrid system satisfies the load demand, it can be an unnecessarily expensive solution. An undersized hybrid system is more cost-effective but may not be able to meet the load demand. The optimum size of the hybrid renewable energy power system depends on several simulations based on specific mathematical models and system components management towards the best solution [16]. This study discusses the implications of a water–energy nexus focusing on renewable energy technologies, a cost-effective and environmentally sustainable supply of energy. It examines a specific case to highlight how renewable energy can address the trade-offs, helping to state the world's pressing water and energy challenges towards an optimization model development and application.

This work presents a consistent framework for optimizing the availability and storage of renewable resources and to evaluate how the system behaves with changing the power capacity, in order to facilitate such complementarity in a flexible and multi-variable process. Different solutions are analyzed, according to the demand and the installed power, for different storage volumes, and different available characteristics for pump/hydro, wind and solar energy sources. The complementarity and the system flexibility to satisfy the consumption is a complex solution with several characteristic parameters that need to be solved simultaneously using different time series for the available sources.

2. Hybrid and Pumped Storage Technologies

2.1. Characterization

Pumped hydropower storage (PHS) accounts over 94% of installed global energy storage capacity and retains several advantages such as lifetime cost, levels of sustainability and scale. The existing 161,000 megawatts (MW) of pumped storage capacity support power grid stability, as significant water batteries, reducing overall system costs and sector emissions [1]. PHS operations and technology are adapting to the changing power system requirements incurred by variable renewable energy (VRE) sources as a new energy transition solution. Variable-speed and ternary PHS systems allow for faster and wider operating ranges, providing additional flexibility, enabling higher penetrations of VRE at lower system costs at high reliable levels. As traditional revenue streams become more unpredictable and markets are volatiles, PHS is seen to secure long-term revenue in order to attract investment, particularly in liberalized energy markets [1,2].

Different energy resources can be combined building an integrated hybrid energy system that complements the drawbacks existing in each individual energy solution. Therefore, the design goals for hybrid power systems are the minimization of power production cost, purchasing energy from the grid (if it is connected), the reduction of emissions, the total life cycle cost and increasing the reliability and flexibility of the power generation system [17–19]. The pumped storage can be seen as the most promising technology to increase renewable energy levels in power systems. Hydro, wind, solar and pumped hydro storage (PHS), as hybrid power solutions, constitute a realistic and feasible option to achieve high renewable levels, considering that their components are properly sized. In some locations, the solar and wind resources have an anti-correlation, complementing each other and giving a combined less variable output than independently [20].

Pumped-storage schemes currently provide the most commercially important means of large-scale grid energy storage and improve the daily capacity factor of the generation system. Pumped hydropower energy storage stores energy in the form of potential energy that is pumped from a lower reservoir to a higher one putting the water source available to turbine to fit the energy demand. In this type of system, low cost electric power (electricity in off-peak time) is used to run the pumps to raise the water from the lower reservoir to the upper one [21]. During the periods of high power demand, the stored water is released through hydro turbines to produce power. Reversible turbine-generator groups act as pump or turbine, when necessary. A typical conceptual pumped hydro storage system with wind and solar power options for transferring water from lower to upper reservoir is represented in Figure 1.

This system is equipped with a photovoltaic (PV) system array, a wind turbine, an energy storage system (pumped-hydro storage), a control station and an end-user (load). This whole system can be isolated from the grid, i.e., a standalone system or in a grid connection where the control station can be the grid inertia capacity. This is currently the most cost-effective means of storing large amounts of renewable energy, based on decisive factors, such as, capital costs, suitable topography and climate changes challenges [22].

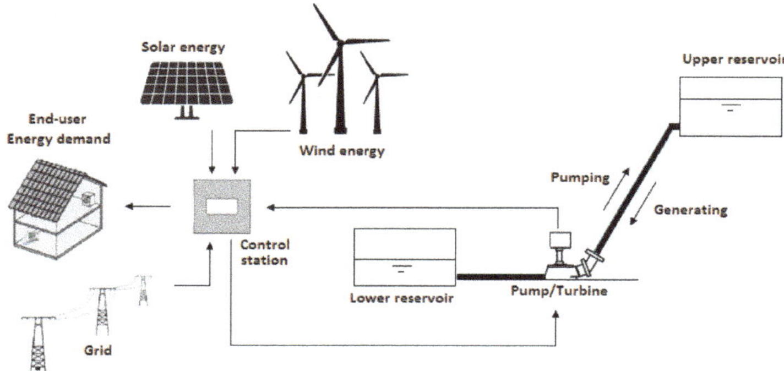

Figure 1. A hybrid hydro-wind-solar system with pumped storage system.

2.2. Optimal Design of a Hybrid System

2.2.1. Dispatch Optimization Model

The purposed mathematical model can predict how much wind, solar power and pumped hydro-storage energy capacity should be installed to satisfy a hybrid renewable solution. Wind is highly fluctuating meteorological parameter changing every hour and annually. Therefore, to connect wind power with the grid and assure quality power supply, large energy storage systems are required. Nevertheless, different types of wind turbines may output different power due to their difference in the power curve characteristics. Therefore, the model used to describe the performance should be different [23–26]. As for solar energy although less fluctuating, it only works during day light hours. It offers more reliable power and can be committed and managed, using relatively smaller energy storage systems to provide continuous and quality power [27]. According to [19], pumped hydropower storage plants have several advantages, such as (1) flexible start/stop and fast response speed, (2) ability to track load variations and adapt to severe load changes, (3) capacity to modulate the frequency and maintain voltage stability and face climate change and reduce footprint effects on an integrated solution.

The optimal design of a hybrid solar–wind-system supported by a pumped-based hydro scheme can significantly enhance the technical and economic performance for efficient energy harnessing. The multi-variable techniques are also known for their accuracy and simplicity when encountering complicated optimization problems. The optimization approaches in hybrid hydro-solar-wind systems is presented in Figure 2. The main objective is to analyze the capacity of such a system to be able to store the excess of wind/solar energy, at times when energy demand is lower, and provide reserves in the form of hydropower at times when consumption exceeds the wind/solar production or, alternatively, making the system self-sufficient. This model was developed for a time scale of one average year, assuming hourly variations, i.e., 8760 iterations, where dimensional data regarding variations in electricity demand and wind/solar energy production.

Figure 2 depicts the proposed framework to coordinate pumped-storage and wind-solar renewable energies, with a closed-loop dispatch process as follows.

The pumped hydro storage (PHS) is the energy storage solution in this study, consisting on a separated pump/motor unit and a turbine/generator unit to manage the other renewable sources inputs to face the energy demand [28]. The turbine generating coefficient (kWh/m^3) in Equation (1) and the water pumping coefficient (m^3/kWh) in Equation (2) are two key parameters of the PHS elements. According to [29,30] the following equation describes the total stored energy E^t (in kWh) in the active volume of a reservoir:

$$E^t = \eta_t \times \rho \times g \times H \times V = c_t \times V \quad (\text{kWh}), \tag{1}$$

where H is the net head (m), η_t the overall efficiency of the turbine/generator unit (%), V the storage capacity (m^3), c_t is the turbine generating coefficient (kWh/m^3), g is the acceleration due to gravity (m/s^2), ρ is the density of the water (kg/m^3).

The energy used to pump the water volume to a specific height, with a specific pumping efficiency is given by

$$E^p = \frac{\rho \times g \times H \times V}{\eta_p} = c_p \times V \qquad \text{(kWh)}, \qquad (2)$$

where H is the pumping head (m), η_p is the overall pumping efficiency (%), and c_p is the water pumping coefficient of the pump/motor unit (m^3/k Wh).

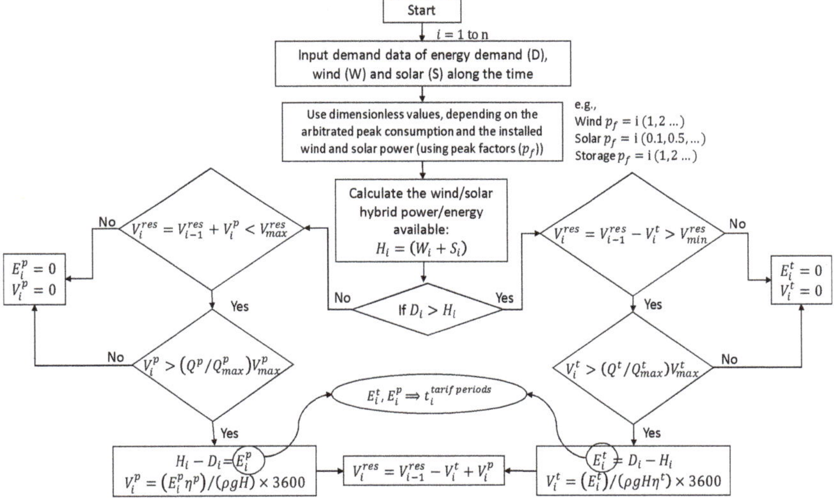

Figure 2. Flowchart to find optimal hybrid system. V is for water volume, E is for energy, D is the demand, H is for hybrid power/energy available, S is the solar energy and W is for wind energy. The superscripts (p, t, res) are assigned for pump, turbine and reservoir.

In the case of energy deficits, water is drawn from the upper reservoir in order to operate the hydro turbines.

Finally, the proposed dispatch model selects the best combination of peak factors to reach the optimal solution in terms of efficiency, energy exploitation, cost, and footprint. The peak factor is the ratio of the total flow to the average daily flow in a water system and is important in the study of a water system to determine potential water consumption values. The ratio of the total flow for large water systems generally varies from 1.2 to 3.0 or even higher for specific small systems [31].

2.2.2. Operating Principles and Important Restrictions

The pumping process is done during the empty hours (i.e., of lower demand) and the hydroelectric generation through peak hours (i.e., for highest demand). Thus, the consumption during the off-peak hours was satisfied exclusively by wind/solar, while in the remaining period, the power generation is complemented by hydro, if insufficient wind/solar production is verified. Excess wind/solar energy that is not used for consumption in the system is used for pumping. In this way, it is possible to reduce the purchase costs of electricity from the grid. The respective operating conditions Equation (3) and restrictions Equation (4) are described below:

Operating conditions:

$$V_{min}^{res} > 0.15 V_{max}^{res}$$
$$V_i^{res} = V_{i-1}^{res} + V_i^p - V_i^t$$
$$V_i^p \geq \frac{Q_{min}^p}{Q_{max}^p} V_{max}^p, \; V_i^t \geq \frac{Q_{min}^t}{Q_{max}^t} V_{max}^t \tag{3}$$
$$\Delta E_i = H_i - D_i$$

Restrictions considered:

$$If \; V_i^{res} \geq V_{max}^{res} \; \rightarrow V_i^p = 0$$
$$Else \; V_i^p = \left(E_i^p \eta^p\right)/(\rho g H) \times 3600$$
$$If \; V_i^{res} \leq V_{min}^{res} \; \rightarrow V_i^t = 0$$
$$Else \; V_i^t = \left(E_i^t\right)/\left(\rho g H \eta^t\right) \times 3600 \tag{4}$$
$$If \; time = off peak \; period \rightarrow E_i^p = Pump \; power \; installed, \; E_i^t = 0$$
$$Else \; E_i^p = 0, \; E_i^t = -\Delta E_i$$

2.2.3. Daily Cycle for Electricity Supply

According to Figure 2, E_i^t, E_i^p depend on the tri-time tariff (Table 1). As the pumping system works in the early hours of the day (0–7 h) the system presents sufficient reserves to be able to assist intermittent renewable energy production failures during the remaining period. The daily cycle for electricity supply, dictated by the tri-time tariff applied in mainland Portugal, is described in Table 1.

Table 1. Time-of-day tariff periods (weekly cycle).

Days	Tarif Period	Winter	Summer
Workdays	Peak	5 h/day	3 h/day
	Half-peak	12 h/day	14 h/day
	Normal off-peak	3 h/day	3 h/day
	Super off-peak	4 h/day	4 h/day
Saturdays	Half-peak	7 h/day	
	Normal off-peak	13 h/day	
	Super off-peak	4 h/day	
Sundays	Normal off-peak	20 h/day	
	Super off-peak	4 h/day	

3. Case Study

3.1. Modelling Assumptions

The benefit in using medium-head pumped-storage plants is to shorten transmission lines from the alternative energy sources to the hydro storage facility, thus minimizing grid overloading due to energy transfer across a country [32]. Moreover, it has the advantage of locating wind or solar farms close to hydropower schemes, which are usually installed in higher topographic zones [33–36].

The proposed solution focuses on a converter connected to a motor/generator. The mostly used is the pump as turbine (PAT) type that allows variable speed to cover the required operating range in both modes of operation. The efficiency considered can vary between 70% and 80% for pump/turbine mode, respectively. The overall energy storage system efficiency is 56%, corresponding to a water pumping and turbine generating coefficients of 1.837 m^3/kWh and 0.305 kWh/m^3, respectively. The energy converter pump as turbine is of variable speed, which allows the exploitation of excess energy produced by PV arrays and wind turbines and also allows covering medium load by a remaining part of hydropower, to improve the overall energy system efficiency.

For both designs, the load profiles and the meteorological data were collected from [37], where the wind power distribution and the solar radiation are presented in Figures 3 and 4, respectively.

Data were collected from meteorological records at a wind and solar power stations located at the geographical coordinates of 38°47′4″ N (latitude) and 9°29′26″ W (longitude), for an average year. Then, dimensionless values were used, depending on the arbitrated peak consumption values and the installed wind and solar energies.

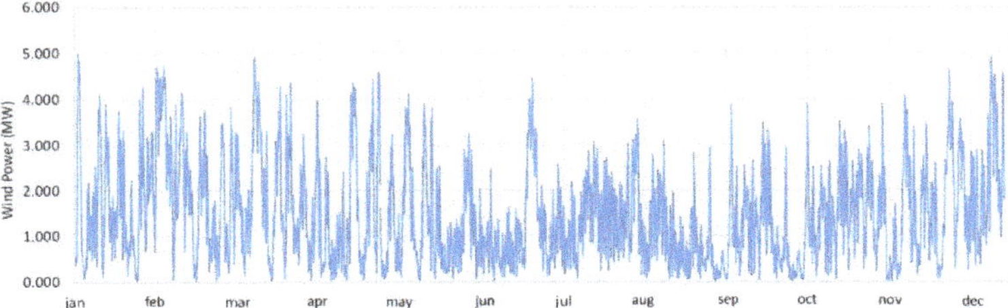

Figure 3. Average wind power distribution during an average year [28].

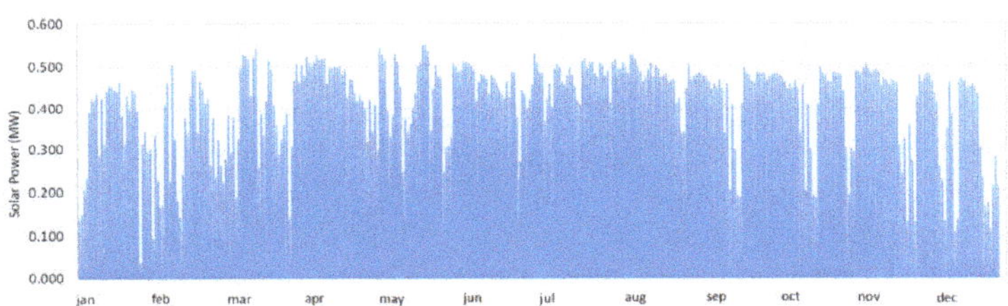

Figure 4. Average solar power distribution during an average year [28].

To demonstrate the contributions when combining pumped hydro storage (PHS) and hydro-wind-solar power system, several analyses were made, considering a peak consumption of 4 MW, and pump/turbine of 2, 4 and 6 MW, respectively. Hence, Figure 5 shows the schematic diagram of all different combinations used: total satisfied consumption and what part is supplied only by the PHS (depending on the installed power for each renewable sources), through wind + hydro and solar + hydro for different storage volumes. Analyzing wind + hydro, it is clear that the storage values considerably influence the satisfaction of demand. For the combination of solar and hydropower, this is not so visible, especially for higher storage values.

The use of wind or solar power as the source of energy supply may not be sufficient to meet the demand without a PHS. On the contrary, the higher the ratio between the installed wind energy and the peak consumption, the higher the satisfied consumption. However, although the wind energy proved to be more effective with a significant increase in the consumption satisfaction, it also translated into a greater surplus and from there the need to integrate a storage system capable of harnessing unused energy [37–39]. Combining the three sources (Figure 6) with an installed PAT of 4 MW allows to reach levels of satisfied consumption of 70 to 85%, with high annual complementarity, against 60–80%, when using P/Ts with 6 MW and 40–60% for P/Ts with 2 MW.

According to Figure 6, four scenarios related to each P/T stand out based on the best combination between the three available renewable sources (Table 2). For these, an estimation of costs and associated profits as a function of the installed power was made where it was assumed that the turbomachines cost corresponds to 35% of the total PHS and that the cost related to the wind presents a linear variation of about 583 €/MW of installed power. The profit from the purchase and sale of electricity to the national grid was accounted through the relative data of the tri-time price rate applied in Portugal mainland (Table 3).

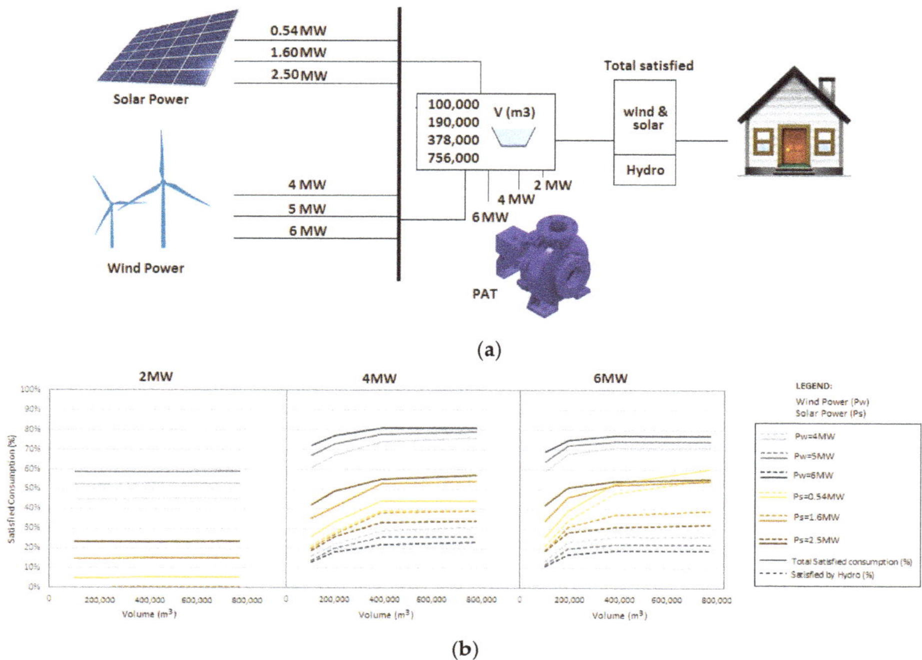

Figure 5. Hybrid solution: shematic diagram of different combinations used (**a**); satisfied demand (wind + hydro; solar + hydro; hydro) for different pump as turbine (PAT)'s power (**b**).

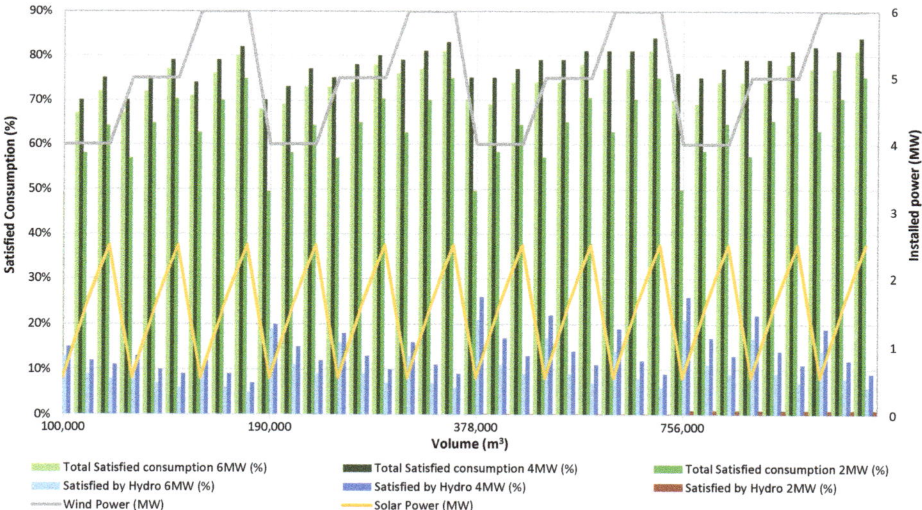

Figure 6. Satisfied consumption (wind + solar + hydro; hydro) using P/Ts with 6, 4 and 2 MW.

Table 2. Estimation of costs and associated profits as a function of the installed power.

V (m³)	Pw (MW)	$P_{P/T}$ (MW)	Cw (€)	$C_{P/T}$ (€)	C_{PHS} (€)	Total (€)	Profit (€)	Return (Years)
378,000	4		2,333,320			3,168,864	282,336	11.2
378,000	5		2,916,650			3,752,194	299,948	12.5
756,000	4	6	2,333,320	303,834	835,544	3,168,864	281,997	11.2
756,000	5		2,916,650			3,752,194	299,601	12.5
378,000	4		2,333,320			2,883,414	357,336	8.1
378,000	5		2,916,650			3,466,744	365,525	9.5
756,000	4	4	2,333,320	200,034	550,094	2,883,414	360,530	8.0
756,000	5		2,916,650			3,466,744	366,695	9.5

Pw—Wind Power; $P_{P/T}$—Pump/Turbine Power.

Table 3. Tariff rate applied in Portugal mainland (€).

Peak	Half-Peak	Off-Peak	Super Off-Peak
0.097	0.0406	0.0115	0.0115

Table 2 shows the investment costs and gross benefit estimation depending on each selected solution presented in Figure 6.

The results from Table 3, reveal that the use of P/Ts with 6 MW do not present any advantage over the P/Ts with 4 MW, using the same wind energies. Concluding, P/TS with 4 MW are economically viable, offering the best technical alternative at the lowest cost, with less return on the investment.

Nevertheless, to understand the relevant aspects of the joint action of these three solutions in consumer satisfaction needs, the energy contribution of four hybrid solutions (based on Table 3) over an average year was studied, assuming the characteristics described in Table 4.

Table 4. Selected scenarios.

Scenarios	Peak Consumption	Wind Power	Solar Power	Pump/ Turbine	Storage
Scenario 1		4 MW	0.54 MW	4 MW	144 MWh (378,000 m³)
Scenario 2	4 MW	5 MW	0.54 MW	4 MW	144 MWh (378,000 m³)
Scenario 3		5 MW	0.54 MW	4 MW	240 MWh (755,685 m³)
Scenario 4		5 MW	1.60 MW	4 MW	144 MWh (755,685 m³)

In all cases, two reservoirs (i.e., upper and lower reservoirs) were considered, with the initial volume of the upper reservoir at 50% of its maximum capacity to allow the system to pump or turbine within the first hour.

3.2. Results

3.2.1. Scenario 1

The power grid and energy storage in Figure 7 (for winter months of February and March) and Figure 8 (for summer months August and September) represent the power and energy variables for the time-line modelled: (i) curves of power demand, wind, solar, hydro and pump (left y-axis); (ii) curve for the storage volume by water pumped into the upper reservoir (right y-axis). Herein, the storage water volume was designed for 1.5 days peak demand power. It can be observed the contribution of renewable sources fulfils the energy demand with more or less effect of pump-storage solution to harmonize the available energy at each instant depending on the summer or winter period analyzed.

For the characteristics defined in Table 4, the system satisfies 15.37 GWh, where 19% comes from hydro and 49% from wind and solar, with an annual maximum wind power of 1.25 × peak demand and solar power of 0.135 × peak demand. The sources pattern varies, and the consumption satisfaction adapts to the sources' availability as function of the power selected to be installed. Even if it seems to have a low share of hydro level source, it is essential to keep flexibility and sustainability performance levels.

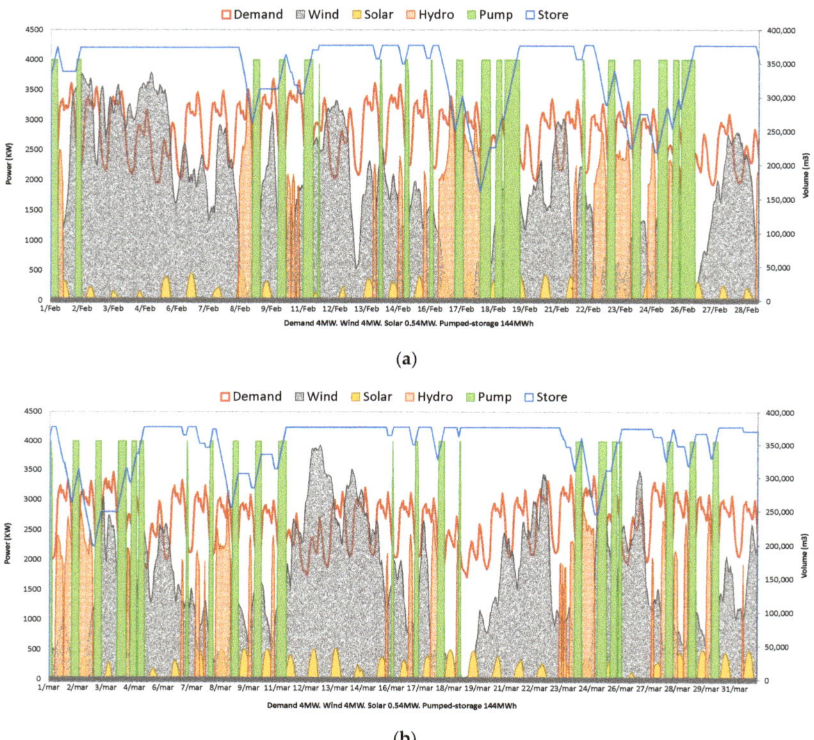

Figure 7. Electricity generation (Wind, PV and Pumped-Storage Hydro) between February (**a**) and March (**b**).

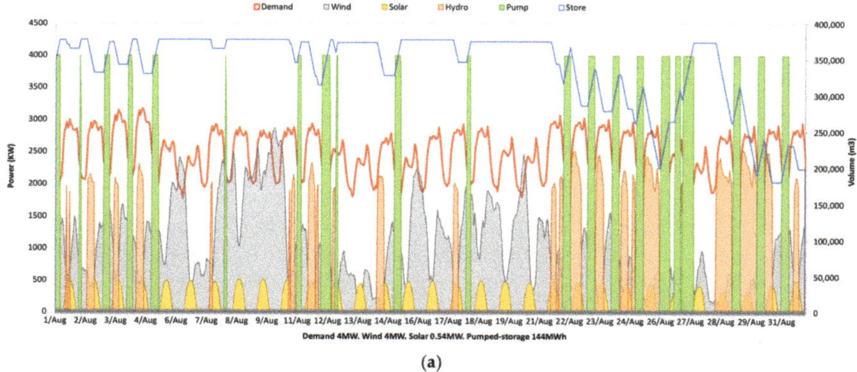

Figure 8. *Cont.*

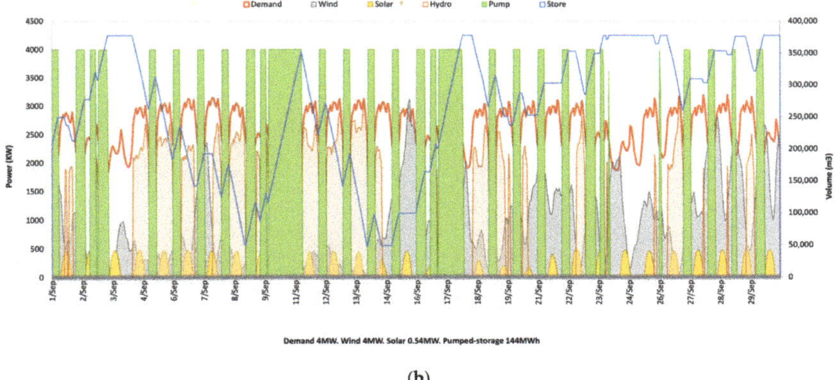

(**b**)

Figure 8. Electricity generation (Wind, PV and Pumped-Storage Hydro) between August (**a**) and September (**b**).

3.2.2. Scenario 2

The next power grid and energy storage timelines (Figures 9 and 10) illustrate when the wind power is 5 MW. Herein, the system produces 3.41 GWh of hydropower responsible for satisfying 15% from the 72% of the total satisfied consumption; the remaining power is guaranteed through wind and solar energies.

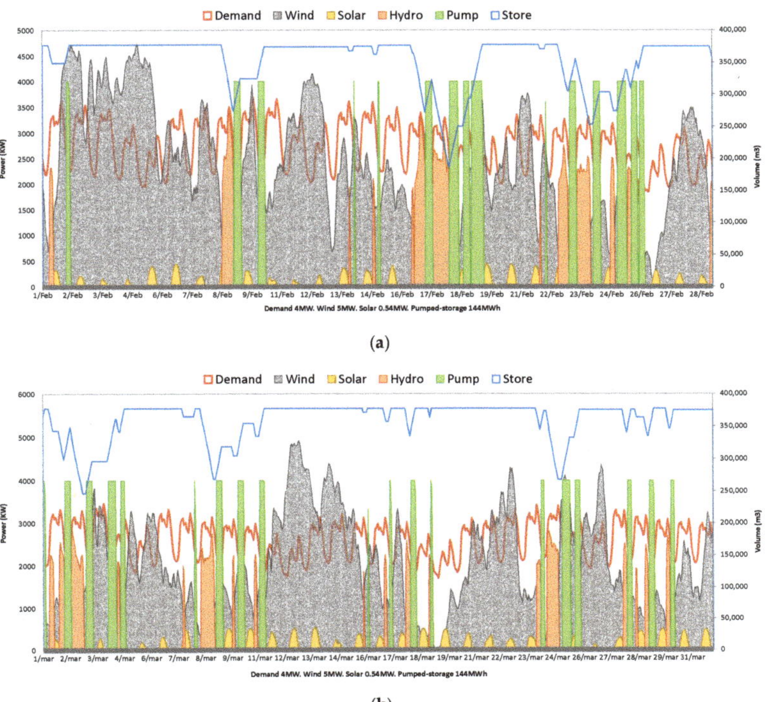

(**a**)

(**b**)

Figure 9. Electricity generation and stored in scenario 2 between February (**a**) and March (**b**).

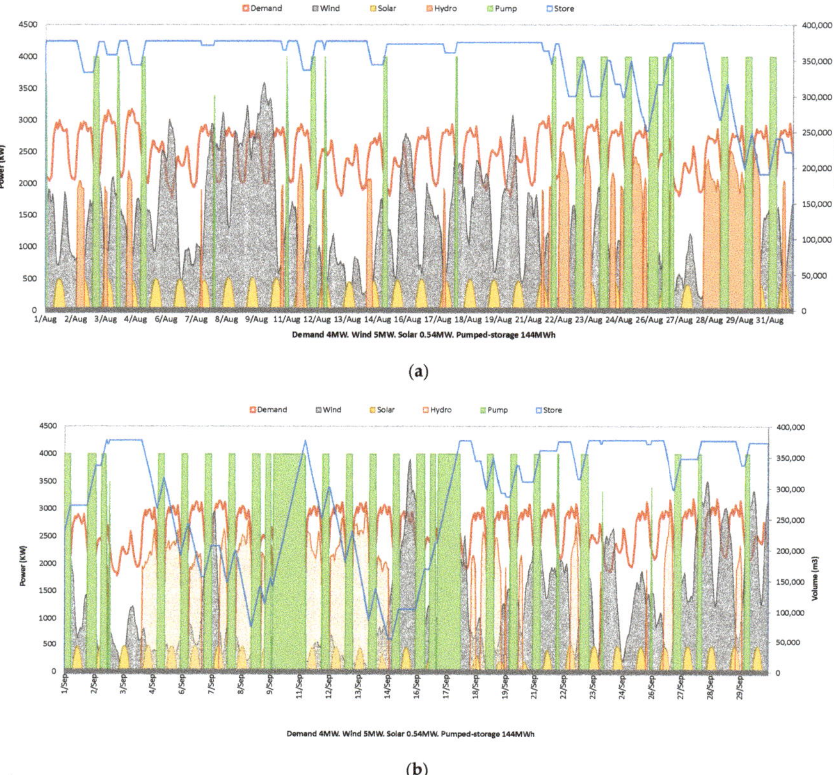

(a)

(b)

Figure 10. Electricity generation and stored in scenario 2 between August (**a**) and September (**b**).

It is notable that increasing the wind power, the hydropower production decreases accordingly. Figure 11 shows the total amount of energy that could be easily harnessed through hydropower in scenario 2, the use of this energy to overcome the wind failures and how excess of wind production in some periods can be used for pumping, creating water reserve for the period of highest consumption. The uniformity of the available sources and storage capacity makes the system more robust.

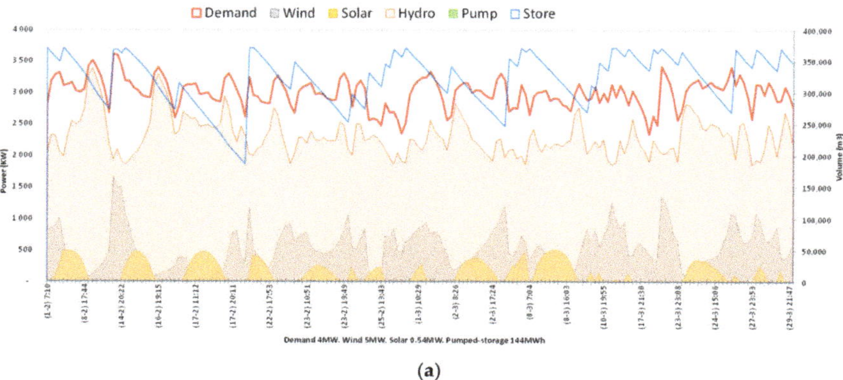

(a)

Figure 11. *Cont.*

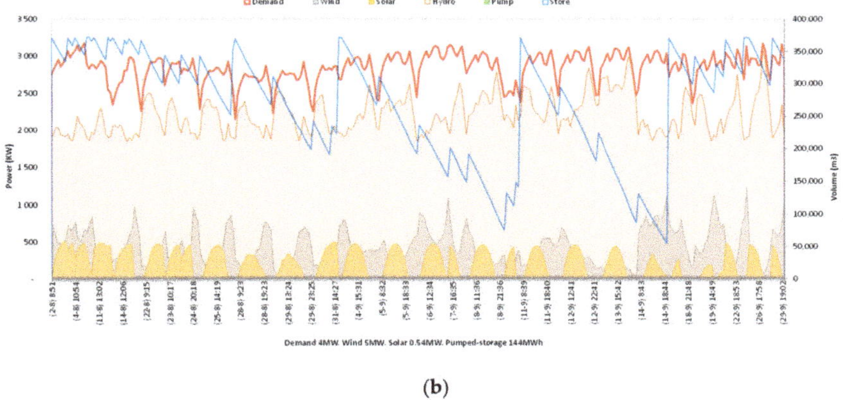

(b)

Figure 11. Total amount of hydropower when using wind energy with 5 MW, between February and March (**a**) and August-September (**b**).

Thus, by increasing the installed maximum wind power from 4 MW up to 5 MW, it is possible to take advantage of a few hours more efficiently, if the power line transmission capacity between the wind generators and the pumps exceeds the peak demand. The storage is adapted to the wind power availability allowing a better compensation between resources.

3.2.3. Scenario 3

In scenario 3, the volume of storage was increased up to 755,685 m^3 (2 times the initial one). The amount of energy that is satisfied by hydro is practically the same, comparing to scenario 2 since the volume used depends on the demand that was kept constant. With this scenario, it is possible to feel the influence or not in some parameters depending on how the intervenient variables are optimized and integrated in a more flexible solution (Figure 12).

Thus, the increasing of the volume storage does not contribute significantly, only 0.56% of the hybrid energy is used to pump. Thus, a balance needs to be made between the total energy that is needed for the best functioning of the PHS, in terms of sources management and the technical level, as well as the investment cost, operation and maintenance cost of this system and also the risk-return profile, in terms of economic performance.

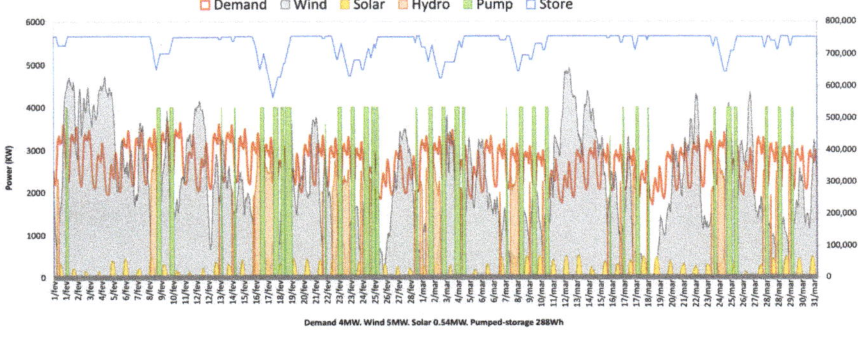

(a)

Figure 12. *Cont.*

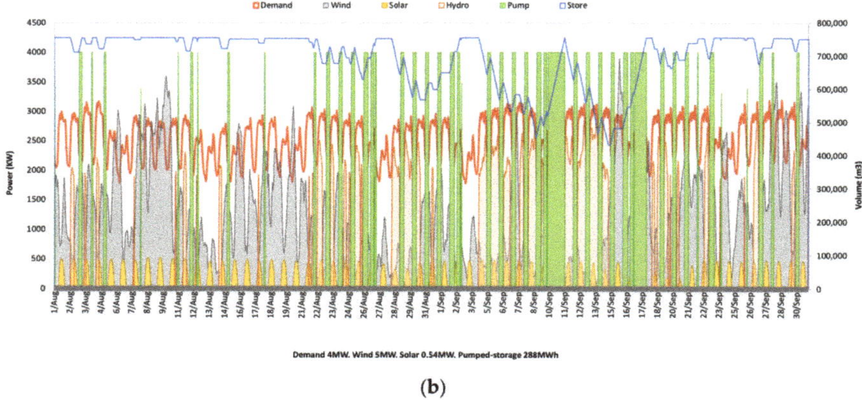

(b)

Figure 12. Electricity generation and stored in scenario 3, between February and March (**a**) and August and September (**b**).

3.2.4. Scenario 4

In Scenario 4 the PV power is increased up to 1.6, i.e., using a peak factor of 2.96 times the previous one (0.135) (Figure 13).

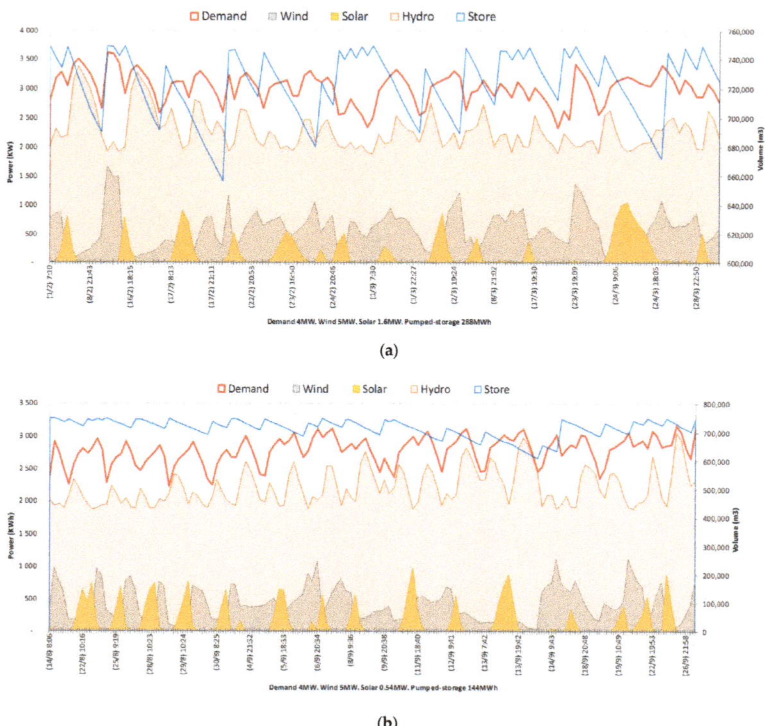

(a)

(b)

Figure 13. Total amount of energy stored vs demand in scenario 4, between February and March (**a**) and August and September (**b**).

The total energy used for consumption was 16.42GWh, representing 73%, where only 8% comes from hydro and the remaining from wind and solar solutions. However, by increasing the solar energy, 11% of this energy is wasted (i.e., 2% more comparing with scenario 2), and only 1% more contributes to the satisfied consumption. Again, the evidence of this sensitivity analysis focused on the increasing power for some resources does not mean a better solution because this multivariate optimization requires several interpretations depending on the main use made of the available sources in each instant and for the analysed period. Having more solar energy imposes a reduction on the use of hydro, because hydro is available when it is more essential not depending on the intermittency of wind and solar, making an interesting balance between the use of sources.

4. Energy Balance

The chosen hybrid hydro-wind and solar power solution with installed capacities of 5 and 0.54 MW, respectively, 4 MW of integrated pumped storage and V = 378,000 m^3 would ensure 72% annual consumption satisfaction. Figures 13 and 14 show for each scenario the energy contribution of the three renewable sources for typical summer and winter days, respectively. Although it is considered a relatively low installed solar power (1/5 wind), this source can be very useful on summer days, especially in the middle of the day, when the wind slows and the solar radiation increases. However, although increasing the PV installed capacity ensures 65% of the consumption through wind + solar (Figures 14d and 15d), comparing with scenario 2 (Figures 14b and 15b), the hydropower can cover that difference with the pump/hydro power solution. For the majority of winter days, there is a surplus of wind production at the beginning of the day, taking advantage of reducing the energy costs for scheduled pumping in the morning. Additionally, for a storage capacity of 144 MWh and 288 MWh, there is practically no significant contribution when comparing to other scenarios (Figures 14c and 15c).

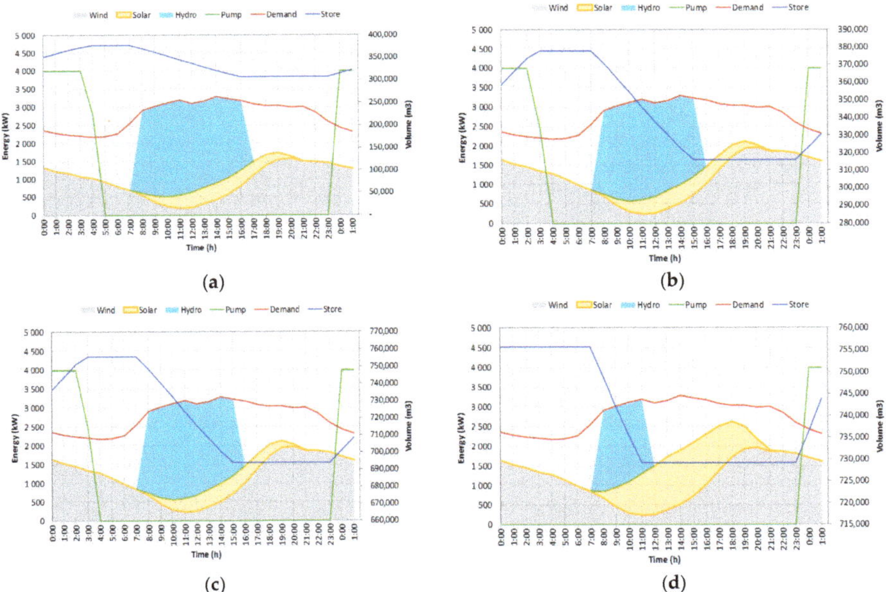

Figure 14. Energy contribution on summer day: (**a**) scenario 1; (**b**) scenario 2; (**c**) scenario 3; (**d**) scenario 4.

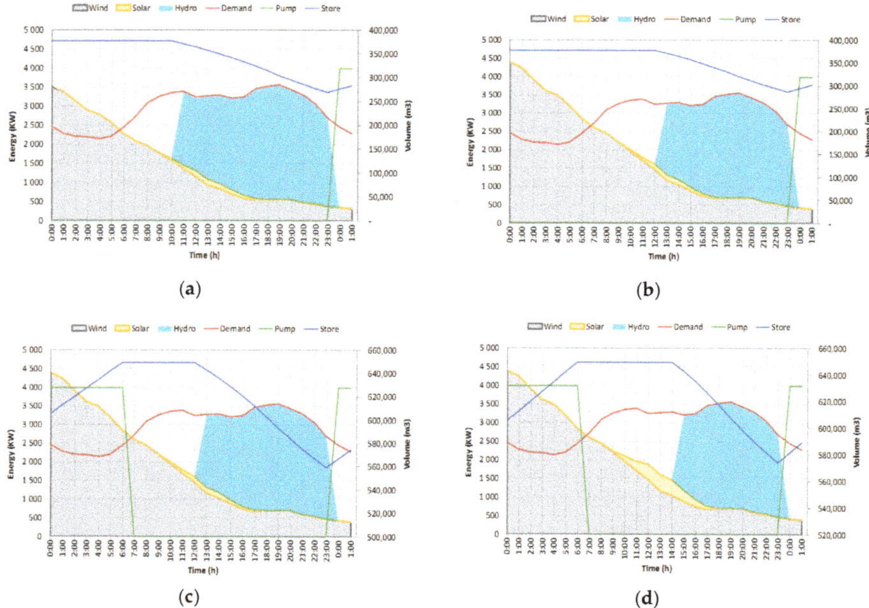

Figure 15. Energy contribution on a winter day: (**a**) scenario 1; (**b**) scenario 2; (**c**) scenario 3; (**d**) scenario 4.

It is important to verify that in all scenarios mostly in Figure 14d, between 12:00–23:00, the turbine volume is below the minimum turbine volume allowed; consequently, the system cannot operate during these hours even with a greater PV energy. Since this system does not depend on the excess of wind and solar energy, a better use of hydropower can be detected in scenario 1 (Figures 14a and 15a). However, it is important to remember that this operating solution needs to be connected to the grid, as the system cannot always be powered by wind/solar energy to perform pumping.

Thus, knowing that the peak demand and the average power is 4 MW and therefore average daily usage only 24 h × 4.0 MW = 96 MWh, for an optimized system, the hydropower energy capacity needed would be 1.5 times the maximum daily energy usage, assuming that the maximum wind and solar power is 0.056 per hour x max daily energy usage. Thus, the annual maximum wind and solar power should be 5.4 MWh and the energy store capacity 144 MWh (i.e. 1.5 × 96 MWh). Increasing the store capacity to 288 MWh, which is still a factor of 2 times more energy storage than the previous one (i.e., 1.5), does not contribute significantly to the operation of the system. This means that the reservoir may be oversized, as a large amount of reserves is not required to satisfy the consumption. Thus, this system depends not only on the maximum daily energy usage of the system but also on the hydropower system. Peak power demand does not completely specify a system's generation and storage requirements since both peak power and maximum daily energy usage are important design considerations.

The contribution of each energy source to the demand satisfaction was assessed (Figure 16) for the four scenarios (i.e., scenario 1, scenario 2, scenarios 3 and 4). It is clear that the storage values (between wind-hydro) considerably influence the satisfaction of consumption. In the combination of solar and hydropower, this is not so visible, especially for higher storage values. In each scenario, the wind energy demonstrates superior efficiency as there is not a large discrepancy between the maximum and minimum generation values. For scenario 1 (Figure 16a), the hydropower contribution presents a greater contribution compared with other scenarios, translating into a bigger unsatisfied demand, with less wind/solar energy impact. In scenario 2, by increasing the wind/solar power, the satisfied consumption through this energy source increases up to 4% compared to the previous one. Still, a water reserve storage system with a capacity twice the initial one brings no advantage (Figure 16c). Nevertheless, comparing with scenario 4 (Figure 16d), increasing the installed solar does not contribute to the balance of the system or to consumer satisfaction. Thus, increasing the energy storage system, capable of holding water reserve, is not advantageous to assist the consumption, it being oversized for matching energy sources to periods of high demand (Figure 16c).

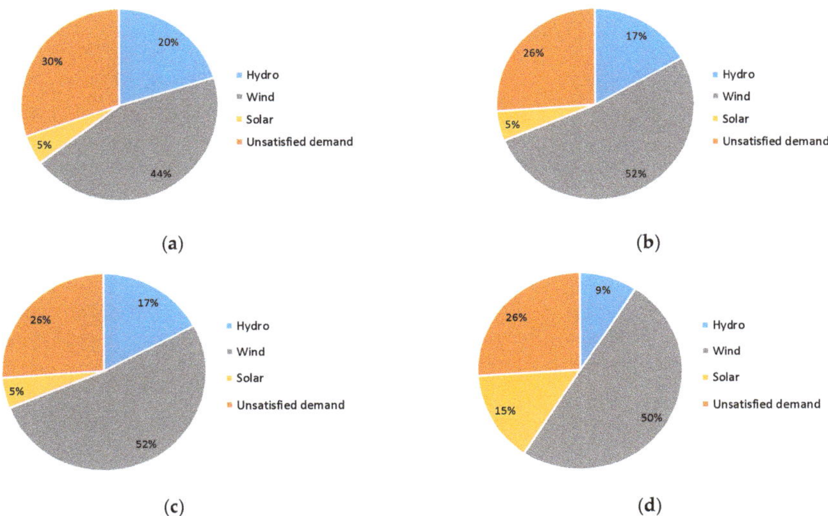

Figure 16. Satisfied consumption by different energy sources and unsatisfied consumption in (**a**) scenario 1; (**b**) scenario 2; (**c**) scenario 3; (**d**) scenario 4.

Thus, scenarios 1 and 2 highlight the major concern of renewables. If some of the surplus energy from renewables produced at times of low demand (e.g., solar power in the summer) could be stored and ready for release when demand raises, many of the problems of renewable supply would be solved. However, from the economic point of view, comparing scenario 2 with scenario 1 (Figure 16b), this solution is more expensive, needing estimation costs for the viability of this system.

Thus, a brief economic analysis was conducted, using key financial parameters such as basic payback period (*BPP*), net present value (*NPV*) and internal rate of return (*IRR*) presented in Equations (5)–(7) [40,41].

$$BPP = \frac{C}{AS} \qquad (\text{years}) \tag{5}$$

$$NPV = \sum \frac{B-C}{(1+r)^n} \qquad (\text{€}) \tag{6}$$

$$IRR = \sum \frac{B}{(1+r)^n} = \sum \frac{C}{(1+r)^n} \qquad (\%), \tag{7}$$

where *BPP* is the minimum time to recover the total investment (in years), *NPV* is the presented value of all future income (€), and expenditure flows and *IRR* is the rate (%) that makes the *NPV* zero [41,42]. Regarding *C*, it is the total investment cost; *AS* is the net annual saving; *B* is the benefit, *C* the cost, *r*—the discount rate and *n*—the lifecycle of the project (years). In this study, as suggested by many turbine manufacturers, the lifespan of the project was assumed to be 20 years, and the overall annual interest rate, *r*, is assumed to be 2.5%. The economic analysis results of the hybrid system are presented in Table 5.

Table 5. Economic analysis results.

Parameters	Scenario 1	Scenario 2	Scenario 3	Scenario 4
Investment Cost (€)	−3,198,412	−3,781,742	−3,781,742	−4,400,072
Total annual cost savings and income (€)	479,762	415,992	415,992	352,006
NPV (€)	4,611,564	5,452,627	5,452,627	6,344,153
IRR (%)	30.63	25.51	25.51	20.24
BPP (years)	6.7	9.1	9.1	12.5

Scenario 1 has a relatively basic pay-back period of approximately 7 years with a lower initial capital cost when compared to the other scenarios. Thus, the results yield that scenario 1 is the most economical and reliable solution.

As renewable energy investments represent a significant subset of the infrastructure sector, where PV and wind power investments have already become well established in the asset allocations of institutional investors, hydropower, by contrast, has a more difficult investment opportunity, since it requires significantly higher up-front investment per capacity unit, making it less scalable than wind power or photovoltaic plants [39–42]. Although, operating pumped-storage power plants is economically viable because electricity can be produced and sold depending on spot market prices, the technical know-how required for hydropower investments is more challenging since the success depends not only on technical and structural components but also on active management.

5. Water–Energy Nexus

This research studied a pumped hydro storage serving for on-grid hybrid energy solutions. The complementary characteristics between solar and wind energy output were presented. Results reveal that the wind turbines have a relatively higher share of energy production than PV since the wind energy resource matches better with the load pattern. Peak factors and power capacity were discussed to calculate the overall energy efficiency of the energy storage system. The case study shows that if wind and solar energies are adopted, with power capacities slightly higher than the peak consumption, a better satisfied demand is guaranteed (Figure 17).

Figure 17. Total satisfied consumption through renewable energy (RE) in each scenario.

In scenarios 2, 3 and 4, the total satisfied consumption is around 74%, against 69% of scenario 1, where the wind/solar capacity of 5/0.54 MW becomes attractive to compensate the load (Scenarios 2 and 3).

Moreover, the potential impact of renewables was estimated (Figure 18), illustrating the greenhouse gas emissions avoided as a result of the renewable energies deployment in each scenario in a complementary way, taking as the base stability the pumped-hydro component. Figure 18b is the result of the mitigation of climate change effects.

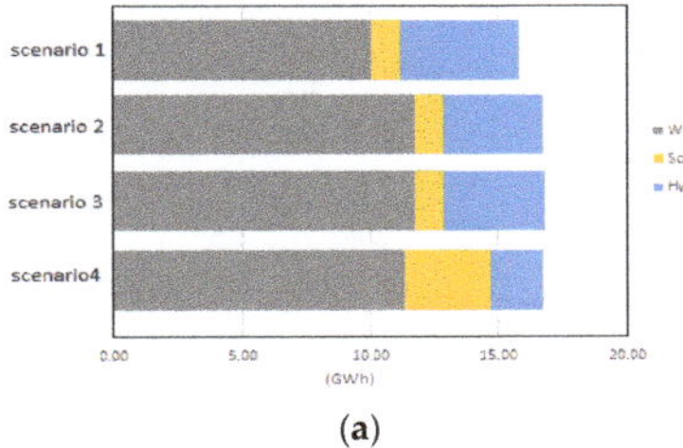

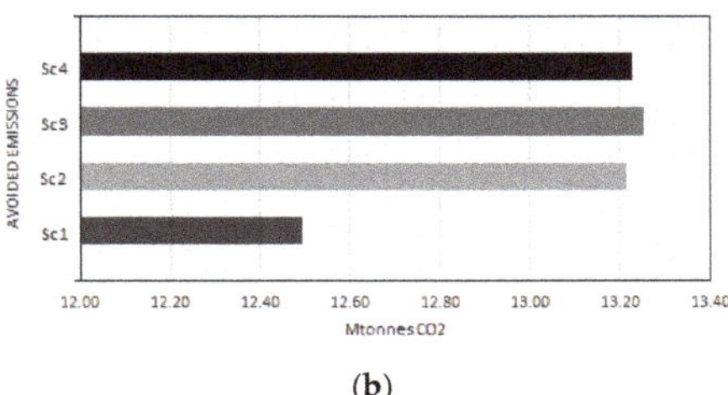

(**a**) (**b**)

Figure 18. Electricity generated from renewables (GWh) in each scenario (**a**); total avoided emissions by each scenario (**b**).

In scenario 1, there is a lower CO_2 emission and water withdrawal, and a cost-optimal market solution also serves to improve the performance (Table 6). Regarding CO_2 and water withdrawal, it remains approximately the same between scenarios 3 and 4. This shows that storage actions taken in the water infrastructure between these scenarios can serve to improve the electric power infrastructure when the two or more sources are coupled together in a water–energy nexus.

Table 6. Cross comparison of cost, CO_2 emissions and water withdrawal in each scenario.

Scenario	Cost (M€)	CO_2 (Mtonnes)	Water Withdrawal (m³)
1	3.20	2.01	169,920
2	3.78	2.72	165,304
3	3.78	2.76	354,304
4	4.40	2.74	357,831

These results suggest an interaction between different renewable sources, which obeys the following pattern: solar energy reduces CO_2 emissions but exacerbates the night ramp in energy demand; in contrast, wind energy may bridge this gap, but it is usually intermittent, unpredictable and weather dependent. By employing an energy storage system, the surplus energy can be stored when power generation exceeds demand and then be released to cover the periods when net load exists, providing a robust back-up to intermittent renewable energy. Thus, water and energy storage presents a promising solution to these two problems, as it allows flattening demand curves and significantly reducing costs.

6. Conclusions

This work demonstrates that technically the pumped-storage hydropower system integrating other renewable sources is an attractive energy solution. The dynamic contribution of individual sources follows different patterns, due to the stability of hydro by pumping and random variability of other energy sources and the energy demand. Employing the three technologies in a complementary and balanced manner, the hybrid system could generate and store electricity at low cost, facing climate changes and reducing the footprint of electricity in a self-sufficient solution. Thus, a consistent multi-criteria framework was developed to optimize the availability and storage of renewable energy, selecting the best combination of peak factors to achieve the optimum solution in terms of efficiency, energy use, costs and footprint. Important considerations are highlighted and summarized from this multi-variable process:

(a) The optimization showed that in a hybrid solution, turbines and pumps can be used at the same time depending on the intermittency, availability and optimized variables, which include different renewable sources, the storage capacity and the load demand. The pumping system can be supplied by intermittent renewable sources when available, and at the same time, can be guaranteed a constant power production by hydraulic turbines. The only one pipe for the P/T solution requires different hours for each operation or the use of separate pipes, which can offer more operating flexibility, where one is kept running when the other is stopped or in operation, depending on the sources' availability, constancy or intermittency, type of storage or type of grid connection;

(b) Three sources were combined considering different pump/turbine installations, wind/solar powers and different water batteries as volume capacities. The analysis revealed that P/Ts with 4 MW are economically viable compared to 6 MW and 2 MW, with 70% to 85% satisfaction of consumption levels;

(c) After selecting the best installation power for P/Ts, four scenarios were tested, changing the wind/solar powers and the water storage capacity. Three types of analyses were performed from the point of view of energy, economy and CO_2 emissions. The results obtained show the process of selecting the best scenario is not straightforward, depending on the final goal. Therefore, this analysis unfolds in important points:

   i.   Scenario 1 stands out from the point of view of reliability and flexibility, where there is a better use of hydropower (Figures 14 and 15), specifically to accommodate the largest shares of other intermittent renewable (solar and wind) energies with a better bridge and compensation between these energy sources;

   ii. Scenario 2 showed a 4% increase in satisfied consumption from an operational point of view, maintaining the same characteristics as scenario 1 but requiring an increase in the installed wind power;

   iii. In scenario 3, increasing storage capacity to 288 MWh does not make a significant contribution to the best operation of the system. As a result, the reservoir is oversized to meet the satisfied consumption, i.e., there is a dependency not only on the maximum daily energy use of the system but also on the hydropower system;

   iv. In both scenarios 1 and 2, surplus energy from renewables produced at times of low demand (e.g., solar power in summer) can be stored and ready for release when demand rises;

   v.   Scenario 1 offers more advantages and greater economic viability also in terms of CO_2 emissions. This hybrid solution is less expensive, with an interesting pay-back period of 7 years, considering the powers installed, with a lower initial capital cost than that of the other scenarios.

Additionally, it can be concluded that replacing fossil fuels by renewable energies requires (i) distributing installations (e.g., wind turbines) widely; (ii) using a range of different intermittent energy sources, especially those that are partially complementary (e.g., sunny weather often means light winds and vice versa); (iii) matching with suitable management of energy sources in periods of high demand. Still, there is clear evidence of how all these modern integrated techniques for complementarity between renewable sources can significantly reduce electricity tariffs and increase the reliability of the energy supply as main targets of hybrid-energy solutions.

Author Contributions: The author M.S. has been involved in the mathematical modelling, in writing the paper, as well as in the analysis of the results. H.M.R. has contributed with the idea, the revision of the document, the supervision of the whole research, and the analyses of the head drop in each system component. All authors have read and agreed to the published version of the manuscript.

Funding: This research is supported by the project REDAWN (Reducing Energy Dependency in Atlantic Area Water Networks) EAPA_198/2016 from INTERREG ATLANTIC AREA PROGRAMME 2014–2020.

Acknowledgments: The authors wish to thank to the project REDAWN (Reducing Energy Dependency in Atlantic Area Water Networks) EAPA_198/2016 from INTERREG ATLANTIC AREA PROGRAMME 2014–2020 and CERIS (CEHIDRO-IST).

Conflicts of Interest: The authors declare no conflict of interest.

Nomenclature

AS	net annual saving (€)
B	benefit (€)
BPP	basic payback period (years)
c_p	coefficient of the pump/motor unit (m^3/kWh)
c_t	turbine generating coefficient (kWh/m^3)
C	total investment cost (€)
$C_{P/T}$	pump/turbine energy solution cost (€)
C_{PHS}	Pump hydro system cost (€)
Cw	wind energy solution cost (€)
E_i^p	Pump power installed (MW)
E_i^t	Turbine power installed (MW)
G	acceleration due to gravity (m/s^2)
H	hybrid power/energy available (MW)
H	net head (m)
IRR	internal rate of return (%)
N	lifecycle of the project (years)
NPV	net present value (€)
p_f	peak factors
Ps	solar power (MW)
PHS	pumped hydro storage
Pw	wind power (MW)
Q^p	pump flow (m^3/s)
Q^t	turbine flow (m^3/s)
R	the discount rate (%)
S	solar energy (MW)
V	storage capacity (m^3)
V^p	pump volume (m^3)
V^{res}	volume of the reservoir (m^3)
V_{max}^{res}	maximum volume of the reservoir (m^3)
V_{min}^{res}	minimum volume of the reservoir (m^3)
V^t	turbine volume (m^3)
W	wind energy (MW)
Greek letters	
η_p	pumping efficiency (%)
η_t	efficiency of the turbine/generator unit (%)
ρ	density of the water (kg/m^3)

References

1. IHA. *The World's Water Battery: Pumped Hydropower Storage and the Clean Energy Transition*; International Hydropower Association, IHA Working Paper: London, UK, 2018.
2. IHA. *Pumped Storage Hydropower Has 'Crucial Role' in Europe's Energy Strategy*; International Hydropower Association, IHA Working Paper: London, UK, 2020.
3. Bhandari, B.; Poudel, S.R.; Lee, K.-T.; Ahn, S.-H. Mathematical modeling of hybrid renewable energy system: A review on small hydro-solar-wind power generation. *Int. J. Precis. Eng. Manuf. Technol.* **2014**, *1*, 157–173. [CrossRef]
4. Hong, C.-M.; Ou, T.-C.; Lu, K.-H. Development of intelligent MPPT (maximum power point tracking) control for a grid-connected hybrid power generation system. *Energy* **2013**, *50*, 270–279. [CrossRef]
5. Kapsali, M.; Kaldellis, J. Combining hydro and variable wind power generation by means of pumped-storage under economically viable terms. *Appl. Energy* **2010**, *87*, 3475–3485. [CrossRef]
6. Xu, B.; Chen, D.; Venkateshkumar, M.; Xiao, Y.; Yue, Y.; Xing, Y.; Li, P. Modeling a pumped storage hydropower integrated to a hybrid power system with solar-wind power and its stability analysis. *Appl. Energy* **2019**, *248*, 446–462. [CrossRef]

7. Vieira, F.; Ramos, H.M. Optimization of operational planning for wind/hydro hybrid water supply systems. *Renew. Energy* **2009**, *34*, 928–936. [CrossRef]

8. Kocaman, A.S.; Modi, V. Value of pumped hydro storage in a hybrid energy generation and allocation system. *Appl. Energy* **2017**, *205*, 1202–1215. [CrossRef]

9. Kapsali, M.; Anagnostopoulos, J.; Kaldellis, J. Wind powered pumped-hydro storage systems for remote islands: A complete sensitivity analysis based on economic perspectives. *Appl. Energy* **2012**, *99*, 430–444. [CrossRef]

10. Deason, W. Comparison of 100% renewable energy system scenarios with a focus on flexibility and cost. *Renew. Sustain. Energy Rev.* **2018**, *82*, 3168–3178. [CrossRef]

11. Jaramillo-Duque, Á.; Castronuovo, E.D.; Sánchez, I.; Usaola, J. Optimal operation of a pumped-storage hydro plant that compensates the imbalances of a wind power producer. *Electr. Power Syst. Res.* **2011**, *81*, 1767–1777. [CrossRef]

12. Yang, W.; Yang, J. Advantage of variable-speed pumped storage plants for mitigating wind power variations: Integrated modelling and performance assessment. *Appl. Energy* **2019**, *237*, 720–732. [CrossRef]

13. Li, J.; Wang, S.; Ye, L.; Fang, J. A coordinated dispatch method with pumped-storage and battery-storage for compensating the variation of wind power. *Prot. Control. Mod. Power Syst.* **2018**, *3*, 2. [CrossRef]

14. Nema, P.; Nema, R.; Rangnekar, S. A current and future state of art development of hybrid energy system using wind and PV-solar: A review. *Renew. Sustain. Energy Rev.* **2009**, *13*, 2096–2103. [CrossRef]

15. Van Meerwijk, A.J.H.; Benders, R.M.J.; Davila-Martinez, A.; Laugs, G.A.H. Swiss pumped hydro storage potential for Germany's electricity system under high penetration of intermittent renewable energy. *J. Mod. Power Syst. Clean Energy* **2016**, *4*, 542–553. [CrossRef]

16. Ramli, M.A.M.; Bouchekara, H.R.E.H.; Alghamdi, A.S. Optimal sizing of PV/wind/diesel hybrid microgrid system using multi-objective self-adaptive differential evolution algorithm. *Renew. Energy* **2018**, *121*, 400–411. [CrossRef]

17. Ashok, S. Optimised model for community-based hybrid energy system. *Renew. Energy* **2007**, *32*, 1155–1164. [CrossRef]

18. Sawle, Y.; Gupta, S.; Bohre, A.K. Review of hybrid renewable energy systems with comparative analysis of off-grid hybrid system. *Renew. Sustain. Energy Rev.* **2018**, *81*, 2217–2235. [CrossRef]

19. Diaf, S.; Notton, G.; Belhamel, M.; Haddadi, M.; Louche, A. Design and techno-economical optimization for hybrid PV/wind system under various meteorological conditions. *Appl. Energy* **2008**, *85*, 968–987. [CrossRef]

20. Kusakana, K. Optimal scheduling for distributed hybrid system with pumped hydro storage. *Energy Convers. Manag.* **2016**, *111*, 253–260. [CrossRef]

21. Ma, T.; Yang, H.; Lu, L.; Peng, J. Technical feasibility study on a standalone hybrid solar-wind system with pumped hydro storage for a remote island in Hong Kong. *Renew. Energy* **2014**, *69*, 7–15. [CrossRef]

22. Ma, T.; Yang, H.; Lu, L.; Peng, J. Optimal design of an autonomous solar–wind-pumped storage power supply system. *Appl. Energy* **2015**, *160*, 728–736. [CrossRef]

23. Bajpai, P.; Dash, V. Hybrid renewable energy systems for power generation in stand-alone applications: A review. *Renew. Sustain. Energy Rev.* **2012**, *16*, 2926–2939. [CrossRef]

24. Hana, A.M.H.A.; Zakaria, A.; Jani, J.; Seyajah, N. Optimization Techniques and Multi-Objective Analysis in Hybrid Solar-Wind Power Systems for Grid-Connected Supply. *IOP Conf. Ser. Mater. Sci. Eng.* **2019**, *538*, 012040. [CrossRef]

25. Rehman, S.; Al-Hadhrami, L.M.; Alam, M. Pumped hydro energy storage system: A technological review. *Renew. Sustain. Energy Rev.* **2015**, *44*, 586–598. [CrossRef]

26. Paska, J.; Biczel, P.; Kłos, M. Hybrid power systems—An effective way of utilising primary energy sources. *Renew. Energy* **2009**, *34*, 2414–2421. [CrossRef]

27. Zhang, L.; Xin, H.; Wu, J.; Ju, L.; Tan, Z. A Multiobjective Robust Scheduling Optimization Mode for Multienergy Hybrid System Integrated by Wind Power, Solar Photovoltaic Power, and Pumped Storage Power. *Math. Probl. Eng.* **2017**, *2017*, 1–15. [CrossRef]

28. Amaro, G.; Ramos, H.M. *Solução Energética Híbrida com Armazenamento por Bombagem: Modelação, Análises de Sensibilidade e caso de Estudo Mestrado Integrado em Engenharia Civil Júri*; Instituto Superior Técnico—Universidade de Lisboa: Lisboa, Portugal, 2018.

29. Mohanraj, K.; Bharathnarayanan, S. Three Level SEPIC For Hybrid Wind-Solar Energy Systems. *Energy Procedia* **2017**, *117*, 120–127. [CrossRef]

30. El-Jamal, G.; Ghandour, M.; Ibrahim, H.; Assi, A. Technical feasibility study of solar-pumped hydro storage in Lebanon. Proceedings of 2014 International Conference on Renewable Energies for Developing Countries, Beirut, Lebanon, 26–27 November 2014; Institute of Electrical and Electronics Engineers (IEEE): Beirut, Lebanon, 2014; pp. 23–28.

31. Canales, F.A.; Beluco, A. Modeling pumped hydro storage with the micropower optimization model (HOMER). *J. Renew. Sustain. Energy* **2014**, *6*, 43131. [CrossRef]

32. Li, J.; Fang, J.; Wen, I.; Pan, Y.; Ding, Q. Optimal trade-off between regulation and wind curtailment in the economic dispatch problem. *CSEE J. Power Energy Syst.* **2015**, *1*, 37–45. [CrossRef]

33. Canales, F.A.; Beluco, A.; Mendes, C.A.B. Modelling a hydropower plant with reservoir with the micropower optimisation model (HOMER). *Int. J. Sustain. Energy* **2015**, *36*, 654–667. [CrossRef]

34. Katsaprakakis, D.A.; Christakis, D.G.; Stefanakis, I.; Spanos, P.; Stefanakis, N. Technical details regarding the design, the construction and the operation of seawater pumped storage systems. *Energy* **2013**, *55*, 619–630. [CrossRef]

35. Li, S.; Sun, F.; He, H.; Chen, Y. Optimization for a Grid-connected Hybrid PV-wind-retired HEV Battery Microgrid System. *Energy Procedia* **2017**, *105*, 1634–1643. [CrossRef]

36. Schmidt, J.; Kemmetmüller, W.; Kugi, A. Modeling and static optimization of a variable speed pumped storage power plant. *Renew. Energy* **2017**, *111*, 38–51. [CrossRef]

37. Carravetta, S.A.; Ramos, H.M.; Houreh, S.D. *Pumps as Turbines, Fundamentals and Applications*; Springer International Publishing: Berlin/Heidelberg, Germany, 2018.

38. Bilal, B.O.; Sambou, V.; Ndiaye, P.; Kébé, C.; Ndongo, M. Optimal design of a hybrid solar–wind-battery system using the minimization of the annualized cost system and the minimization of the loss of power supply probability (LPSP). *Renew. Energy* **2010**, *35*, 2388–2390. [CrossRef]

39. Goyena, R. Summary for Policymakers. In *Climate Change 2013—The Physical Science Basis*; Intergovernmental Panel on Climate Change, Ed.; Cambridge University Press: Cambridge, UK, 2019; Volume 53, pp. 1–30.

40. Luo, W.; Jiang, J.; Liu, H. Frequency-adaptive modified comb-filter-based phase-locked loop for a doubly-fed adjustable-speed pumped-storage hydropower plant under distorted grid conditions. *Energies* **2017**, *10*, 737. [CrossRef]

41. Kose, F.; Kaya, M.N.; Ozgoren, M. Use of Pumped Hydro Energy Storage to Complement Wind Energy—A case study. *Therm. Sci.* **2020**, *24*, 777–785. [CrossRef]

42. Nikolaou, T.; Stavrakakis, G.; Tsamoudalis, K. Modeling and optimal dimensioning of a pumped- 2 hydro energy storage system for the exploitation of 3 the rejected wind energy in the non—Interconnected 4 electrical power system of the Crete island, Greece. *Energies* **2020**, *13*, 2705. [CrossRef]

Article

Transient-Flow Induced Compressed Air Energy Storage (TI-CAES) System towards New Energy Concept

Mohsen Besharat [1,*], Avin Dadfar [1], Maria Teresa Viseu [2], Bruno Brunone [3] and Helena M. Ramos [1]

[1] Department of Civil Engineering and Architecture, CERIS, Instituto Superior Técnico, Universidade de Lisboa, 1049-001 Lisbon, Portugal; avin.dadfar@tecnico.ulisboa.pt (A.D.); helena.ramos@tecnico.ulisboa.pt or hramos.ist@gmail.com (H.M.R.)

[2] Hydraulics and Environment Department, Laboratório Nacional de Engenharia Civil (LNEC), 1049-001 Lisbon, Portugal; tviseu@lnec.pt

[3] Department of Civil and Environmental Engineering, University of Perugia, G. Duranti 93, 06125 Perugia, Italy; bruno.brunone@unipg.it

* Correspondence: mohsen.besharat@tecnico.ulisboa.pt

Received: 24 January 2020; Accepted: 19 February 2020; Published: 22 February 2020

Abstract: In recent years, interest has increased in new renewable energy solutions for climate change mitigation and increasing the efficiency and sustainability of water systems. Hydropower still has the biggest share due to its compatibility, reliability and flexibility. This study presents one such technology recently examined at Instituto Superior Técnico based on a transient-flow induced compressed air energy storage (TI-CAES) system, which takes advantage of a compressed air vessel (CAV). The CAV can produce extra required pressure head, by compressing air, to be used for either hydropower generation using a water turbine in a gravity system or to be exploited in a pumping system. The results show a controlled behaviour of the system in storing the pressure surge as compressed air inside a vessel. Considerable power values are achieved as well, while the input work is practically neglected. Higher power values are attained for bigger air volumes. The TI-CAES offers an efficient and flexible solution that can be exploited in exiting water systems without putting the system at risk. The induced transients in the compressed air allow a constant outflow discharge characteristic, making the energy storage available in the CAV to be used as a pump storage hydropower solution.

Keywords: hydro-energy; CAES; transient flow; energy concept; energy storage; similarity law

1. Introduction

The history of using the energy of water is comparable with the history of human civilization. In recent decades, the generated power from water (hydropower) has been increased considerably as a significant source of renewable energy. Worldwide generation from hydropower, which varies each year with shifts in weather patterns and other local conditions, was an estimated 1292 GW (4200 TWh) in 2018, which is about 25.6% of the global electricity generation and 15.9% of renewable electricity (Figure 1) [1]. Pumped-storage hydropower (PSH) systems have played an essential role in the flexibility of renewable energy sources such as the integration of wind and solar energies. The global PSH installed capacity in 2018 was 160.3 GW, which increased by about 1% during the year [2]. The hydropower capacity has been growing quite rapidly during the last few decades and the forecasts show that it will remain the world's primary source of renewable power in 2024. Hydropower will account for 10% of the total increase in renewable capacity based on a forecast for the period of

2019 to 2024 by the International Energy Agency (IEA) [3]. The main source of renewable electricity in Europe is provided by hydropower, with an estimated electricity generation of 643 TWh (pump-storage not included), which accounts for 17% of the total generation [1]. Despite the fact that most of the European countries do not provide good potential in the hydropower generation field due to their flat landscape, some countries, like Norway, Russia and Portugal, can provide a good contribution in hydropower energy generation [4].

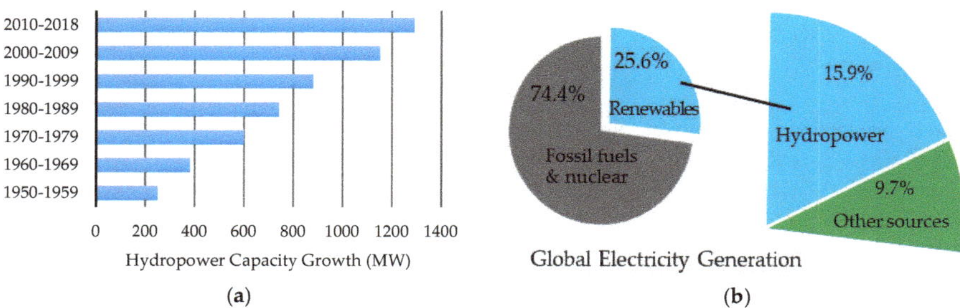

Figure 1. (**a**) Hydropower capacity growth, (**b**) hydropower rank in the global electricity generation [1]

Traditionally, hydropower facilities can appear in different types, namely impoundment (storage), pumped-storage, run-of-river, offshore hydropower plants and inline solutions in the water sector. Hydropower potential can be categorized in terms of the installed capacity as large (>10 MW), small (<10 MW), mini (<1 MW), micro (<0.1 MW) and pico (<0.005 MW) [4]. Indeed, many of the potential hydropower spots cannot be exploited as a result of technical and economic reasons, but this estimation proves the huge potential behind the hydropower energy generation [4]. Traditional and large hydropower plants are mostly of the impoundment type, which calls for large financial investment. In addition, the development of large hydropower plants can trigger big environmental impacts and requires specific topographical conditions. For this reason, this study mostly focuses on less expensive technologies.

Apart from the traditional hydropower types, which are mostly of large potential types, some new lower potential types have been introduced and examined recently. Among them, PSH has been very promising. This system takes advantage of off-peak electricity to transfer water to the upper reservoir using a pump and then produces electricity from the potential energy of water during peak hours [5–7]. PSH is currently the most common form of on-grid electricity storage, with an operating condition which is very well fitted to reversible turbomachines, namely Francis or pumps as turbines (PATs) [8]. The recent increasing influence of variable renewable energy sources forces innovations in energy storage technologies [9]. This has been the main motivation behind other PSH types, such as underwater PSH (UPSH), which some studies have shown can be competitive with conventional PSH and compressed air energy storage (CAES) systems. In UPSH, seawater acts as an upper reservoir and an underwater tank has the role of a lower reservoir working with the same concept of a conventional PSH [9].

Accordingly, the novel renewable energies should be designed on the basis of new technologies and small decentralized hydropower plants to decrease the investment costs and increase the reliability and flexibility of the renewable source by making it less dependent on climate factors [10–12]. There is a huge untouched potential in the micro-hydropower category with low and medium heads available in different water conveyance systems [13]. In that sense, energy recovery in potable water networks (urban, rural), irrigation water networks and wastewater and process industry systems received a lot of attention. For this purpose, the pump as turbine (PAT) systems have been recently under consideration and increasing studies have come out addressing different aspects of that [14,15]. PATs are hydraulic pumps that operate in reverse mode for energy production purposes by means of a connected induction motor working as a generator [9]. The operational cost of a PAT is five to 10 times lower than that of a conventional turbine [16]. However, the contribution to renewable energy sources is very slight, which may come from the lack of knowledge, some institutional barrier for implementation and interest. But lately, several studies have been published, mostly examining the PAT application in the water sector through the REDAWN project. In a study [17], a hybrid solution of leakage reduction and energy recovery using PATs was analysed in rural water distribution networks in Ireland, showing the significant contribution of generated power in local energy demand. Another study [18] focused on the off-grid PAT combined with a self-excited induction generator, which led to a boost in system efficiency of 20%.

Besides rural and urban water networks, irrigation water networks also have provided great potential in energy recovery, where a substantial amount of energy used for water distribution can be recovered [19]. The energy harvesting in irrigation water networks is relatively a new area, with very challenging concerns because of the flow variation. Still, studies show that high potential exists in this area [20–22]. Although studies have shown that conventional turbines (Francis, Pelton, Kaplan, Turgo, and crossflow) can still compete with new emerging technologies, the new technologies might provide more economic solutions due to a modular design [23]. However, new unconventional hydropower solutions are putting a new spin on hydropower generation for future energy generation [24].

Recently, the CAES systems were considered as a proper solution for geographical conditions in which PSH and PATs cannot be implemented. The first test of a CAES system was in 1949 by Stal Laval, when he registered the first patent of CAES to store energy in the form of compressed air in underground caverns. The basic concept of a CAES is based on stored compressed air which is released to generate energy when needed [25]. Currently, two large capacity plants exist in Germany and the USA, but a lot of studies are appearing in the literature focusing on different solutions, such as the thermodynamic behaviour and hybrid solutions, along with some novel conceptual ideas. A lot of studies have focused on the thermodynamic inspection of CAES systems which have led to the development of different types of CAES system, such as advanced adiabatic CAES (AA-CAES), isothermal CAES (I-CAES), adiabatic CAES (A-CAES) or others [26,27].

In two recent studies, a hybrid solution was proposed to assess the combination of PSH and CAES (pumped hydro CAES or PH-CAES). These studies [28,29] presented an experimental investigation of a PH-CAES system operating using a Pelton turbine. The reported efficiency for the PH-CAES system was 45% and the authors claimed that, unlike conventional CAES systems, this system works with constant power [28,29]. In another study, a new concept for the CAES system was a geographically independent system which was proposed and tested in the laboratory. The feasibility of the concept was proved while reporting the experimental electrical roundtrip efficiency and hydraulic (thermodynamic) efficiency of 24% and 97%, respectively. The authors mentioned the energy loss in the machinery as the main reason for a low electrical roundtrip efficiency [30]. Also, some new concepts for using compressed air have been tackled by some researchers, either in underground caverns or underwater inflatable balloons [31]. Also, another study was published to evaluate the dynamic behaviour of air-water for pressure storage and recovery [32].

Correspondingly, this study aims to present a new conceptual idea of energy storage/recovery using Transient Flow-Induced Compressed Air Energy Storage (TI-CAES). The transient flow is induced in the system by creating a waterhammer event, which has been widely studied and addressed in the literature from different aspects [33,34], giving enough knowledge to employ it for this purpose. For that, experimental data were collected in a system composed of a water pipeline, a compressed air vessel (CAV) and some valves. The results of experimental investigations are used to assess two ideas: (a) generating energy by means of a water turbine; (b) pumping water using the stored pressure inside the entrapped air.

2. System Description

A schematic drawing of the experimental system is presented in Figure 2, showing the auxiliary components for the purpose of this study. The system is composed of PVC pipes with a nominal diameter of 63 mm and the nominal pressure of 16 bar connecting two hydro-pneumatic tanks. By means of recirculating the flow, a steady-state flow condition is established. The data acquisition system records the pressure values and also the velocity profiles. The pressure sampling is done using S-10 WIKA pressure transducers, with the maximum sampling error of ±0.5% as reported by the manufacturer [35]. The velocity profiles are measured by an ultrasonic doppler velocimetry (UDV) system from MET-FLOW offering an accuracy of ±1 ms [36]. The sampling rates for pressure and velocity measurements were 910 and 10 Hz, respectively. Three pressure transducers were installed in the system, as shown in Figure 2. In the measurements, the charging/discharging flow rate and pressure were recorded for different steady-state velocities in the main pipeline (represented by Reynolds (Re) number in the current paper) and the volumes of air inside the CAV. A volume fraction ratio (VFR) has been defined as the volume of air over the volume of CAV in percentage to express the volume of air inside the CAV. The CAV can operate through two different approaches, i.e., the electricity dispatch system by means of a hydro-turbine (Figure 2a) and the pumping system (Figure 2b). The transient flow is induced into the system by the action of BV-01, a ball valve manoeuvred by an electro-pneumatic actuator from +GF+ company [37].

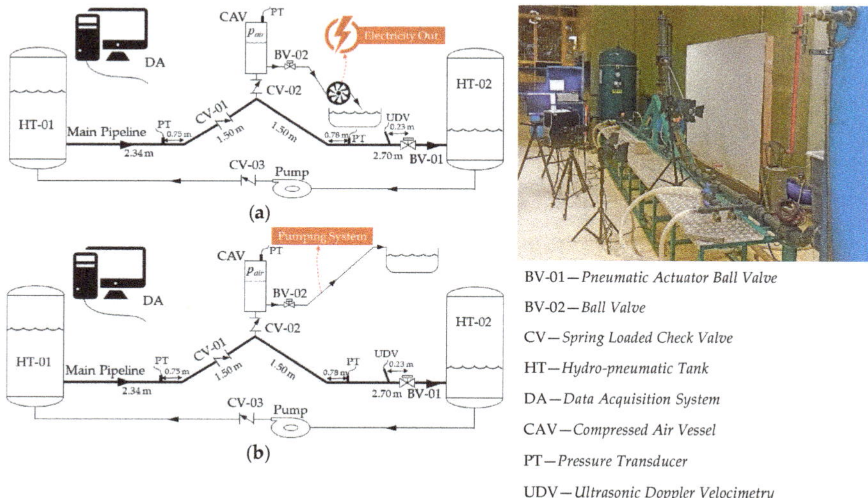

Figure 2. Experimental apparatus as part of the transient flow-induced compressed air energy storage (TI-CAES) system and energy conceptual ideas: (**a**) electricity dispatch; (**b**) pumping system.

The system can be used to generate energy using a hydro turbine, when the CAV works in two stages, i.e., water charge and discharge (Figure 2a). In the water charge stage, the BV-01 is actuated in a closed/open action to induce a waterhammer into the system. A pressure surge will be propagated in the system by the upsurge occurrence travelling in the upstream direction. As a result of the pressure surge and the special arrangement of CV-01 and CV-02 in the system, the backflow will be directed towards the CAV. The pressure of air in the CAV (p_{air}), which initially was equal to the atmospheric pressure, will increase due to the compression of the air. In the water discharge stage, since the BV-02 is in open position, the compressed air starts to expand and consequently expels water through BV-02. The charge and discharge stages can be performed frequently. In this study, 9 charge/discharge stages were examined, meaning 9 waterhammer event occurrences.

On the other hand, the system might work for pumping (Figure 2b). For that, the CAV operates as well in the charge and discharge stages. In the charge stage, the BV-01 closes and opens to create upsurges. The pressure surge will be stored in the CAV (because the BV-02 is closed). The waterhammer action is repeated until the desired or maximum pressure is reached inside the CAV. Then, the BV-02 will be opened, allowing compressed air to expand and drive water out of the CAV. Depending on the pressure of the air, it can elevate water to a significant differential level or compensate head losses in a pumping system.

The current study focuses on the energy generation idea (Figure 2a), in which the transient effect includes 9 consecutive waterhammer events taking a total time of 18 s. This effect is created by the closing/opening action of BV-01, taking 0.4 s (0.2 s for each opening or closing action). The major measuring parameters in this system are the main pipeline pressure, the air pressure, the air volume (displayed by VFR), the flow velocity and the CAV outlet flow rate. The measured range for major parameters of TI-CAES system in the current experimental tests is presented in Table 1. The initial pressure of the whole system has been set to the atmospheric pressure. The pressure in the main pipeline starts fluctuating upon starting the transient event. When the transient event finishes, the pressure in the main pipeline returns to the initial condition. The flow condition before and after the transient event should be the same.

Table 1. System parameters summary.

Parameter	Measuring Range
Pipe pressure [bar]	0.00 to 6.62
Air pressure [bar]	0.00 to 5.08
VFR [%]	3.33 to 66.67
Flow Velocity [m/s]	1.36 to 5.13
Flow Rate [L/s]	1.76 to 7.43
Re number [–]	36,000 to 155,000

3. Parametric Analysis

The pressure variation in the entrapped air within the CAV is recorded, and demonstrates the charge and discharge stages through pressure spikes (Figure 3). The pressure spikes caused by the compression of the air under the waterhammer action shows different values and a fluctuation behaviour based on the air size (represented by VFR) and Re of the pipeline flow. Two pressure variation results are presented in Figure 3 for a constant Re = 155,000, showing quite different fluctuation and minimum pressure values when compared to the initial pressure (p_0).

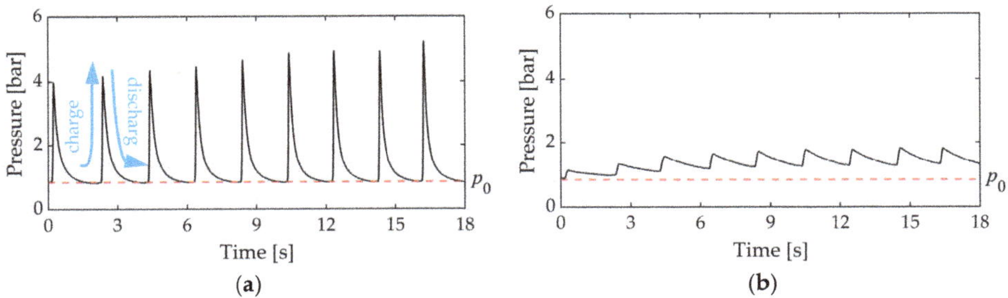

Figure 3. Experimental pressure variation in charge/discharge stages of the CAV, for Re = 155,000: (**a**) volume fraction ratio (VFR) = 8.33%; (**b**) VFR = 66.67%.

It was analysed how the mentioned different pressure consequences can affect the functionality of the TI-CAES system. For that, a thermodynamic approach was used to analyse the pressure data. Thermodynamic behaviour in closed air vessels is traditionally modelled using the polytropic equation:

$$p_{air} \cdot \forall_{air}^n = \text{constant}, \tag{1}$$

where p_{air} and $\forall_{air}$ represent the air-phase pressure and volume, respectively [38]. n is a real number known as the polytropic exponent which includes the heat transfer and depends on the type of thermodynamic process (usually for isothermal process $n = 1$ and for adiabatic process $n = 1.4$) [39]. In order to define the n value, one should first determine the type of the thermodynamic process. Previous studies proved that, for rapid transitions, the process is close to adiabatic with zero heat transfer [39–46]. However, in this kind of transient, an average value of $n = 1.2$ is suggested [35,43], because of the different behaviour of the transient phenomenon from the beginning of the process to the end. In the study [39], it was demonstrated that, for medium and big air volumes, the polytropic exponent $n = 1.2$ gives very good agreement with the measured pressures. According to this, the new volume of air, changing due to transient (V_{tr}) and corresponding to the new pressure surge (p_{tr}), is calculated using Equation (2).

$$V_{tr} = V_0 \cdot \left(\frac{p_0}{p_{tr}}\right)^{1/n}, \tag{2}$$

where p_0 and V_0 stand for the initial pressure and volume, respectively. Hence, V_{tr} corresponds to a new volume due to changing the pressure from p_0 to p_{tr}. Based on that, the pressure–volume diagram for different volumes of air inside the CAV is plotted in Figure 4. It demonstrates that bigger changes in the air volume correspond to higher VFRs.

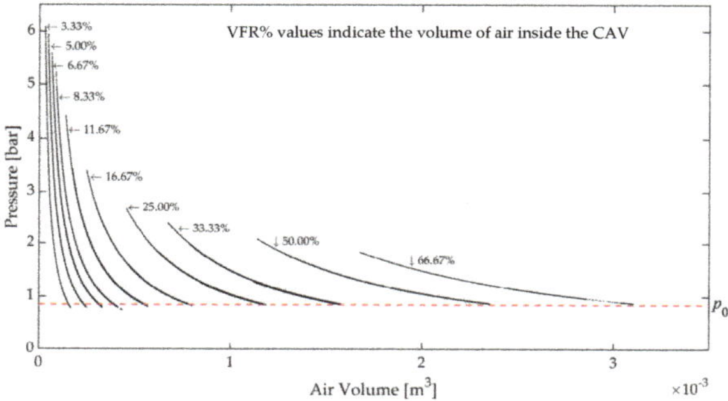

Figure 4. The pressure–volume diagram.

The area below each pressure–volume curve shows the work done by the expansion of air for a specific VFR. In Figure 5, the works done for bigger VFRs are greater than those with smaller VFRs, promising a higher hydraulic power generation.

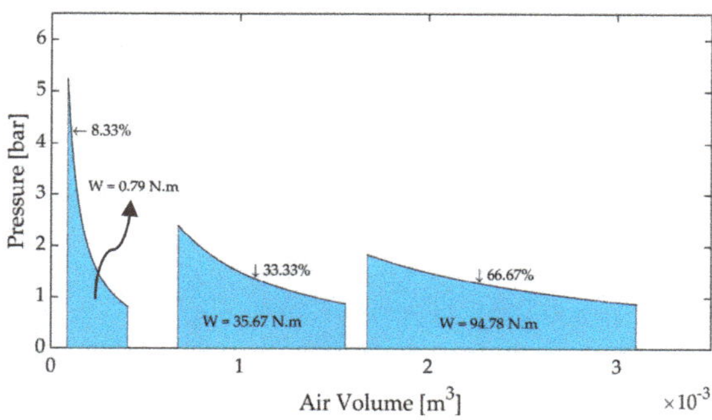

Figure 5. The work done by the air expansion is the area below the pressure–volume graph in different VFRs.

The hydraulic power can be determined by integrating the work over the charge and discharge time. It can be calculated using Equation (3).

$$P_{hyd} = \int W \, dt = Q \cdot p_{air}, \tag{3}$$

where W is the work and Q is the flow rate. Two approaches can be tackled in the calculation of p_{air} since it is fluctuating during the time as shown in Figure 3. Either p_{air} can be considered as the integration of the pressure variation graph over time or a function of time. For the first approach, the integral of p_{air} along the time was calculated for different tests using Equation (4), in which p_{int} stands for integrated pressure values.

$$p_{int} = \frac{1}{t_{tr}} \int_0^{t_{tr}} p_{air} \, dt, \tag{4}$$

The resulting integrated pressure values were divided by t_{tr} (the total transient events' duration) to achieve an average pressure. The attained average pressures are presented in Figure 6a for different VFRs and Re numbers. Despite very high peaks for smaller VFRs, as shown in Figure 6b, the average pressure is lower, and the highest average pressure belongs to 10% < VFR < 60%. For each VFR, pressure results of seven different Re numbers are shown in Figure 6a, starting from low Re to high Re, including 36×10^3, 56×10^3, 75×10^3, 93×10^3, 115×10^3, 132×10^3, and 155×10^3.

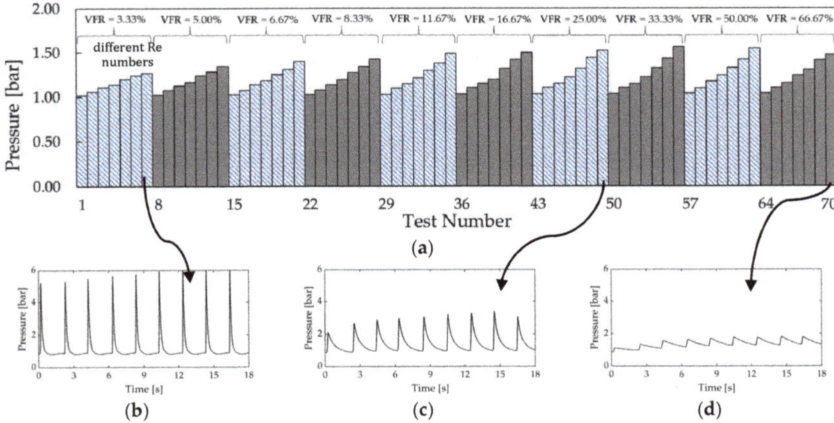

Figure 6. The concept of average and real-time air pressure: (**a**) average pressure for different VFR and Re numbers; (**b**) p_{air} for VFR = 3.33% and Re = 155,000; (**c**) p_{air} for VFR = 25.00% and Re = 155,000; (**d**) p_{air} for VFR = 66.67% and Re = 155,000.

In the second approach, the hydraulic power can be calculated using Equation (3) considering p_{air} as a function of time (Figure 7). The resulting power will be a function of time, as presented in Figure 7. As is evident in this figure, by increasing the VFR (i.e., increasing the air volume) the maximum spikes of achieved power are decreased. But, on the other hand, the minimum attained power with time is higher. The maximum attained hydraulic power is 90.08 W, which is a power related to VFR = 8.33% and Re = 155,000.

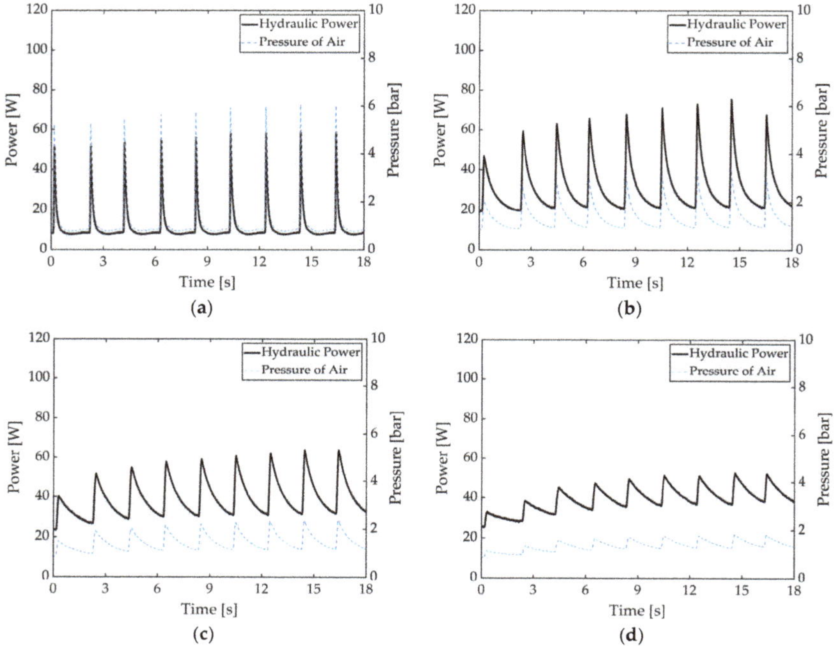

Figure 7. The hydraulic power and p_{air} changing during transient for Re = 155,000 and: (**a**) VFR = 3.33%; (**b**) VFR = 16.67%; (**c**) VFR = 33.33%; (**d**) VFR = 66.67%.

The hydraulic power of CAV was estimated using Equation (3) when the p_{air} is calculated from both abovementioned approaches. Both calculation approaches were found, as expected, to give the same values of power as in Figure 8. Considering the small volume of CAV in these experiments (4.7 L), and also the very short time associated to the transient events (only 18 s), the achieved power can be a significant and promising value for higher Re numbers.

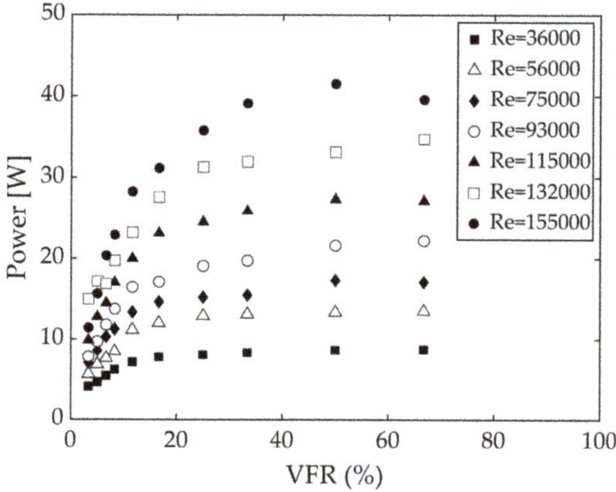

Figure 8. The power production for experimental CAV volume based on integrated pressure values.

4. Recovery Flow Behaviour

To approach the energy recovery concept in the TI-CAES system, it is important to verify the consistency of the main pipeline flow with less perturbation and possible dissipative effects when inducing waterhammer events. For this purpose, the velocity profile in the transient was recorded by means of ultrasonic doppler velocimetry (UDV) [47] to see how the waterhammer action affects the flow characteristics. As studied by Brunone and Berni [48], for a constant flows rate, the shape of velocity profile in a fast transient condition is very different from the steady-state one. Figure 9 shows the plot of the variation of velocity profiles based on a 0.2 s interval. The time intervals that were selected were small enough to show the variations of the velocity profiles in detail during the transient. In the majority of the cases, the flow returns to the initial condition in around 1.2 s.

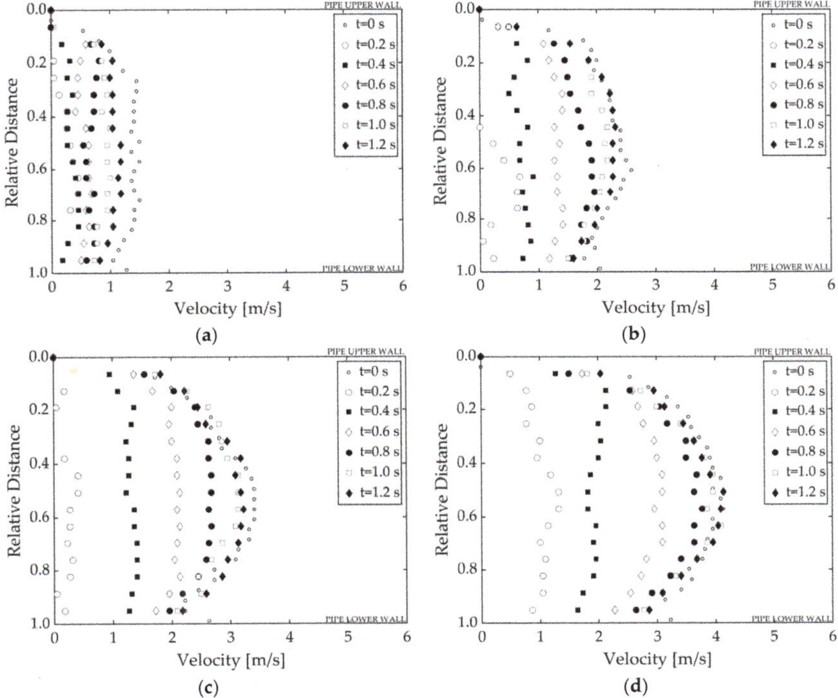

Figure 9. Velocity profiles along time: (**a**) Re = 58,000; (**b**) Re = 93,000; (**c**) Re = 132,000; (**d**) Re = 155,000.

By calculating the velocity differences between a transient mean velocity (V_1) and the initial mean velocity (V_0) and dividing it by maximum initial velocity as $\left|(V_1 - V_0)/(V_{\max})_0\right|$, an index for the velocity change can be achieved for further evaluation. It can help us to understand how much the velocity is going to change with time and provide information about the dissipative effects associated to flow velocity. This parameter is plotted in Figure 10 for all the tests and time steps of 0.8, 1.0, 1.2 and 1.4 s from the beginning of the transient event. The selected time steps are the closest time steps to the initial conditions, making the comparison easier. As can be seen in Figure 10a–d, the velocity index is very small after 1 s from the beginning of the transient event, proving the recovery of initial flow conditions. Also, it can be deduced that the initial condition recovery occurs better for higher Re numbers.

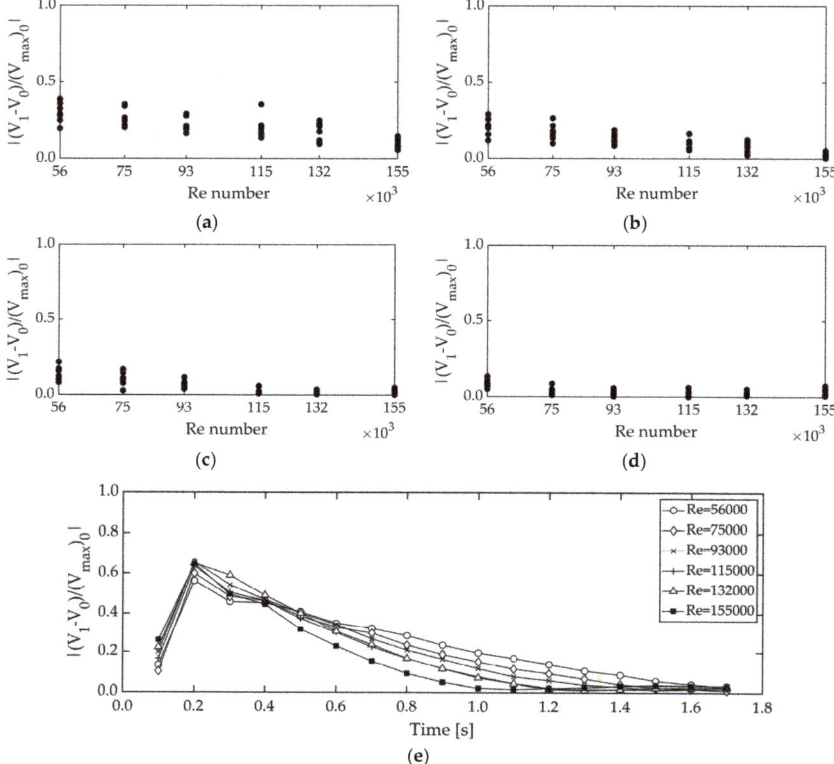

Figure 10. Difference between the transient mean velocity (V_1) and the initial mean velocity (V_0) over the maximum initial velocity $(V_{max})_0$ for all the tested conditions and time steps after starting the transient event as: (**a**) 0.8; (**b**) 1.0; (**c**) 1.2; (**d**) 1.4; (**e**) for more time.

As a general conclusion, the initial flow velocity recovery with time is plotted in Figure 10e, showing the highest difference in the time of valve closure (0.2 s) and recovering the flow very fast upon re-opening the valve. As a result, the flow change during the current transient event can be recuperated with almost neglected dissipative effects for the main flow conditions.

5. Dimensional Analysis

In this section, the Buckingham pi theorem is exploited to study the transient hydraulic behaviour caused in the CAV. The flow is based on the pressure with the same liquid (water) being adopted in the model and prototype. So, on the basis of dimensional analysis, a scale factor is defined by:

$$\lambda_L = \frac{L_m \, (\text{variable in the model})}{L_p \, (\text{variable in the prototype})}, \tag{5}$$

where λ is defined as a scale parameter of model dimension over prototype dimension. The subscripts m and p stand for model and prototype, respectively.

The CAV (as an air vessel) can be reproduced similarly to a reservoir, where the gravity force is the main predominant force. The parameters, demonstrated in Table 2, are important in the operation of the CAV in the TI-CAES model functionality (represented by subscript: m):

Table 2. Major parameters for operation of CAV in TI-CAES model.

Parameters	Definition
$(t_{tr})_m$	time of transient (waterhammer) event
$(∀_{CAV})_m$	volume of CAV
$(∀_{air})_m$	volume of air inside CAV
$(D)_m$	exit pipe diameter
$(p_{air})_m$	air pressure inside CAV
$(Q_{out})_m$	outlet flow from CAV
$(P_{hyd})_m$	hydraulic power of CAV
$(E)_m$	available energy in CAV

Applying dimensional analysis is possible to forecast the characteristic parameters for different length-scales, supposing the similarity law applied only to the CAV operation, since dimensions and characteristics of the model were obtained based on the Froude and Allievi similarity, and the effect of viscosity forces on the combined friction effect of the flow inside the tank and inertia forces were neglected. Hence, from the basic equations of dimensionless variables, the compatibility between the Froude and Allievi criteria is obtained by imposing that the elastic wave velocity scale (λ_c), and the liquid velocity scale (λ_V); both are equal to $\lambda_L^{1/2}$. This criterion corresponds to the assumption that the Mach number (V/c) is constant [49]. Therefore, for example, with $\lambda_L = 1/10$, it was necessary to ensure, in the experimental installation, a slower speed than the prototype. Thus, for the wave speed of 1000 m/s in a prototype, the equivalent in the experimental installation should be 316 m/s. This reduction has been achieved by selecting PVC material for the pipe. Failure to comply with Reynolds' criterion leads to the unit pressure drop not being similar. However, given that the objective of the study is the analysis of the total integrated pressure (p_{int}) that occurs in the transient time, it is assumed that the effect of friction losses are not relevant for this type of study, as proved in former research [49].

In this context, by putting different variable scales as a function of the size, one can get the following scales for different parameters based on Froude similarity:

Velocity scale:
$$\frac{V_m}{(gL_m)^{1/2}} = \frac{V_p}{(gL_p)^{1/2}} \Rightarrow \frac{V_m}{V_p} = \frac{(gL_m)^{1/2}}{(gL_p)^{1/2}} \Rightarrow \lambda_V = \lambda_L^{1/2}, \tag{6}$$

Time scale:
$$L = V.t \Rightarrow \lambda_L = \lambda_V.\lambda_t \Rightarrow \lambda_t = \frac{\lambda_L}{\lambda_V} \Rightarrow \lambda_t = \lambda_L^{1/2}, \tag{7}$$

Acceleration scale:
$$A = \frac{dV}{dt} \Rightarrow \lambda_a = \frac{\lambda_V}{\lambda_t} \Rightarrow \lambda_a = 1, \tag{8}$$

Flow rate scale:
$$Q = V.A \Rightarrow \lambda_Q = \lambda_V.\lambda_L^2 \Rightarrow \lambda_Q = \lambda_L^{5/2}, \tag{9}$$

Mass scale:
$$m = \rho.∀ \Rightarrow \lambda_m = \lambda_\rho.\lambda_L^3 \Rightarrow \lambda_m = \lambda_L^3, \tag{10}$$

Force scale:
$$F = m.a \Rightarrow \lambda_F = \lambda_m.\lambda_a \Rightarrow \lambda_F = \lambda_L^3, \tag{11}$$

Pressure scale:
$$p = \frac{F}{S} \Rightarrow \lambda_p = \frac{\lambda_F}{\lambda_L^2} \Rightarrow \lambda_p = \lambda_L, \tag{12}$$

Energy scale:
$$E = F.L \Rightarrow \lambda_E = \lambda_F.\lambda_L \Rightarrow \lambda_E = \lambda_L^4, \tag{13}$$

Power scale:
$$P = \frac{E}{t} \Rightarrow \lambda_P = \frac{\lambda_E}{\lambda_t} \Rightarrow \lambda_P = \lambda_L^{7/2}, \tag{14}$$

As a result, the transient or waterhammer event time, flow rate, power and energy in the prototype can be calculated using the following equations:

$$(t_{tr})_p = \frac{(t_{tr})_m}{\lambda_t} = \frac{(t_{tr})_m}{\lambda_L^{1/2}}, \tag{15}$$

$$(p_{air})_p = \frac{(p_{air})_m}{\lambda_p} = \frac{(p_{air})_m}{\lambda_L}, \tag{16}$$

$$(Q_{out})_p = \frac{(Q_{out})_m}{\lambda_Q} = \frac{(Q_{out})_m}{\lambda_L^{5/2}}, \tag{17}$$

$$(P_{hyd})_p = \frac{(P_{hyd})_m}{\lambda_P} = \frac{(P_{hyd})_m}{\lambda_L^{7/2}}, \tag{18}$$

$$(E)_p = \frac{(E)_m}{\lambda_E} = \frac{(E)_m}{\lambda_L^4}, \tag{19}$$

The measured values of pressure and flow rate related to Re = 155,000 (corresponding to highest achieved power) along with physical specifications of the experimental system were used to calculate the hydraulic power (the same data as Figure 8) and energy of the system, as shown in Table 3. The data for different air volumes (VFRs) are presented in Table 3 to be comparable with later scaled-up data of prototypes.

Table 3. Measured and calculated parameters in the laboratory model apparatus for different VFRs.

Parameters	$(t_{tr})_m$ [s]	$(\forall_{CAV})_m$ [m³]	VFR [%]	$(\forall_{air})_m$ [m³]	$(D)_m$ [m]	$(p_{int})_m$ [kPa]	$(Q_{out})_m$ [m³/s]	$(P_{hyd})_m$ [kW]	$(E)_m$ [kWh]
Dimensions	T	L³	-	L³	L	M L⁻¹ T⁻²	L³ T⁻¹	M L² T⁻³	M L² T⁻²
			3.33	0.00016		117.27	0.00010	0.01140	0.00006
			5.00	0.00023		125.04	0.00013	0.01563	0.00008
			6.67	0.00031		130.75	0.00016	0.02034	0.00010
			8.33	0.00039		133.09	0.00017	0.02292	0.00011
Model	18	0.0047	11.67	0.00054	0.02	139.49	0.00020	0.02828	0.00014
Values			16.67	0.00078		140.21	0.00022	0.03116	0.00016
			25.00	0.00116		142.98	0.00025	0.03574	0.00018
			33.33	0.00155		146.70	0.00027	0.03912	0.00020
			50.00	0.00233		145.15	0.00029	0.04153	0.00021
			66.67	0.00310		137.86	0.00029	0.03960	0.00020

The experimental values shown in Table 3 are used to calculate the scaled-up values for three length-scales of 1/10, 1/20 and 1/25 presented in Tables 4–6 for the same Re = 155,000, demonstrating an estimation of the installed power and available energy for the respective prototypes.

Table 4. Scaled-up parameters for a prototype with length-scale 1/10 and different VFRs.

Parameters	λ_L	$(t_{tr})_p$ [s]	$(\forall_{CAV})_p$ [m³]	VFR [%]	$(\forall_{air})_m$ [m³]	$(D)_m$ [m]	$(p_{air})_m$ [kPa]	$(Q_{out})_m$ [m³/s]	$(P_{hyd})_m$ [kW]	$(E)_m$ [kWh]
				3.33	0.16		1172.65	0.03	36.05	0.57
				5.00	0.23		1250.37	0.04	49.43	0.78
				6.67	0.31		1307.46	0.05	64.32	1.02
				8.33	0.39		1330.94	0.05	72.49	1.15
Prototype	1/10	56.92	4.70	11.67	0.54	0.20	1394.85	0.06	89.44	1.41
Values				16.67	0.78		1402.08	0.07	98.53	1.56
				25.00	1.16		1429.77	0.08	113.03	1.79
				33.33	1.55		1467.00	0.08	123.71	1.96
				50.00	2.33		1451.50	0.09	131.33	2.08
				66.67	3.10		1378.60	0.09	125.22	1.98

Table 5. Scaled-up parameters for a prototype with length-scale 1/20 and different VFRs.

Parameters	λ_L	$(t_{tr})_p$ [s]	$(V_{CAV})_p$ [m³]	VFR [%]	$(V_{air})_m$ [m³]	$(D)_m$ [m]	$(p_{air})_m$ [kPa]	$(Q_{out})_m$ [m³/s]	$(P_{hyd})_m$ [kW]	$(E)_m$ [kWh]
				3.33	1.24		2345.31	0.17	407.89	9.12
				5.00	1.86		2500.74	0.22	559.18	12.50
				6.67	2.48		2614.92	0.28	727.64	16.27
				8.33	3.10		2661.88	0.31	820.07	18.34
Prototype	1/20	80.50	37.60	11.67	4.35	0.40	2789.71	0.36	1011.94	22.63
Values				16.67	6.21		2804.16	0.40	1114.72	24.93
				25.00	9.31		2859.54	0.45	1278.83	28.60
				33.33	12.42		2934.00	0.48	1399.60	31.30
				50.00	18.62		2903.01	0.51	1485.79	33.22
				66.67	24.83		2757.21	0.51	1416.65	31.68

Table 6. Scaled-up parameters for a prototype with length-scale 1/25 and different VFRs.

Parameters	λ_L	$(t_{tr})_p$ [s]	$(V_{CAV})_p$ [m³]	VFR [%]	$(V_{air})_m$ [m³]	$(D)_m$ [m]	$(p_{air})_m$ [kPa]	$(Q_{out})_m$ [m³/s]	$(P_{hyd})_m$ [kW]	$(E)_m$ [kWh]
				3.33	2.43		2931.63	0.30	890.69	22.27
				5.00	3.64		3125.92	0.39	1221.06	30.53
				6.67	4.85		3268.64	0.49	1588.92	39.72
				8.33	6.06		3327.35	0.54	1790.76	44.77
Prototype	1/25	90.00	73.44	11.67	8.49	0.50	3487.14	0.63	2209.73	55.24
Values				16.67	12.13		3505.20	0.69	2434.16	60.85
				25.00	18.19		3574.43	0.78	2792.52	69.81
				33.33	24.25		3667.50	0.83	3056.25	76.41
				50.00	36.38		3628.76	0.89	3244.46	81.11
				66.67	48.50		3446.51	0.90	3093.48	77.34

6. Discussion

The results presented in the current study are based on measurements in an experimental model and a set of properly developed analyses. This study inspected the TI-CAES idea from different aspects and reported different achievements, most importantly the amount of attainable hydraulic power. The presented data for three different prototypes demonstrate encouraging values of hydraulic power having the maximum values of 131.33, 1485.79 and 3244.46 in kW for the length-scales of 1/10, 1/20 and 1/25, respectively, while the maximum power of the model was equal to 41 W. The power grew considerably by increasing the length-scale, as can be seen in the comparison of the length-scales of 1/20 and 1/25 in Figure 11, when the maximum power growth is more than double, changing from 1485.79 to 3244.46.

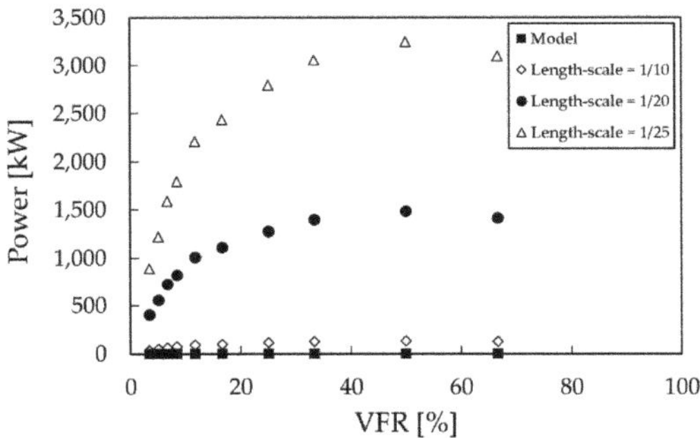

Figure 11. Hydraulic prototype power for different length-scales and VFRs.

Isolating the CAV from the main pipeline, to observe the behaviour of air in charge/discharge phases, it can be concluded the CAV behaviour does not depend on the Re number. This has been justified earlier in the discussions of Section 5, since the friction factor effect in the CAV can be omitted. However, the CAV receives influence from the main pipeline in each waterhammer event. So, as presented in Figure 8, the Re number in the main pipeline will affect the output hydraulic power. The above results showed the outcomes only for a Re = 155,000, this being the highest tested one. In order to have an idea of how the hydraulic power varies for different Re numbers (i.e., different flow velocities in the main pipeline), a dimensionless parameter is defined as Equation (20) by removing the effect of CAV volume and maximum pressure:

$$(P_{\text{dim}})_m = \frac{(P_{\text{hyd}})_m}{\forall_{\text{CAV}} \times \frac{1}{(t_{wh})_m}(p_{\text{max}})_m},\tag{20}$$

where $(p_{\text{max}})_m$ is the maximum recorded p_{air} for each individual test in the experimental system. The dimensionless graph in Figure 12 shows that for VFRs < 20%, the power in different prototype scales does not depend on the Re numbers of the main pipeline flow. Nevertheless, bigger VFRs are more impressionable and therefore for VFRs > 20%, the power will be more dependent on the Re number of the main pipeline flow.

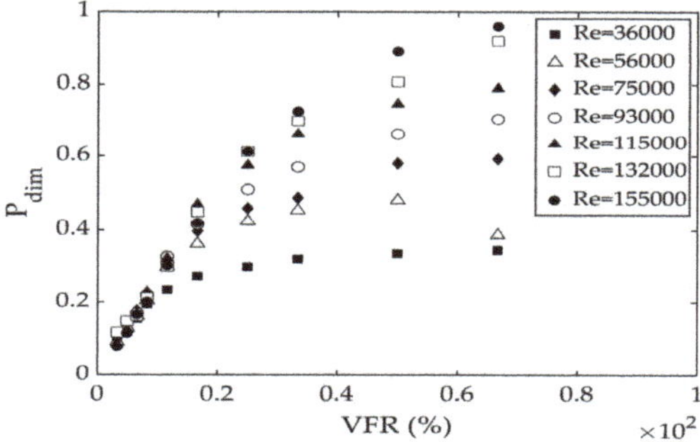

Figure 12. Dimensionless power for different VFR and Re numbers in TI-CAES system.

The results show that the CAV is working in charge/discharge cycles by the compression/expansion of the air after each waterhammer occurrence. The charge/discharge action of the CAV includes two pressure values of p_0 (the initial pressure) and p_{tr} (the transient pressure in each cycle). Small VFRs produce very high peaks, as presented in Figure 3, creating the expectation to have higher powers for small VFRs. In fact, due to the direct relation between hydraulic power and pressure, the variation of power during the transient time will also include high power peaks for small VFRs, as presented in Figure 7. Regardless, the total hydraulic power during the test time calculated based on an integrated pressure value (see Figure 6) will be higher for bigger VFRs. In fact, for big VFRs, the air–water interface encounters a big displacement upon receiving the first pressure surge. Since the time interval between waterhammer repetitions is small, the available time is not enough for the air–water interface to return to the initial position and as a result, the initial pressure (p_0) is not recovered (see Figure 3b). This will be repeated in the next entrance of the pressure surges into the CAV as well. So, after a while, the pressure of the whole CAV increases, leading to less displacement but a higher pressurized CAV system that generates greater outlet flow rate and consequently hydraulic power. This issue has been proved also via the expansion work analysis in Figure 5.

In another part of the study, the consistency of the main pipeline flow was studied as another significant parameter in the feasibility of the TI-CAES, verifying a very short-lasting change in the velocity profile of the main pipeline. In addition, the maximum flow rate taken from the main pipeline was equal to 0.29 L/s (as shown in Table 3), indicating a total volume of about 5 L for the tested transient time (i.e., 18 s). These values are less than 4% of the main pipeline flow and the corresponding volume for Re = 155,000, which confirms the flow consistency and limited perturbation again.

7. Conclusions

The current research offers a new idea of a transientflow-induced compressed air energy storage system called TI-CAES. The main motivation behind this idea was decreasing the energy dependency in pressurized infrastructures and water systems by taking advantage of transient flow and compressibility nature of the air. Currently, hydro-power renewable energy sources hold the biggest share in renewable sources. Existing water systems supply the relevant capacity to increase the current share of hydro-power sources, diminishing the energy dependency and environmental impacts associated.

In this context, the feasibility of the TI-CAES system has been examined using an experimental system. The valve closure is always longer than critical closure time to avoid high values of pressure surges. This research presented an extensive analysis of different parameters to introduce the behaviour of the system and also verify the feasibility. Consequently, the major achievements of this study can be stated as:

1. The system showed a very controlled behaviour under the pressure surge occurrence, where the pressure surge can be stored in the form of compressed air in a vessel. The maximum pressure in the main pipeline is slightly higher than the maximum pressure of the air that provides a reliable calculation and prediction of the system.
2. The attained hydraulic power consists of considerable values, while the amount of input work to induce pressure surge into the system is negligible. This fact promises a high-efficiency system in which most of the parts usually exist in advance in a water system. Hence, the initial investment will not be large to set-up a TI-CAES system in an existing water system.
3. The analysis proved that the perturbation created in the main pipeline flow will be short-lasting and insignificant as changes in velocity profiles proved. The results show that the initial velocity of the main flow can be recovered in a very short time. Also, it was found that the velocity recovery and perturbation elimination is better for higher Reynolds numbers, making the TI-CAES system a better solution for water system with high Reynolds numbers.
4. It was shown that higher VFRs will provide considerably higher power values. In addition, higher VFRs will not provide extreme pressure peaks, which is an important advantage increasing the safety measures of the system. So, bigger air volumes provide a safe and economic energy recovery attitude due to very controlled pressure spikes while offering significant work done by air expansion.
5. The results show the TI-CAES concept offers a flexible, geography independent and efficient system that can provide high hydraulic power due to suppling a very high extra pressure head in the form of pressurized air.

No technical challenge regarding the implementation of TI-CAES in a real system has been encountered. However, further implementation in technical scale is required to verify the achieved values.

Author Contributions: The authors M.B. and H.M.R. contributed to the development of the idea and writing the paper; M.B. did the experimental tests with A.D.'s support; Results are analysed with the contribution of M.B., H.M.R., M.T.V. and B.B. All authors have read and agreed to the published version of the manuscript.

Funding: This work was supported by the Fundação para a Ciência e Tecnologia (FCT), Portugal under grant number PD/BD/114459/2016.

Acknowledgments: Authors would like to thank CERIS (Civil Engineering, Research, and Innovation for Sustainability) Centre from Instituto Superior Técnico, Universidade de Lisboa, Portugal, for providing the experimental facilities and also the European project REDAWN (Reducing Energy Dependency in Atlantic Area Water Networks) EAPA_198/2016 from INTERREG ATLANTIC AREA PROGRAMME 2014–2020.

Nomenclature

Symbols

a	acceleration [m/s^2]
c	wave speed [m/s]
D	pipe diameter [m]
E	energy [N.m]
F	force [N]
g	gravitational acceleration [m/s^2]
L	length [m]
m	mass [kg]
n	polytropic exponent [–]
p	pressure [bar] or [kPa]
P	hydraulic power [W] or [kW]
$\forall$	volume [m^3]
W	work [N.m]
t	time [s]
Q	flow rate [L/s] or [m^3/s]
V	flow velocity [m/s]

Indices

air	air inside CAV
dim	dimensionless value
0	initial (at time zero)
hyd	hydraulic
int	integrated value
L	length
m	model
max	maximum value
out	outlet
p	prototype
tr	related to transient condition

Greek letters

γ	specific weight [N/m^3]
λ	scale factor
ρ	fluid density [kg/m^3]

Abbreviations

AA-CAES	Advanced Adiabatic Compressed Air Energy Storage
BV	Ball Valve
CAES	Compressed Air Energy Storage
CAV	Compressed Air Vessel
CV	Check Valve
DA	Data Acquisition system
HT	Hydro-pneumatic Tank
PAT	Pump as Turbine
PH-CAES	Pumped Hydro Compressed Air Energy Storage
PSH	Pumped-Storage Hydropower
PT	Pressure Transducer
Re	Reynolds number
TI-CAES	Transient-flow Induced Compressed Air Energy Storage
UDV	Ultrasonic Doppler Velocimetry
UPSH	Underwater Pumped-storage Hydropower
VFR	Volume Fraction Ratio

References

1. IHA (International Hydropower Association). Hydropower Status Report: Sector Trends and Insights. 2019. Available online: https://www.hydropower.org/status2019 (accessed on 13 May 2019).
2. REN21. Renewables Global Status Report. 2019. Available online: https://www.ren21.net/wp-content/uploads/2019/05/gsr_2019_full_report_en.pdf (accessed on 13 January 2020).
3. IEA (International Energy Agency). Renewables: market analysis and forecast from 2019 to 2024, Paris. 2019. Available online: https://www.iea.org/reports/renewables-2019 (accessed on 13 January 2020).
4. Hoes, O.A.C.; Meijer, L.J.J.; Van der Ent, R.J.; Van de Giesen, N.C. Systematic high-resolution assessment of global hydropower potential. *PLoS ONE* **2017**, *12*, e0171844. [CrossRef] [PubMed]
5. Vieira, F.; Ramos, H.M. Hybrid solution and pump-storage optimization in water supply system efficiency: A case study. *Energy Policy* **2008**, *36*, 4142–4148. [CrossRef]
6. Vieira, F.; Ramos, H.M.; Covas, D.I.C.; Almeida, A.B. Pump-Storage optimization with renewable energy production in water supply systems. In Proceedings of the 4th International Conference on Water and Wastewater Pumping Stations, BHR Group, Cranfield, UK, 17–18 June 2008.
7. Ramos, H.M.; Amaral, M.P.; Covas, D.I.C. Pumped-storage solution towards energy efficiency and sustainability: Portugal contribution and real case studies. *J. Water Resour. Prot.* **2014**, *6*, 1099. [CrossRef]
8. Morabito, A.; Oliveira e Silva, G.; Hendrick, P. Deriaz pump-turbine for pumped hydro energy storage and micro applications. *J. Energy Storage* **2019**, *24*. [CrossRef]
9. Kougias, I.; Aggidis, G.; Avellan, F.; Deniz, S.; Lundin, U.; Moro, A.; Muntean, S.; Novara, D.; Pérez-Díaz, J.I.; Quaranta, E.; et al. Analysis of emerging technologies in the hydropower sector. *Renew. Sustain. Energy Rev.* **2019**, *113*. [CrossRef]
10. Ramos, H.M. *Guidelines for Design of Small Hydropower Plants*, 1st ed.; WREAN (Western Regional Energy Agency & Network) and DED (Department of Economic Development): Belfast, North Ireland, 2000.
11. Vieira, F.; Ramos, H.M. Optimization of operational planning for wind/hydro hybrid water supply systems. *Renew. Energy* **2009**, *34*, 928–936. [CrossRef]
12. Pérez-Sánchez, M.; Sánchez-Romero, F.J.; Ramos, H.M.; López-Jiménez, P.A. Energy recovery in existing water networks: Towards greater sustainability. *Water* **2017**, *9*, 97. [CrossRef]
13. Ramos, H.M.; Zilhao, M.; López-Jiménez, P.A.; Pérez-Sánchez, M. Sustainable water-energy nexus in the optimization of the BBC golf-course using renewable energies. *Urban Water J.* **2019**, *16*, 215–224. [CrossRef]
14. Pérez-Sánchez, M.; Sánchez-Romero, F.J.; López-Jiménez, P.A.; Ramos, H.M. PATs selection towards sustainability in irrigation networks: Simulated annealing as a water management tool. *Renew. Energy* **2018**, *116*, 234–249. [CrossRef]
15. Carravetta, A.; Derakhshan Houreh, S.; Ramos, H.M. *Pumps as Turbines*; Springer Tracts in Mechanical Engineering; Springer: Cham, Switzerland, 2018. [CrossRef]
16. Novara, D.; Derakhshan, S.; McNabola, A.; Ramos, H.M. Estimation of unit cost and maximum efficiency for pumps as turbines. In Proceedings of the 9th Eastern European IWA Young Water Professionals, Budapest, Hungary, 24–27 May 2017.
17. García, I.F.; Ferras, D.; McNabola, A. Potential of energy recovery and water saving using micro-hydropower in rural water distribution networks. *J. Water Resour. Plan. Manag.* **2019**, *145*. [CrossRef]
18. Fernandes, J.F.P.; Pérez-Sánchez, M.; Ferreira da Silva, F.; López-Jiménez, P.A.; Ramos, H.M.; Branco, P.J.C. Optimal energy efficiency of isolated PAT systems by SEIG excitation tuning. *Energy Convers. Manag.* **2019**, *183*, 391–405. [CrossRef]
19. Morillo, J.G.; McNabola, A.; Camacho, E.; Montesinos, P.; Rodríguez Díaz, J.A. Hydro-power energy recovery in pressurized irrigation networks: A case study of an Irrigation District in the South of Spain. *Agric. Water Manag.* **2018**, *204*, 17–27. [CrossRef]
20. Chacón, M.C.; Rodríguez Díaz, J.A.; Morillo, J.G.; McNabola, A. Hydropower energy recovery in irrigation networks: Validation of a methodology for flow prediction and pump as turbine selection. *Renew. Energy* **2020**, *147*, 1728–1738. [CrossRef]
21. Pérez-Sánchez, M.; Sánchez-Romero, F.J.; Ramos, H.M.; López-Jiménez, P.A. Modeling Irrigation Networks for the Quantification of Potential Energy Recovering: A Case Study. *Water* **2016**, *8*, 234. [CrossRef]

22. Pérez-Sánchez, M.; Sánchez-Romero, F.J.; Ramos, H.M.; López-Jiménez, P.A. Optimization Strategy for Improving the Energy Efficiency of Irrigation Systems by Micro Hydropower: Practical Application. *Water* **2017**, *9*, 799. [CrossRef]

23. Sari, M.A.; Badruzzaman, M.; Cherchi, C.; Swindle, M.; Ajami, N.; Jacangelo, G. Recent innovations and trends in in-conduit hydropower technologies and their applications in water distribution systems. *J. Environ. Manag.* **2018**, *228*, 416–428. [CrossRef]

24. Timilsina, A.B.; Mulligan, S.; Bajracharya, T.R. Water vortex hydropower technology: A state-of-the-art review of developmental trends. *Clean Technol. Environ. Policy* **2018**, *20*, 1737–1760. [CrossRef]

25. Chen, H.; Zhang, X.; Liu, J.; Tan, C. Compressed Air Energy Storage, Energy Storage - Technologies and Applications, Ahmed Faheem Zobaa, IntechOpen. Available online: https://www.intechopen.com/books/energy-storage-technologies-and-applications/compressed-air-energy-storage (accessed on 13 January 2020). [CrossRef]

26. Zhang, Y.; Yang, K.; Li, X.; Xu, J. The thermodynamic effect of thermal energy storage on compressed air energy storage system. *Renew. Energy* **2013**, *50*, 227–235. [CrossRef]

27. Grazzini, G.; Milazzo, A. Thermodynamic analysis of CAES/TES systems for renewable energy plants. *Renew. Energy* **2008**, *33*, 1998–2006. [CrossRef]

28. Camargos, P.L.T.; Pottie, D.L.F.; Ferreira, R.A.M.; Maia, T.A.C.; Porto, M.P. Experimental study of a PH-CAES system: Proof of concept. *Energy* **2018**, *165*, 630–638. [CrossRef]

29. Pottie, D.L.F.; Ferreira, R.A.M.; Maia, T.A.C.; Porto, M.P. An alternative sequence of operation for Pumped-Hydro Compressed Air Energy Storage (PH-CAES) systems. *Energy* **2019**. [CrossRef]

30. Odukomaiya, A.; Abu-Heiba, A.; Graham, S.; Momen, A.M. Experimental and analytical evaluation of a hydro-pneumatic compressed-air Ground-Level Integrated Diverse Energy Storage (GLIDES) system. *Appl. Energy* **2018**, *221*, 75–85. [CrossRef]

31. Storli, P.-T.; Lundström, T.S. A New Technical Concept for Water Management and Possible Uses in Future Water Systems. *Water* **2019**, *11*, 2528. [CrossRef]

32. Besharat, M.; Viseu, M.T.; Ramos, H.M. Experimental study of air vessel behavior for energy storage or system protection in waterhammer events. *Water* **2017**, *9*, 63. [CrossRef]

33. Triki, A.; Fersi, M. Further investigation on the water-hammer control branching strategy in pressurized steel-piping systems. *Int. J. Press. Vessels Pip.* **2018**, *165*, 135–144. [CrossRef]

34. Triki, A.; Chaker, M.A. Compound technique -based inline design strategy for water-hammer control in steel pressurized-piping systems. *Int. J. Press. Vessels Pip.* **2019**, *169*, 188–203. [CrossRef]

35. WIKA Data Sheet PE 81.01. 2018. Available online: https://en.wika.com/upload/DS_PE8101_en_co_1392.pdf (accessed on 13 January 2020).

36. *UVP-DUO User's Guide*; MET-FLOW. Release 5.2; Met-Flow SA: Lausanne, Switzerland, 2014; Available online: https://met-flow.com/support/download (accessed on 13 January 2020).

37. George Fischer (GF) Piping Systems Datasheet Pneumatic Actuators PA 30–PA 90. Available online: https://www.gfps.com/appgate/ecat/common_flow/100022/AT/en/109564/109576/P120047/product.html (accessed on 13 January 2020).

38. Wylie, E.B.; Streeter, V.L. *Fluid Transients in Systems*; Prentice Hall: New York, NY, USA, 1993.

39. Besharat, M.; Ramos, H.M. Theorical and experimental analysis of pressure surge in a two-phase compressed air vessel. In Proceedings of the 12th International Conference on Pressure Surges, BHR Group, Dubline, Ireland, 18–20 November 2015; Available online: https://www.scopus.com/inward/record.uri?eid=2-s2.0-84966351254&partnerID=40&md5=4c831ebf8171ef3543b8add5064aa020 (accessed on 13 January 2020).

40. Vereide, K.; Tekle, T.; Nielsen, T. Thermodynamic behavior and heat transfer in closed surge tanks for hydropower plants. *J. Hydraul. Eng.* **2015**, *141*. [CrossRef]

41. Goodall, D.C.; Kjørholt, H.; Tekle, T.; Broch, E. Air cushion surge chambers for underground power plants. *Int. Water Power Dam Constr.* **1988**, *40*, 29–34.

42. Steward, E.H.; Borg, J.E. Moose river air chamber design and performance. In *Waterpower'89*; U.S. Army Corps of Engineers; ASCE: Buffalo District, NY, USA, 1989; pp. 567–576.

43. Svee, R. Surge chamber with enclosed compressed air cushion. In Proceedings of the First International Conference on Pressure Surges, Canterbury, UK, 10 September 1972; pp. 15–24.

44. Zhou, L.; Liu, D.; Karney, B. Investigation of hydraulic transients of two entrapped air pockets in a water pipeline. *J. Hydraul. Eng.* **2013**, *139*. [CrossRef]

45. Zhou, L.; Liu, D.; Karney, B.; Wang, P. Phenomenon of white mist in water rapidly filling pipeline with entrapped air pocket. *J. Hydraul. Eng.* **2013**, *139*. [CrossRef]
46. Chaudhry, M.H. *Applied Hydraulic Transients*; Springer: New York, NY, USA, 2014.
47. Simão, M.; Besharat, M.; Carravetta, A.; Ramos, H.M. Flow Velocity Distribution towards Flowmeter Accuracy: CFD, UDV, and Field Tests. *Water* **2018**, *10*, 1807. [CrossRef]
48. Brunone, B.; Berni, A. Wall shear stress in transient turbulent pipe flow by local velocity measurement. *J. Hydraul. Eng.* **2010**, *136*. [CrossRef]
49. Ramos, H.M. Simulação e Controlo de Transitórios Hidráulicos em Pequenos Aproveitamentos Hidroelétricos. Ph.D. Thesis, Civil Engineering, Instituto Superior Técnico, Universidade de Lisboa, Lisboa, Portugal, 1995. (In Portuguese).

MDPI

St. Alban-Anlage 66

4052 Basel

Switzerland

Tel. +41 61 683 77 34

Fax +41 61 302 89 18

www.mdpi.com

Water Editorial Office

E-mail: water@mdpi.com

www.mdpi.com/journal/water